v a d e m e c u m

Tropical Dermatology

Roberto Arenas

Department of Dermatology
Dr. Manuel Gea Gonzalez General Hospital
Mexico City, Mexico

Roberto Estrada

Dermatology Service of the General Hospital of Acapulco
Acapulco, Mexico

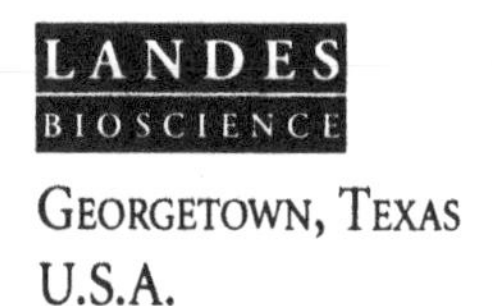

GEORGETOWN, TEXAS
U.S.A.

Tropical Dermatology
Vademecum
LANDES BIOSCIENCE
Georgetown, Texas

Please address all inquiries to the Publisher:
Landes Bioscience, 810 S. Church Street, Georgetown, Texas, U.S.A. 78626
Phone: 512/ 863 7762; FAX: 512/ 863 0081

ISBN 13: 978-1-57059-493-9.(pbk)

Library of Congress Cataloging-in-Publication Data

Tropical Dermatology/ [edited by] Roberto Arenas, Roberto Estrada.
p.cm.
"Vademecum."
Includes bibliographical references and index.
ISBN 1-57059-493-7 (alk. paper)
1. Skin--Diseases--Handbooks, manuals, etc. 2. Tropical Medicine--Handbooks, manuals, etc. I. Arenas, Roberto. II. Estrada, Roberto, 1943- .
[DNLM: 1. Skin diseases handbooks. 2. Tropical Medicine handbooks. WR 39 T856 2001]
RD598.H286 2001
616.5'00913--dc21
DNLM/DLC 98-10905
for Library of Congress CIP

While the authors, editors, sponsor and publisher believe that drug selection and dosage and the specifications and usage of equipment and devices, as set forth in this book, are in accord with current recommendations and practice at the time of publication, they make no warranty, expressed or implied, with respect to material described in this book. In view of the ongoing research, equipment development, changes in governmental regulations and the rapid accumulation of information relating to the biomedical sciences, the reader is urged to carefully review and evaluate the information provided herein.

Dedication

To The Hospital General Dr. Manuel Gea Gonzalez (Mexico); to The Hospital General de Acapulco y Universidad de Guerrero (Mexico); and to our patients all over the world.

Contents

H. Treponematosis and Genital Ulcers

Editors

Roberto Arenas, M.D.
Department of Dermatology, Division of Mycology
Dr. Manuel Gea Gonzalez General Hospital
Mexico City, Mexico
Chapters 1, 2, 3, 5-8, 11, 12, 22, 23, 28, 30, 31, 33, 35-37, 46, 49, 53-60, 65, 67, 69

Roberto Estrada, M.D.
Professor of Dermatology
University of Guerrero, Mexico
Acapulco, Mexico
Chapters 29, 41-43, 45, 47, 48, 51, 60, 61, 70

Contributors

Giancarlo Albanese, M.D.
Division of Dermatology, Mycology, and Tropical Dermatology
Ospedale S. Gerardo di Monza
Milano, Italy
Chapters 44, 63

Marin Arce, M.D.
Dr. Manuel Gea Gonzalez General Hospital
Mexico City, Mexico
Chapters 5, 6

Esperanza Avalos-Diaz, M.D.
Immunotechnology Unit
Autonomous University of Zacatecas
Zacatecas, Mexico
Chapter 50

Alexandro Bonifaz, M.B.
Department of Mycology, Dermatology Service
General Hospital of Mexico
Mexico
Chapters 4, 10.1, 10.2, 13

Josefina Carbajosa, M.D.
Instituto Nacional de la Nutricion
Mexico
Chapter 69

Hector Caceres-Rios, M.D.
Professor, Department of Dermatology and Pediatrics, Universidad Peruana Cayetano Heredia
Assistant Pediatric Dermatology Service, Instituto de Salud del Nino
Lima, Peru
Chapter 70

Guadalupe Chavez, M.D.
General Hospital of Acapulco
Acapulco, Mexico
Chapter 59

Roberto Cortes-Franco, M.D.
Department of Dermatology
Dr. Manuel Gea Gonzalez General Hospital
Mexico City, Mexico
Chapters 24-27, 66, 68

Luciano Dominguez-Soto, M.D.
Department of Dermatology
University of Guerrero
Mexico
Chapter 68

J. Octavio Flores, M.D.
Pascua Dermatology Center
Mexico City, Mexico
Chapter 21

Sergio Eduardo Gonzalez-Gonzalez, M.D.
Medicine Faculty of the Autonomous University of Nuevo Leon
Dermatology Service
Dr. Jose E. Gonzalez de Monterrey University Hospital
Monterrey, Nuevo Leon, Mexico
Chapter 34, 38

Esther Guevara, M.D.
Department of Dermatology
Dr. Manuel Gea Gonzalez General Hospital
Mexico City, Mexico
Chapter 68

Roderick J. Hay, D.M. F.R.C.P.
Department of Clinical Sciences
London School of Hygiene and Tropical Medicine
London, England
Chapter 62

Rafael Herrera, M.D.
Immunotechnology Unit
Autonomous University of Zacatecas
Zacatecas, Mexico
Chapter 50

Ma Teresa Hojyo, M.D.
Department of Dermatology
Dr. Manuel Gea Gonzalez General Hospital
Mexico City, Mexico
Chapter 68

Pedro Lavalle, M.D.
Pascua Dermatology Center
Mexico City, Mexico
Chapter 53

Ruben Lopez-Martinez, M.D.
Department of Microbiology and Parasitology, UNAM
Mexico
Chapter 19

Jorge A. Mayorga-Rodriguez, Ph.D.
Mycology Department
Dr. Jose Barba Rubio Dermatology Institute of Jalisco
Jalisco, Mexico
Chapters 9, 15

Clemente Moreno-Collado, M.D.
Dermatology Department
Military Medical School
Mexico
Chapters 10.2, 39, 40

Fernando Munoz-Estrada, M.D.
Dermatology Department
School of Medicine
Autonomous University of Sinaloa
Culiacan, Sinaloa, Mexico
Chapters 9, 15

Gisela Navarrete, M.D.
Department of Dermatopathology
Pascua Dermatology Center
Mexico City, Mexico
Chapters 32, 52

Rataporn Ungpakorn, M.D.
Mycology Department
Institute of Dermatology
Phyathai, Bangkok, Thailand
Chapter 20

Ma Elisa Vega-Memije, M.D.
Department of Dermatology
Dr. Manuel Gea Gonzalez
Mexico City, Mexico
Chapter 68

Jorge Vega-Nunez, M.D.
Department of Dermatology
University of Michoacan
Morelia, Michoacan, Mexico
Chapter 16

Edmundo Velazquez-Gonzalez, M.D.
Dermatology Department
Pediatric Hospital
Siglo XXI National Medical Center
Mexico City, Mexico
Chapters 64.1, 64.2

Rodolfo Vick, M.D.
UNAM Faculty of Medicine
Department of Internal Medicine
Dr. Manuel Gea Gonzalez General Hospital
Mexico City, Mexico
Chapter 58

Oliverio Welsh, M.D.
Department of Dermatology
Jose Eleuterio Gonzalez University Hospital
Monterrey, Nuevo Leon, Mexico
Chapter 17

Clarisse Zaitz, M.D., Ph.D.
Faculty of Medical Sciences
Santa Casa de Sao Paulo
Sao Paulo, Brazil
Chapters 14, 18

Preface

This handbook presents the geographical distribution, etiology, clinical picture, and treatment of dermatoses in the tropics. The tropical diseases have been known as exotic pathology, colonial medicine, or tropical public health. In some developed countries they are called imported diseases.

European doctors, soldiers and missionaries were the first to study these diseases in the 17th and 18th centuries. The English doctor, Patrick Manson (1844-1922), is considered the Father of the tropical medicine. Together with Joseph Chamberlain, Manson founded the School of Tropical Medicine in London. In 1907 the School of Tropical Medicine in Liverpool founded *Annals of Tropical Medicine and Parasitology.*

The tropics and subtropics comprise about 75% of the world population. The tropical diseases are not merely a group of nosologic diseases indigenous to the intertropical zone. Many are diseases of poor public health originating from poverty, ignorance, and population upheaval. Tropical dermatoses represent a public health problem in 127 countries with a population of 3 billion people who do not have access to health care. In rural areas these diseases represent 30% of doctor visits.

Originally, infectious diseases predominated, but some have been eradicated by sanitary and hygienic measures and others have decreased considerably due to antibiotics. Now infectious and parasitic diseases along with emergent diseases such as AIDS, or old re-emergent, drug-resistant diseases constitute the majority of tropical dermatoses.

The basic dermatologic problems are mycosis, parasitosis, mycobacteriosis, treponematosis and pyodermas. These differ in their clinical manifestation, distribution, and incidence due mainly to racial and environmental factors.

Due to the social, environmental and economic impact of the tropical diseases, multidisciplinary organizations have been created in the world to control them, especially to the seven most important diseases: malaria, filariasis, leishmaniasis, leprosy, Chagas disease, schistosomiasis and trypanosomiasis. Most have prominent skin manifestations.

Almost all tropical dermatoses are found in Mexico. Since some tropical diseases are more prevalent in the rest of Latin America or in other parts of the world, we have invited international authorities to contribute to this handbook.

Each disease is treated in accordance with concise format. We succinctly describe the geographic distribution of the disease, the clinical and laboratory diagnosis and treatment. This handbook is for students and physicians throughout the world. We hope it will be a valuable resource.

Roberto Arenas, M.D.
Roberto Estrada, M.D.

Acknowledgments

We wish to express our gratitude to Julieta Ruiz-Esmenjaud, M.D. for helping us in editing this book.

A. Superficial Mycosis

Dermatophytosis

Pityriasis Versicolor

Candidiasis

Tinea Nigra

Trichosporonosis

Dermatophytosis

Roberto Arenas

Superficial mycoses are caused by parasitic fungi of the keratin. They are called dermatophytes and belong to the gender *Trichophyton, Microsporum* and *Epidermophyton.* They infect the skin, the hair and nails, and occasionally, they involve deep tissue.

GEOGRAPHIC DISTRIBUTON

Dermatophytosis, especially tinea capitis, tinea corporis and tinea cruris, occurs worldwide but is very common in tropical countries. Ninety-eight percent of tinea capitis is seen in children; there are occasional cases in adult women. From 3-28% of children from low social-economic groups have tinea capitis. Favus is a tinea found only in Africa and South America. Tinea imbricata is endemic in the Pacific Islands, some parts of Malaysia, India and Latin America. Tinea cruris and tinea pedis are frequently seen in adult men with a range of 17%-20% and 70%, respectively. Tinea pedis is not common in children (1.5%). In som reports onychomycosis occurs in 54-70% of adult men. It occurs in 18%-60% of onychopathies and in 30% of dermatophytosis.

The most frequent causal agent is *T. rubrum;* it is reported in 36-52% and even up to 80% of dermatophitic infections. In tinea pedis it has been found in 79% and in onychomycosis in 76.2% of cases. *M. canis* is isolated in 14-24% and *T. tonsurans* in 15-18% (present in 88.1% of tinea capitis and 40.6% of tinea corporis in the United States). *T. mentagrophytes* and *E. floccosum* occur in 3-8% of cases. *M. gypseum, T. violaceum* and *T. verrucosum* are rare in Mexico. *T. soudanense* is found in Africa. *M. audouinii* has practically been eradicated in America and is rare in Europe where a similar species, *M. langeroni,* is found. *T. tonsurans* is not found in Africa, but its frequency in the USA has been increasing lately.

ETIOLOGY

There are three groups of dermatophytes: *Trichophyton, Microsporum* and *Epidermophyton* (there are 41 species, just 11 are common). The infections they cause are restricted to keratin-containing structures such as skin, hair and nails. There is a natural host defense to dermatophytes which depends on an antifungal serum factor whose existence is controversial, sebaceous secretion and acquired immune resistance.

Tropical Dermatology, edited by Roberto Arenas and Roberto Estrada.

Dermatophytic infections can be transmissible from the environment (geophilic, *M. gypseum*), from infected animals [zoophilic, *M.canis* (dogs and cats), rodents (*T. mentagrophytes*), bovine cattle (*T. verrucosum*), monkeys (*T. simii*)], or from infected people (anthropophilic). The severity and the course of the infection depends on the species of the dermatophyte and the host response. The granular colonies (zoophilic) generally cause an acute tinea and the anthropophilic cause mild inflammation and a chronic course. Dermatophytes can infect humans when exposed to a contagious source. Factors include a genetic predisposition to infection, humidity, heat, diabetes, the prolonged use of glucocorticoids, unventilated shoes, poor hygiene, or wet feet. Dermatophytes in the hair can be transmitted from contaminated combs, spray, or oils.

Dermatophytes infect keratin of the hair without involving the keratinagenous zone. On the nails the infection can start with involvement across the distal edge (hypoonychia) or on its lateral borders (distal and lateral subungual onychomycosis), and rarely, inflammation surrounds the cuticle and the eponychia affecting the proximal portion of the nail (proximal subungual onychomycosis). Also, infection can involve the dorsal surface of the ungual lamina (leukonychia tricophytica).

Mannans produced by *T. rubrum* inhibit cellular immunity by blocking monocytes. Two cytokines participate: IL-4 and IFN-γ. The first increases production of IgE and the second enhances delayed hypersensitivity. The humoral response mediated by IgG and IgM is related to acute inflammatory processes characterized by the presence of an infectious focus (usually tinea pedis in adults, or kerion in children), and evidence of distant clinical lesions—usually vesicles, lichenoid reaction, or erythema nodosum and a positive hypersensitivity trichophytin reaction (*J Am Acad Dermatol* 1993:28:S19-S23; *Monografias de Dermatologia. Madrid. Aula Medica* 1993: VI(6):384-390). Another pathogenic factor is the production of extracellular enzymes, proteases, which cause thickening of the keratin. Deoxyribonuclease and elastase are also involved in acute and chronic tineas. Elastase (in *T. tonsurans* and *T. mentagrophytes* infections) is associated with marked inflammation and lipase (in *T. rubrum* infection) with a mild inflammation and a chronic course (*Rev Lat Amer Microbiol* 1994; 36:17-20).

Dermatophytes have species-specific antigens. The carbohydrate fraction is related to immediate hypersensitivity (type I, IgE mediated), the protein fraction to delayed hypersensitivity (type IV) and the peptide-galactomannan complex is related to the allergic reaction. The low frequency of relapses in tinea capitis is explained by acquired immunity. Chronic tinea pedis and tinea cruris are due to highly congenial conditions (humidity and maceration favor fungal hyperproliferation), immune deficiency, atopy (IgE blocks receptors in antigen presenting cells).

CLASSIFICATION

Superficial dermatophytosis includes tinea capitis, tinea cruris, tinea manuum, tinea pedis, onychomycosis, and tinea imbricata. Deep dermatophytosis includes inflammatory dermatophytosis, tinea barbae, kerion celsi, favus, trichophytic granuloma, mycetoma, and dermatophytic disease.

CLINICAL PICTURE

The incubation of dermatophytes takes days or weeks. Tinea capitis occurs almost exclusively in children and is most common in kindergarten and elementary school ages; it usually is causeD by *M. canis* (40%-80%) and *T. tonsurans* (15%-60%). The dry type presents with skin scaling and short (2-3 mm), thick, split, deformed hairs and sometimes hair with a whitish layer.

Trichophytic tinea (Fig. 1.1) causes diffuse alopecia with small and irregular plaques mixed with healthy hairs and scales, "ringworm." Microsporic tinea presents as one or a few pseudoalopecic round plaques with short and regular affected hairs that seem to have been trimmed with a lawn mower (Fig. 1.2).

The inflammatory type, or kerion, is the most obvious manifestation of acquired immunity. It is more often due to *M. canis* and *T. mentagrophytes* (Fig. 1.3). This type can be localized to any area in the skin, but it is common on the scalp. The inflammatory plaque is painful and has abscesses, pustules, ulcers, and crusts. Adenopathy without fever is frequently present. In the early stages, it is a dermatophytic folliculitis, and in late stages the kerion is evident, so named from the Greek word "honey comb" because of its appearance. *T. verrucosum* (*T. ochraceum*) can cause a wide ulceration. The alopecia is quite evident, but parasitized hairs are hard to find. With proper treatment, it can be cured in 2-5 months, but without therapy, alopecia may be permanent.

Favus or tinea favosa is due to *T. schoenleinii* or *M. gypseum*, and it is characterized by disk-like crusts, called scutula, a group of filaments with a fetid odor similar to that of a wet mouse.

Tinea corporis, herpes circinatus, or tinea of the glabrous skin, can be caused by, *M. canis*, *T. rubrum*, *T. tonsurans*, *T. mentagrophytes* and *E. floccosum*. There is pruriginous, erythematous plaque with scaling and a vesicular border. *T. tonsurans* is more common in children and *T. rubrum* in adults (Figs.1.4 and 1.5). The microsporic type produces small plaques of 0.5-2 cm in diameter. This type is frequently seen in epidemics with a common source, usually an infected cat or dog.

A type of dermatophytosis which is not very well known is dermatophytic gluteae or epidermatophytosis of the diaper area. It affects children under 3 years of age. The diaper area and adjacent skin are involved, with characteristic annular, erythematous, squamous plaques with a few vesicules with areas of spared healthy skin. This type is caused by *E. floccosum* and rarely by *T. rubrum*.

Tinea cruris or Hebra's eczema can involve the inguino-crural and perineal areas. Sometimes it extends to the abdomen, buttocks, and rarely to the scrotum and penis (Fig. 1.6) as erythematous, squamous plaques with a vesicular border. Lesions are pruritic and chronic. Tinea cruris is common in warm weather and in people who remain sitting for long periods of time. Causal agents are *T. rubrum*, *E. floccosum*, and *T. mentagrophytes*. The same fungi cause tinea pedis or "athlete's foot" which usually affects adults. Its interdigital localization is frequently associated with other microorganisms and it is known as a dermatophytosis complex. On

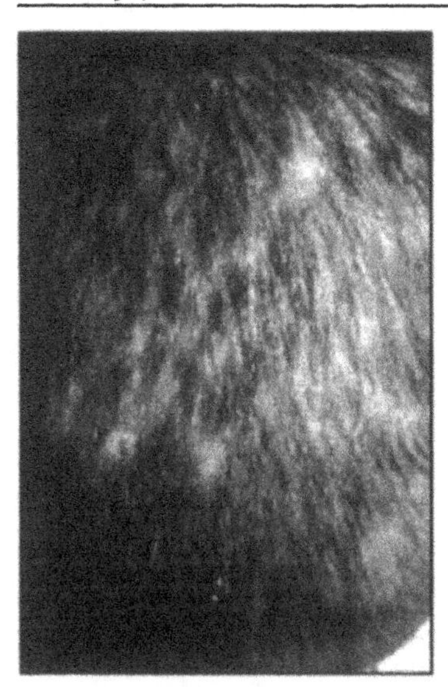

Fig . 1.1. Tinea capitis due to *T. tonsurans*.

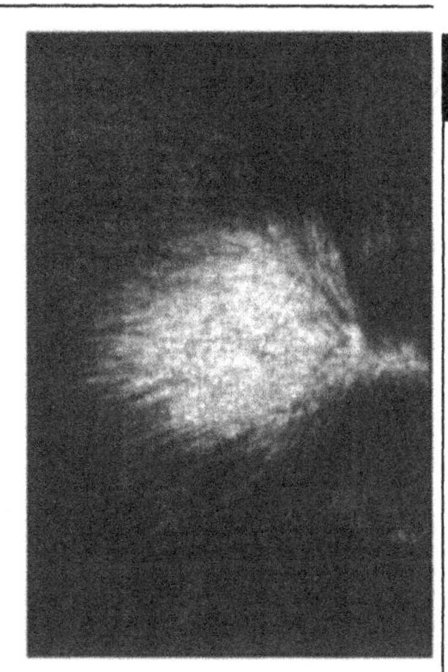

Fig. 1.2. Tinea capitis due to *M. canis*.

the foot, the interdigital folds, the sole and the borders are affected. It is characterized by fissures, scaling, vesicles, blisters, and crusting (dyshidrotic or eczematous), or it can be hyperkeratotic. It can be complicated by secondary impetigo, erysipelas and contact dermatitis.

Tinea manuum can affect one or both palms, and it is usually caused by *T. rubrum* (90%). There is diffuse hyperkeratosis, desquamation, anhydrosis, erythema, and sometimes an inflammatory reaction with vesicles and pustules. If it spreads to the dorsum of the hand it looks like tinea corporis (Fig. 1.7).

The onychomycosis due to dermatophytes or tinea unguium affects fingernails in 27%, toenails in 70% and both in 3% of cases. Nail thickening, fragility, striae, and a yellow-brown discoloration are observed in distal subungual onychomycosis. In severe cases, dystrophy is observed (Fig. 1.8). The common causal agent is *T. rubrum*. Predisposing factors are trauma, and now AIDS and organ transplants in which the trichophytic leukonychia or superficial white onychomycosis, and the proximal subungual white onychomycosis, are more frequently seen (*Int J Dermatol* 1995:34(8):591).

The chronic dermatophytic syndrome due to *T. rubrum* has been described as an early plantar infection that can arise together with groin involvement, with distal subungual onychomycosis, and in some cases with dermatophytic granuloma (*Int Dermatol* 1996:35(9):614-17).

The id phenomenon or dermatophytid (trichophytid) is seen in the hands. It is frequently related to tinea pedis and vesicles (dyshidrosis) and/or scaling are often present, but a lichenoid reaction or erythema nodosum and polymorphous erythema polimorphous also can occur.

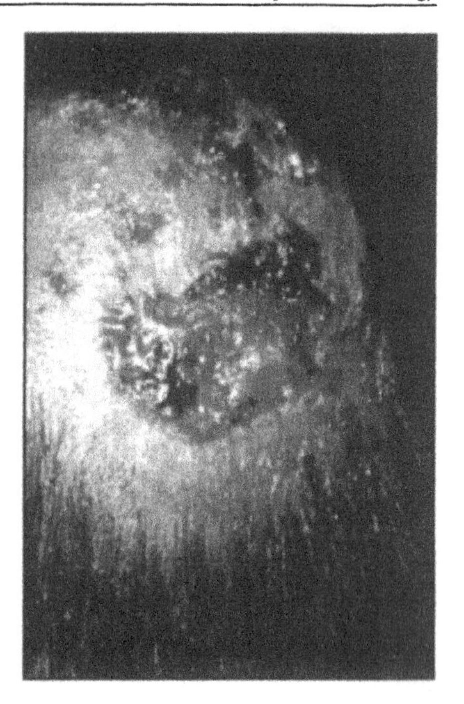

Fig. 1.3. Celsi's kerion.

Inflammation in tinea originates from the dermatophyte itself, or nowadays, from glucocorticoids (tachyphylasis). This feature of dermatophytosis is called tinea incognito and is characterized by exacerbation of the erythema, enlargement of the lesion, satellite plaques, atrophic marks, and isolation of one or two dermatophytes, and sometimes also *C. albicans.*

Tinea barbae is characterized by follicular pustules with a chronic course that can cause scarring and alopecia.

Tokelau or tinea imbricata is uncommon and caused by *T. concentricum* which only affects certain ethnic groups. It is characterized by a fine concentric scaling in a lace-like fashion or in a diffused pattern with lichenification (Fig. 1.9).

The trichophytic or dermatophytic granuloma is usually caused by *T. rubrum* and affects lower limbs. It is characterized by a hard nodule amd mild pain. It can be solitary or confluent and sometimes it can be seen as erythematosquamous plaque with a chronic course. There is almost always a history of prolonged use of glucocorticoids use and/or leg shaving. Cellular immunity is normal, and the trichophytin test is positive; in the disseminated types there is often a certain degree of immunosuppression.

Infection caused by *Scytalidium* (*Hendersonula toruloidea*) *hyalinum* or *dimidiatum*, has been observed in tropical and subtropical areas. It has been reported in the East and West of Africa, India, Pakistan, Thailand, Hong Kong, and in some places of Latin America.

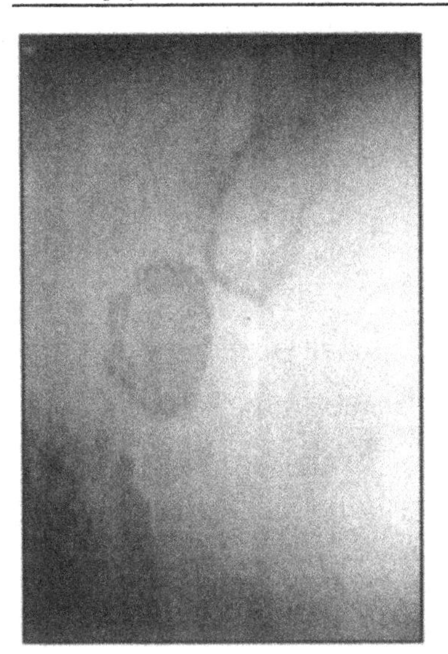

Fig. 1.4. Tinea corporis due to *T. rubrum*.

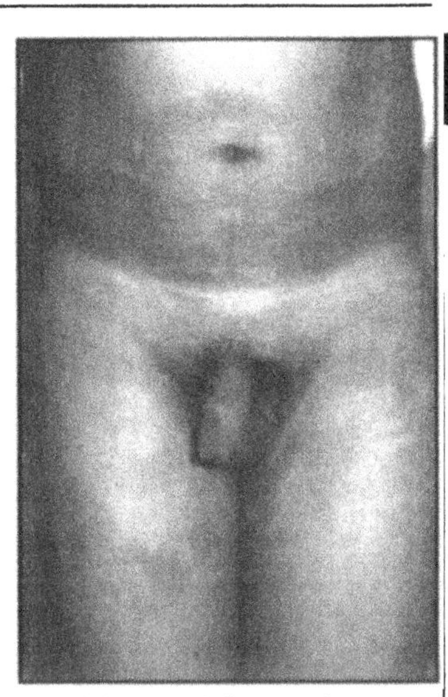

Fig. 1.5. Tinea corporis due to *T. rubrum*.

The clinical appearance of *Scytalidium* infection is similar to the dry infections caused by *T. rubrum* with scaling and fissures on the palms and soles. The infections are usually asymptomatic. There can be ungual dystrophy and onycholysis. Sinuous and irregular filaments are found on direct examination of the plaques. Culture should be done on Sabouraud's medium without cycloheximide. *Scytalidium* infection usually does not respond to systemic antimycotics.

LABORATORY DATA

Biopsy is not necessary except in deep mycosis. Superficial types show hyperkeratosis and keratinous follicular plugs. PAS or Gomori-Grocott stains show filaments in the horny layer. Neutrophils, vasodilatation and a mild to dense lymphohistiocytic infiltrate are seen. In kerion there are pustules or abscesses, and spores and/or filaments can be found in the dermis. In the trichophytic granuloma, besides the presence of the parasite, giant and epithelioid cells are observed.

A Wood's lamp is useful in the microsporic tinea capitis. There is a green fluorescence. Direct examination with KOH plus dimethylsulfoxide or with black chlorazol shows filaments and spores. In trichophytic tinea capitis, hair reveals endothrix spores (trichophytic and favic type) and on microscopy, ectoendothrix spores (microsporic, microide and megasporic types), which indicates filaments and spores are found inside and/or outside the affected hair. An easy way to collect the parasitized hair is to rub the affected hairy skin area with saline-soaked gauze. Culture can be done with a sterile swab, a piece of floor carpet or with a tooth

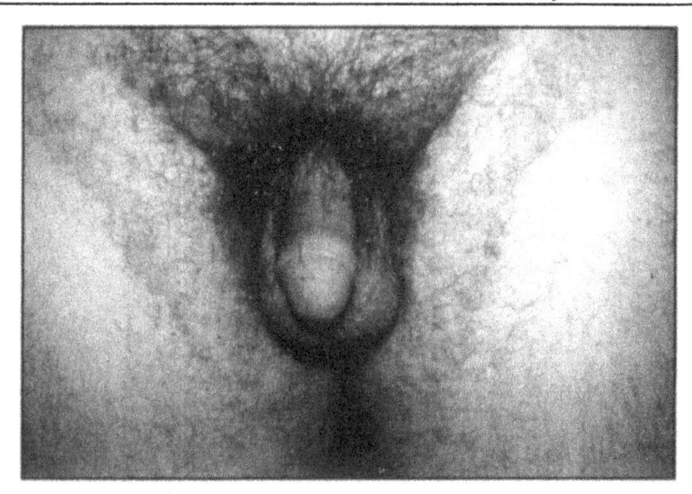

Fig. 1.6. Tinea cruris due to *T. rubrum.*

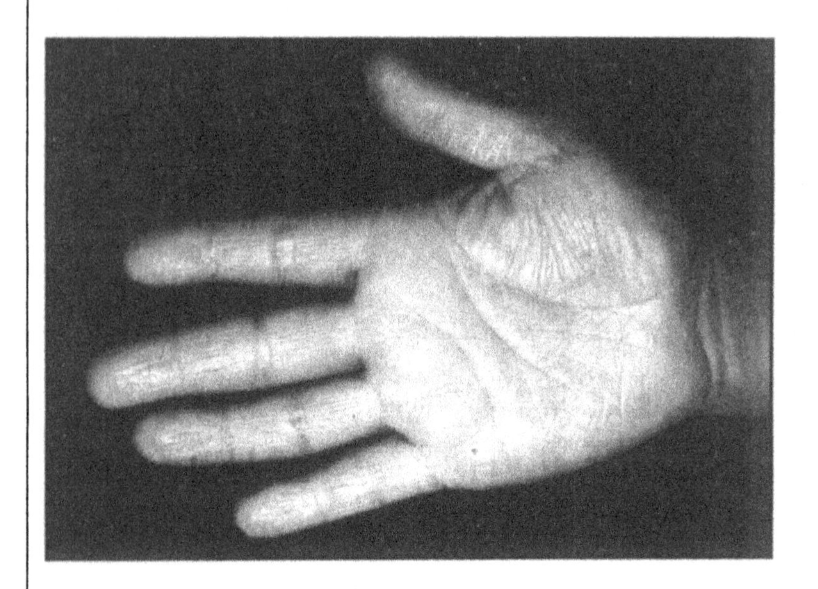

Fig. 1.7. Tinea manuum.

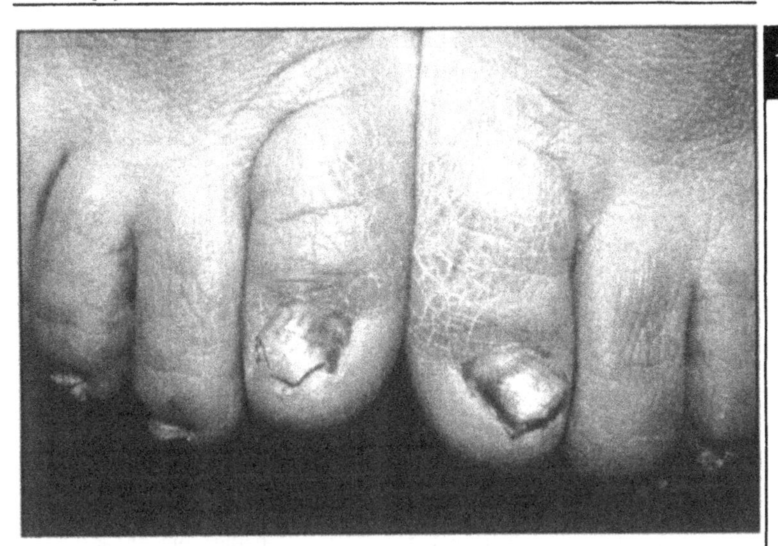

Fig. 1.8. Dermatophytic onychomycosis.

brush. Culture on Sabouraud's medium with or without antibiotics allows growth (1-2 weeks, or more) and identification of the causal agent. The trichophytin test is not of practical use.

TREATMENT

Treatment of tinea capitis is oral griseofulvin, 10 to 30 mg/kg/day (20-25 mg of the micronized form and 15-20 mg of the ultramicronized) for 2-3 months. A practical regimen is to administer 125 mg in children younger than 3 years old, 250 mg in children 4 to 7 years old, 375 mg in 8 to 12 years of age and 500-1000 mg/day in adults. The main side effects are nausea, headache, and photosensitivity. In kerion, some authors recommend 0.5 mg/kg/day of prednisone for two weeks along with an antimycotic (*Pediatr Dermatol* 1994; 11:69-71). In tinea capitis, terbinafin is also effective, 3-6 or even 10 mg/kg/day for 4-8 weeks. A practical regimen consists in 62.5 mg/day in children weighing less than 20 kg, or 125 mg/day in children that weigh 20-40 kg or are older than 5 years, respectively. In individuals of 40 kg and in adults, 250 mg/day is recommended (*Br J Dermatol* 1995:132:683-89; *Br J Dermatol* 1995: 132:98-105). In children who are able to ingest capsules, itraconazole is administrated, 5 mg/kg/day (*Int J Dermatol* 1994; 33:743-47); it is being tested as pulse-therapy for one week per month. There is some experience with weekly doses of fluconazole, although it has the advantage of being in an oral suspension. It is considered that an ideal antimycotic for children should be a liquid with a secure cap and a pleasant flavor; it must be effective and have few drugs interactions.

On the head, lightly rubbing the affected areas while bathing eliminates the parasitized hair. As an adjuvant, shampoos with 2.5% selenium sulfide, 2% ketoconazole or iodide-povidone, are used.

In tinea corporis the following topical compounds can be applied locally for 1-2 months: 0.5-1% iodide solution, Whitfield ointment (vaseline with 3% salicylic acid and 6% benzoic acid), tolnaftate 1% solution, cream or powder, tolciclate, pyrolnitrine or undecylenic acid, 1%-2% imidazolic cream or solutions of omoconazole, clotrimazole, or isoconazole twice a day, or ketoconazole, sertaconazole, oxiconazole, flutrimazole, tioconazole, croconazole, eberconazole or bifonazole once a day. Sometimes the following are administered: oral or topical griseofulvin, alillamines as naphtifin and terbinafin lotions (*J Am Acad Dermatl* 1992;26:956-90). In children > 40 kg or in adults, terbinafin 250 mg/day for two weeks if tinea corporis, or four weeks if tinea pedis is recommended. In tinea imbricata oral griseofluvin is effective.

In tinea cruris the predisposing factors, such as the use of synthetic and tight clothing and excessive sweating, should be avoided. In tinea pedis if secondary infection is present, it must be treated with topical antiseptics or topical and systemic antibiotics. Feet should be kept dry and patients should avoid the use of plastic, unventilated shoes. Antimycotic powders may help prevent relapse. Preventive measures should be taken as reinfection is observed.

Onychomycosis responds partially to topical treatments. Occlusion increases drug penetration and the infected nail keratin can be eliminated through partial surgical or chemical removal with 40% urea (25 g vaseline, 25 g lanolin, 10 g white wax and 40 g urea), also available as bifonazole 2% and urea 40%. This treatment is recommended when just a few nails are affected, or in children and pregnant women. The first, occlusive phase lasts 1-4 weeks, until the infected portion of the nail falls. The second phase, until complete cure (usually several months), involves the daily application of 1% bifonazole cream. Also, ticonazole 28%, cyclopirox 8%, or amorolfin 5% (nail polish) once or twice a week can be applied. The cyclopirox and amorolfin penetrate well. They are water resistent and are esthetically acceptable. Before each new application, the nail should be cleaned with acetone. Nail polish is useful for distal, superficial forms after complete cure as prophylaxis.

The oral treatment of onychomycosis with griseofulvin 50-100 mg/day for 6-12 months has a low cure rate and is more effective in fingernails than in toenails (simultaneous use of vitamin E enhances its absorption). Itraconazole 200 mg/day (*J Am Acad Dermatol* 1996; 35(1): 110-11) or terbinafin 250 mg/day for three months are used. Itraconazole 400mg or terbinafin 500 mg/day can be administered as pulse-therapy one week per month until cure. Fluconazole, 150 mg, is recommended weekly for eight months or 300 mg for a shorter length of time. Sometimes it is necessary to combine local and systemic treatments.

For inflammatory dermatophytosis, trichophytic granuloma, or tinea in immunocompromised patients, ketoconazole can be used orally 200 mg/day in adults, and 5 mg/kg/day in children. Prolonged therapy required for onychohomycosis has a risk of hepatotoxicity.

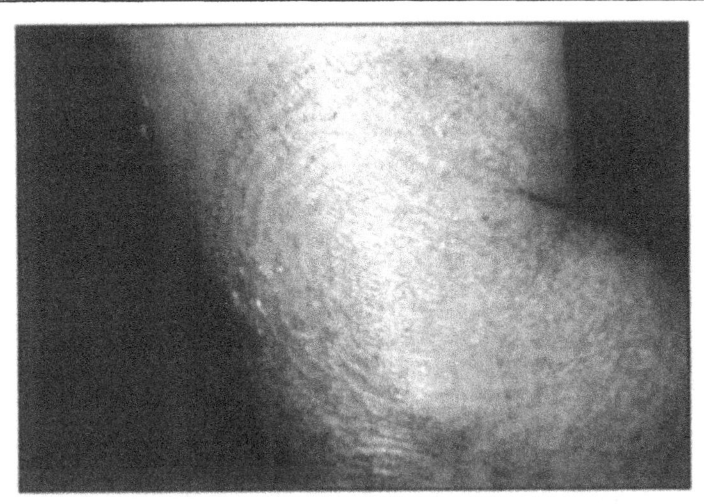

Fig. 1.9. Tinea imbricata (Tokelau).

Resistance to antimycotics is rare; it is usually relative, and it disappears by increasing the oral dose or the topical frequency. Therapeutic failures are due to incorrect diagnosis, the wrong dose, short duration of treatment, poor absorption, drug interactions or poor compliance.

Selected Readings

1 Arenas R. Tinea manuum. Datos epidemiologicos and micologicos en 366 casos. Gac Med Mex 1991; 127(5):435-38.
2 Chang P, Logeman H. Onychomycosis in children. Int J Derm 1994: 33(8):550-51.
3 Elweski BE. Cutaneous mycoses in children. Br J Dermatol 1996: 134(Suppl 46):7-11.
4 Frieden IJ, Howard R. Tinea capitis: Epidemiology, diagnosis, and control. J Am Acad Dermatol 1994; 31:S42-S46.
5 Gupta AK, Sauder DN, Shear NH. Antifungal agents: an overview. Part I and II. J Am Acad Dermatol 1994; 30:677-98, 911-33.
6 Kemna ME, Elewski BE. A U.S. epidemiologic survey of superficial fungal diseases. J Am Acad Dermatol 1996; 35:539-42.
7 Schwartz RA, Janniger CK. Tinea capitis. Pediatr Dermatol 1995; 55:29-33.

Pityriasis Versicolor

2

Roberto Arenas

Pityriasis versicolor, a superficial mycosis caused by the lipophilic *Malassezia furfur*, is characterized by hypochromic or hyperchromic macules (tinea versicolor) covered by fine scales affecting the trunk, neck, and the proximal part of theupper extremities. The course is chronic with frequent relapses.

GEOGRAPHIC DISTRIBUTION

Pityriasis versicolor is endemic in the tropics. It is most common in Samoa, Fiji, Mexico, Central America, South America, Cuba, Antilles, Mediterranean, and in some regions of Africa.

ETIOLOGY

The casual agent is *Malassezia furfur* (*Pityrosporum*). Three serotypes have been characterized A, B, and C, but the clinical presentation is identical for all serotypes (*Br J Dermatol* 1993; 129:533-540). The micelial pathogenic form (*M. furfur*) is frequently seen associated with excessive sweating, increased sebum secretion, synthetic clothing, the application of oils, and topical or systemic glucocorticoids. Also constitutional factors such as genetic predisposition, Cushing's syndrome, diabetes and immunodeficiency are important. But it is not usually present in AIDS where seborrheic dermatitis is frequently seen. *M. furfur* is capable of inducing an inflammatory response with a mild lymphocytic $CD4^+$ infiltrate. Specific antibodies and the capacity to activate complement have been demonstrated.

Scales seem to be due to the keratolytic effect of the fungus or to the transformation of triglycerides to irritant fatty acids. Color changes have been related to alterations in melanosome size, as well as to a cytotoxic effect on melanocytes produced by a dicarboxylic acid that inhibits tyrosinase and also has an antibacterial effect. Dicarboxylic acid is a product of fungal metabolism. (*Exp Dermatol* 1996; 5(1):49-56).

Taxonomy and nomenclature of this fungus have always been controversial. Not too long ago it was called *Pityrosporum* (*P. ovale-P. orbicular*) in the yeast phase present on healthy skin and in seborrheic dermatitis. Now the term *Malassezia* is commonly used when referring to both forms. As a result of morphologic studies and molecular biological studies, seven species of *Malassezia* have been identified: *M. globosa*, *M. sympodialis*, *M. restricta*, *M. slooffiae*, *M. obtusa*, and

Tropical Dermatology, edited by Roberto Arenas and Roberto Estrada. ©2001 Landes Bioscience.

M. pachydermatis. This last, which is isolated from animals, is not lipodependent, and, with the exception of *M. furfur,* they all have stable morphologic characteristics (*J Mycol Med* 1996;6 103-110).

CLINICAL PICTURE

It can affect any age and both sexes. It predominates in men 2:1. It is more frequent in men between 20 to 45 years of age, and it also has been observed in children and even in nursing infants. The frequency under 10 years of age varies from 1% to 11% (*Br J Dermatol.* 1996; 134(Suppl 46): 7-11). There is a family history in 19%. The lesion has a centripetal distribution on the thorax, back and proximal parts of the extremities. Infrequently it appears on the neck, forearms, or distal portion of extremities (Fig. 2.1). In children it can affect the face, forehead, and preauricular areas (Fig. 2.2), and in nursing infants the diaper area (*Pediatr Dermatol* 1991; 8:9-12).

Clinically abundant, lenticular hypochromic and brown or pink macules are present. Their size varies from 2-4 mm in diameter, or even up to 1 or 2 cm. They are covered by furfuraceous scaling, and they can be confluent forming large plaques (Figs. 2.1-2.3). Sometimes there are follicular lesions. Lesions are often chronic and asymptomatic. Occasionally there is mild pruritus. In immune-compromised hosts, it affects unusual areas like the head and genitals. If the lesions are rubbed with a curette or a fingernail, a mark is seen on the skin created by the detachment of the scales (Besnier's or scratch sign).

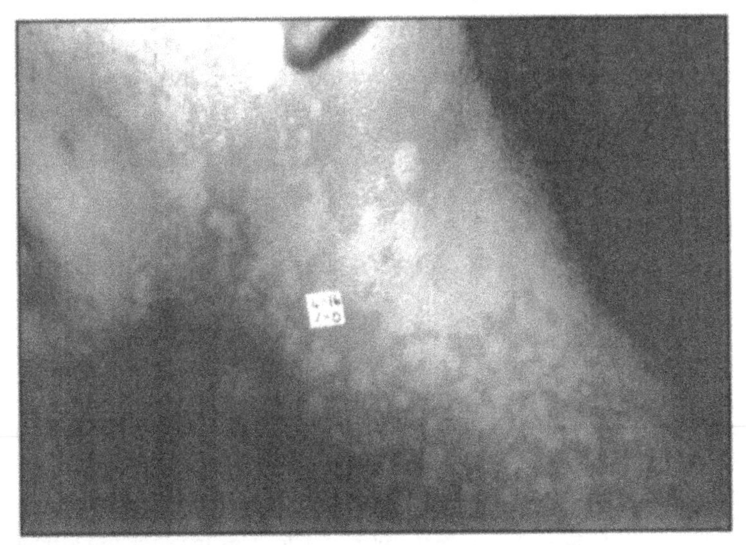

Fig. 2.1. Hypochromic pityriasis versicolor.

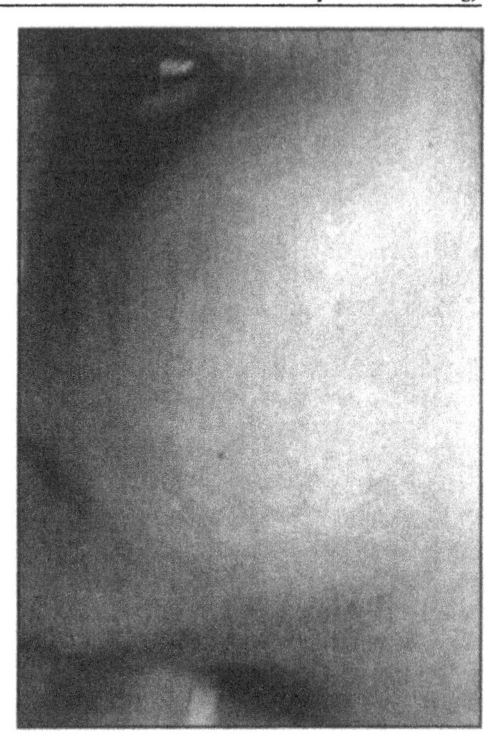

Fig. 2.2. Infantile pityriasis versicolor.

The folliculitis caused by *Malassezia* (*Pityrosporum*) affects young people taking glucocorticoids or systemic antibiotics (tetracycline), diabetics and AIDS patients. It can be seen just with the use of occlusive clothing. It presents as pruritic follicular papules with keratotic plugs and pustules. Systemic infection is unusual. It has been observed in newborns and in patients with IV catheters. The relationship between confluent and reticulated papillomatosis is controversial and may be due to the hyperkeratosis.

LABORATORY DATA

With a Wood's lamp a golden-yellowish color is seen, especially in mild cases. The direct exam with KOH, the scotch tape test, black chlorazol and Parker blue ink show spores 3-6 µm and short filaments, "spaghetti and meatballs" (Fig. 2.4). Culture is not necessary but it has been performed in enriched agar with lipids like 10% olive oil. The intradermal skin test is not of practical use. Biopsy is not necessary, but with hematoxylin-eosin, PAS and Gomori-Grocott stains, yeasts and filaments can be seen in the horny layer or in the pilar infundibulum. In folliculitis a granulomatous or lymphohistiocytic infiltrate is seen in the involved hairs.

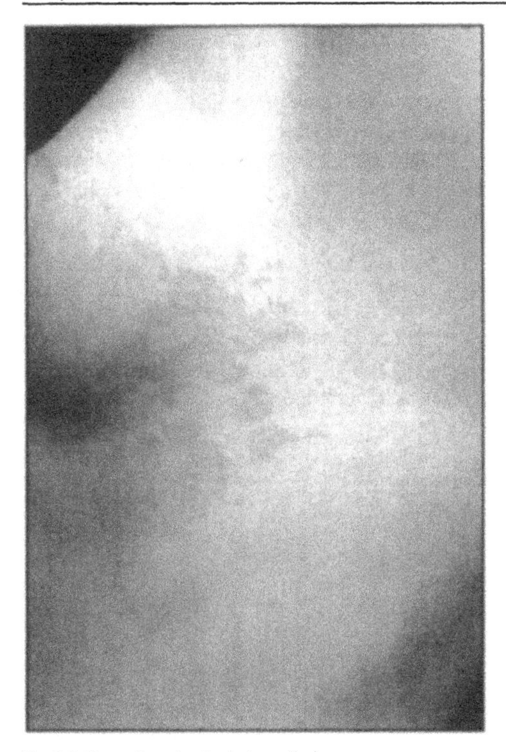

Fig. 2.3. Hyperchromic pityriasis versicolor.

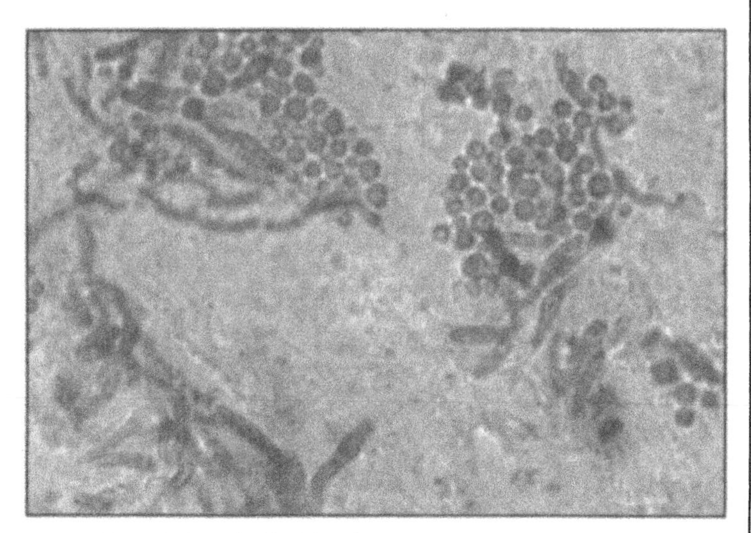

Fig. 2.4. *Malassezia furfur* (Parker blue ink 40X).

TREATMENT

Most local treatments are effective, but the nonresponding cases are explained by the difficulty in eliminating the predisposing factors.

Lotions, creams or soaps with 1-3% salicylic acid and sulfur, Whitfield ointment, 1% iodine solution, tolnaphtate, zinc pirythione, 2.5% selenium sulfur, 2% ketoconazole, both in cream solution or shampoo, 20% sodium hyposulfite or 50% propyleneglycol, both in aqeuous solution, 0.05% retinoic acid in solution or cream, topical 1-2% imidazoles in cream or solution (miconazole, clotrimazole, econazole, ketoconazole, tioconazole, oxiconazole, bifonazole, and sertaconazole), topical alillamines like terbinafin and naphtifin, as well as amorolfin, cyclopiroxolamine and griseofulvin—all of them are applied for several weeks.

Oral treatment: ketoconazole 200mg/day in the morning for 10 days to 1 month, or 400 mg in a single dose; itraconazole 100mg/day for 15 days to 1 month, or 200 mg/day for 5 to 7days (*J Am Acad Dermatol* 1996; 34:785-7); fluconazole 50 mg/ day, or 150 mg weekly for 1 month, or 400 mg in single dose (*Acta Derm Venereol* 1992; 72:74-75; *Acta Derm Venereol* 1996; 76:444-446).

Selected Readings

1 Arenas R. Dermatology. Atlas, Diagnostico y Tratamiento. Mexico. Interamericana/ McGraw Hill. 1996:77-81.
2 Borelli D, Jacobs PH, Nall L. Tinea versicolor. Epidemiologic, clinical and therapeutic aspectsl J Am Acad Dermatol 1991; 25:300-305.
3 Faergemann J. Jones TC, Hettler O, Loria Y. Pityrosporum ovale (Malassezia furfur) as the causative agent of seborrhoeic dermatitis; new treatment options. Br J Dermatol 1996: 134 (Suppl 46): 12-15.
4 Moreno-Gimenez JC. Pityriasis versicolor and other related processes. Monographs in dermatology. Madrid. Group Aula Medica 1995: VIII(4): 611-617.
5 Silva-Lizama E. Tinea versicolor. Int J Dermatol 1995; 34(9):611-617.

Candidiasis

Roberto Arenas

Candidiasis or moniliasis is caused by opportunistic yeasts from the genus Candida, most frequently, *C. albicans*. Infections may be superficial or deep, involving of the skin, mucosa, or the internal organs. The course may be acute, subacute or chronic.

GEOGRAPHIC DISTRIBUTION

Candidiasis occurs worldwide; it causes up to 25% of the superficial mycoses involving nails (35%), skin (30%), and the mucosae (20%). There is no predilection for race, sex, or age. It affects 4-8% of newborns. Candida vulvovaginitis is more common between age 20-30 years of age, and is present in 20-30% of gynecological infections. Balanitis occurs in adult and elderly men. Oral candidiasis occurs in persons with immunoincompetence, in children under 10, and in the elderly. The intertrigous and onychomycotic forms are more frequent in women, and interdigital candiasisis is more common in the tropics. Deep and systemic forms are uncommon; they may occur after cardiovascular surgery, with the use of intravenous catheters or in intravenous drug abusers.

ETIOLOGY

Candida is a saprophytic fungus. It is found on the mucous membranes: the digestive tract (24%) and the vagina (5-11%). The most common pathogenic species is *C. albicans* which has two serotypes (A and B). Other species are found: *C. parapsilosis, C. guilliermondi, C. tropicalis, C. kefyr, C. stellatoidea*, and *C. krusei.* Candida is opportunistic and becomes a pathogen when factors favor its growth such as antibiotics, glucocorticoids, cytotoxic drugs, or sex hormone imbalances caused by contraceptives and pregnancy, in diabetes or thyroid insufficiency, poor general health, or poor hygiene. Local factors such as poorly fitted dentures, poor dentition, direct contact with products high in sugars (bakers, fruit-packers), in finger sucking, fingernail biting, poor manicure and pedicure, and with synthetic clothing (plastic boots, disposable diapers).

Candida infection involves epithelial adherence, colonization, epithelial penetration, and vascular invasion followed by, dissemination, adherence to endothelium, and penetration into tissue (*Features* 1994: 60(6):313-318). There are nine virulence factors in *C. albicans*: Also many features of Candida characterize infection: the formation of hyphae; thigmotropism, is a tactile sensitivity that directs hyphae to follow tissue surfaces; hydrophobicity, which facilitates non-specific adherence; adhesion molecules; the microorganism has the capacity to produce or acquire a molecular surface that mimics that of the host; proteinases (*J Mycol Med* 1995; 5:145-166) and (Table 3.1).

CLASSIFICATION

1) Localized forms: mouth, big folds, small folds, diaper area, genitals, nails and periungual region.
2) Disseminated and deep forms: chronic mucocutaneous candidiasis and granuloma.
3) Systemic forms: septicemia from Candida, iatrogenic candidemia and fungal invasive dermatitis.

CLINICAL PICTURE

Candida affects any tissue or organ. Infection in the mouth, thrush, is manifested by redness and whitish mucosal plaques (Fig. 3.1). Lesions can be diffuse or affect a single region like the palate, the buccal mucosa, gums, or tongue (glossitis). Plaques are asymptomatic or accompanied by a burning sensation. The following forms have been described: erosive, hyperplastic, pseudomembranous, erythematous (atrophic), acute or chronic, as well as forms with plaques and nodules. Lip involvement is exceptional, but frequently in the corners of the mouth, angular cheilitis, a triangular area of fissures and erythema, is seen. Black-hairy tongue may be due to Candida or Geotrichum strains. Esophageal involvement occurs mainly in patients with AIDS and in leukemic patients.

Intertriginous candidiasis is characterized by erythema, scaling, maceration, peeling borders and satellite papules (Fig. 3.2). These lesions are seen in interdigital

Table 3.1 Virulence factors in C. albicans (modified from Odds, FC. Candida species and virulence. Features 1994; 60(6):313-318).

Phenotype switching
Hypha formation
Thigmotropism
Hydrophobicity
Surface virulent molecules
Molecular mimicry
Lytic enzymes production
Growth rate
Nutrient requirements

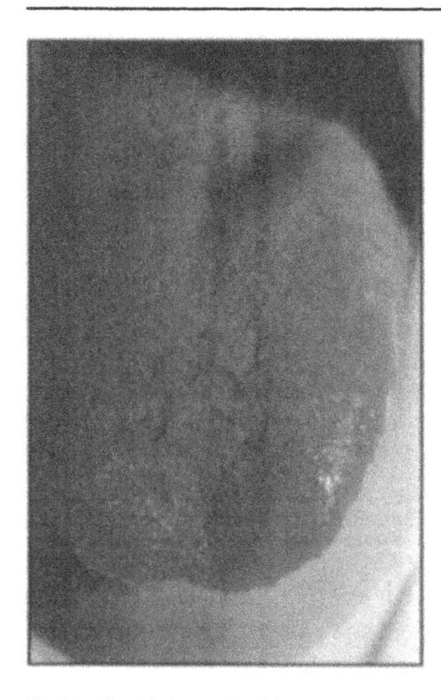

Fig. 3.1. Glossitis due to Candida.

spaces in housekeepers or in people whose hands are frequently wet. It also affects the feet or large intertriginous areas: axillary, inguinal, submammary or intergluteal (Figs. 3.3, 3.4). In the diaper area, erythema, scaling vesicles, blisters, pustules, and ulcers (usually associated with a secondary dermatitis) are seen, usually secondary to a previous dermatitis. Neonatal candidiasis manifests as thrush or disseminated pustular or vesicular lesions.

In vaginitis, inflammation, grumous leukorrhea and pruritus are present, as well as dyspareunia when lesions involve the vulva and perineum. There are erythematous and pseudomembranous forms associated with frequent and

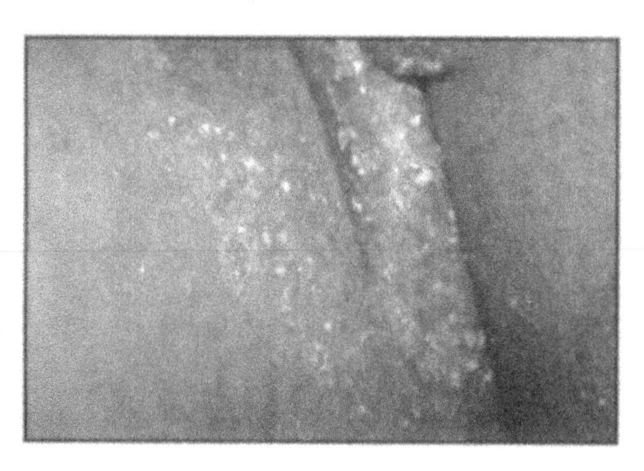

Fig. 3.2. Candidiasis with erythema, scaling, and pustules.

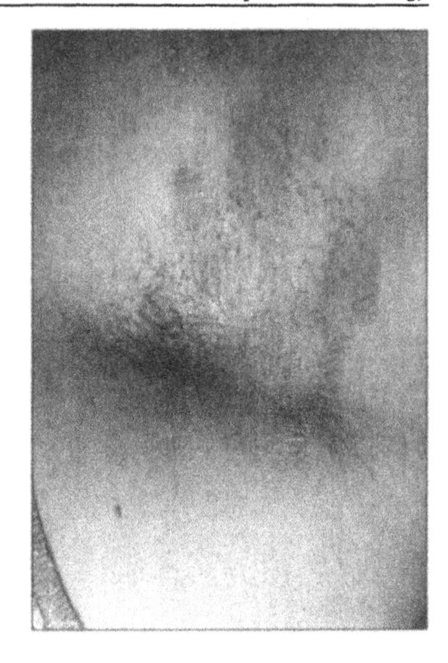

Fig. 3.3. Candidiasis of the axillary folds.

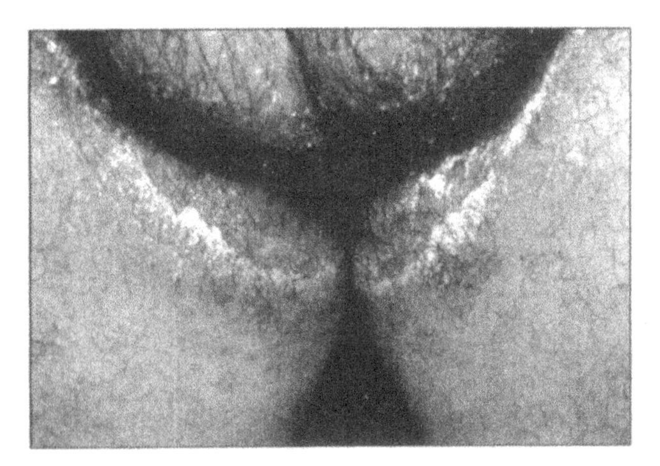

Fig. 3.4. Inguinal candidiasis.

persistent recurrence. Vaginitis is caused mainly by *C. albicans*, but *C. tropicalis* and *C. glabrata* (*T. glabrata*) are also cultured. On the penis (balanitis or balanopostitis), the skin is macerated, has white plaques, and erosions. Sometimes vesicles and pustules appear; dysuria and polyuria may be present. In onychomycosis, the ungual lamina is engorged, mainly in the base, and has transverse striae. There is depigmentation, or the nail acquires a yellow, green, or black discoloration (Fig. 3.5). Perionyxis and pain can be severe. Periungual inflammation is thecause of nail damage. Sometimes pus can be expressed from the affected area, and onycholysis may also be present.

Mucocutaneous candidiasis is a severe problem. It begins in nursing infants and is related to thymic abnormalities, functional defects in lymphocytes and leukocytes, and endocrinopathies or, rarely, is idiopathic. In adults it is associated with malignancy. Lesions appear on the skin, mucosae and nails, or also within the deep, nodular and scaling Candida granuloma.

Systemic candidiasis can involve any organ but is found mainly in the esophagus and heart. Candida septicemia usually has an intestinal origin. Iatrogenic candidemia is associated with parenteral nutrition.

Invasive fungal dermatitis has been described recently in premature newborns (< 24 weeks) after vaginal delivery and postnatal treatment with glucocorticoids (81%) who develop of hyperglycemia (*Pediatrics* 1995: 95(5):682-686). It is signalled around the ninth day by the appearance of erosive and scaly lesions which provide an access for Candida causing a systemic infection (69%).

LABORATORY DATA

Biopsy is not necessary, but when performed, the superficial forms show, in the thick stratum corneum, the microorganism as filaments and yeasts, that are more evident with PAS or Gomori-Grocott. In the dermis there is mild edema. With deep involvement, there are abscesses and a granulomatous reaction.

The intradermal reaction to candidin is not useful because it is positive in all

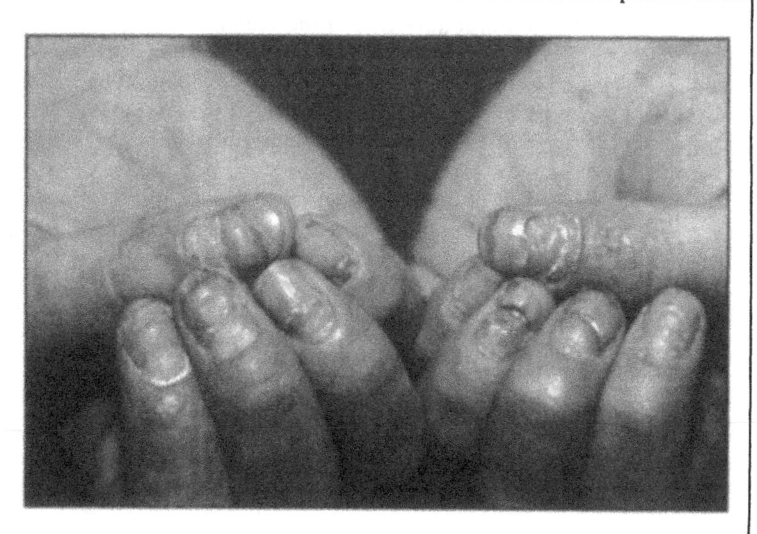

Fig. 3.5. Onychomycosis caused by Candida.

the individuals who have had previous contact with the fungus (60%). Direct examination with KOH, Lugol's solution or distilled water reveals pseudohypha, hypha and blastospores 2-4 µm in diameter. Also a smear, stained with Gram, PAS or methylene blue is useful. Cultures in Sabouraud, with or without addition of antibiotics, support abundant colonies of *Candida*. These species are sensitive to cycloheximide (Actidione): *C. tropicalis, C. krusei*, and *C. parapsilosis*. Chlamydospores in potato-carrot agar or corn meal indicates *C. albicans*. It is possible to make physiologic determinations (auxonogram and zymogram) through systems available commercially. Immunological and serum filamentation are more specialized. Serological tests involve immunodiffusion, latex agglutination, complement fixation, ELISA or fluorescent antibodies. They are useful in deep and systemic forms.

TREATMENT

Above all, it is important to eliminate factors favorable to Candida. Bicarbonate mouthwash and miconazole gel improve oral candidiasis. Vaginal miconazole or nystatin tablets, given their slow dissolution, are also recommended. Treatment with Gentian violet 1% and Castellanis stain are effective but messy and have been abandoned. In the genital and diaper areas, vinegar or acetic acid are applied (5-10 ml in a liter of water or Burow's solution). Nystatin cream (100,000 U), capsules (500,000 U), powder for oral suspension or drops (100,000 U), topical powder, or vaginal tablets TID for seven days to several weeks. This antimycotic is not absorbed through the gastrointestinal tract. Ketoconazole orally, 200 mg/day, can be used for chronic and deep infection of the skin, mucosa and nails. The mucocutaneous forms improve in days or weeks, and the others in several months. Secondary hepatic and antiandrogenic effects with the prolonged administration must be kept in mind. For vaginitis, 400 mg/day for 5 days or terconazole cream or vaginal tablets for 5 days are recommended. In superficial forms the following are useful: clioquinol 3%, clotrimazole, econazole, isoconazole, ticonazole, ketoconazole, sulconazole, bifonazole, and miconazole; topical terbinafin, amorolfin, and cyclopirox. In mucocutaneous forms, itraconazole, 100 mg/day, is administrated until the symptoms disappear; in onychomycosis of fingernails itraconazote 200 mg/day, for six months; and in toenails 200 mg/day for 3 months plus observation without treatment for 4-6 months is recommended. In vaginitis, 400-600 mg in a single dose, or 200 mg/day for 3 days is indicated. In chronic vaginitis the dose is repeated the first day of the menstrual period for 5 months. Fluconazole 150 mg in a single dose or weekly for 4 weeks can be used. In chronic cases it can be administered daily. In deep and systemic forms, amphotericin B, 0.6 mg/kg/day, is used either in its standard form or in liposomes, or 5-fluocytosine 150 mg/kg/day, or both.

Selected Readings

1 Arenas R. Dermatologia. Atlas, Diagnostico, y Tratamiento. Mexico. Interamericana/McGraw Hill. 1997:315-20.
2 Odds FC. Candida species and virulence. Features 1994; 60(6): 313-318.
3 Rowen JL, Atkins JT, Levy ML, et at. Invasive fungal dermatitis in the >1000 gram neonate. Pediatrics 1995; 95(5):682-686.
4 Senet JM, Robert R. Physiopathologie des candidoses. J Mycol Med 1995; 5:145-166.
5 Sobel JB. New insights into the pathogenesis of vaginal candidiasis. Int J Exp Clin Chem 1989; 2(suppl 1): 9-18.

Tinea nigra

Alexandro Bonifaz

The superficial mycosis caused by the fungus, *Exophiala werneckii* (*Phaeoanellomyces werneckii*), is characterized by very limited hyperpigmented, asymptomatic macules covered with fine flakes that present mainly on the palms, rarely on the soles or elsewhere.

GEOGRAPHIC DISTRIBUTION

This mycosis occurs frequently in tropical areas. Areas of the highest incidence are: Central and South America and the Caribbean (Panama, Colombia, Venezuela, Brazil and Cuba); Asia (India, Sri Lanka, and Myanamar); Polynesia, and the African coast. In Europe and the United States, it is rare; in the USA 100 cases have been reported, most of them in the Southeast (Louisiana, Alabama, and North Carolina).

ETIOLOGY

The causal agent is a pleomorphic demiatiaceous fungus called *Exophiala werneckii.* Its taxonomy and classification have always been controversial. It was originally classified within the genus Cladosporium, and nowdays it is called *Phaeoannellomyces werneckii* (*J Med Vet Mycol* 1985:23:179-88). Another species, *Stenella araguata* (*Cladosporium castellanii*), has been reported. The fungus that causes tinea nigra thrives in the soil and on vegetation in humid climates. Transmission in humans has not yet been reported. It is probable that the fungus enters through the skin after minor trauma which explains why the hands are the most frequent site of involvement. *Exophiala werneckii* culture starts as black yeast-like colonies; later the colonies become hairy. In the first phase, blastoconidia with septae are observed, and in the second abundant septated hyphae with a great amount of annelloconidium are noted.

CLINICAL PICTURE

It affects all ages and both sexes equally. Nevertheless, most authors report that it is more frequent in people under 20 years of age with a slight preference for

Tropical Dermatology, edited by Roberto Arenas and Roberto Estrada.

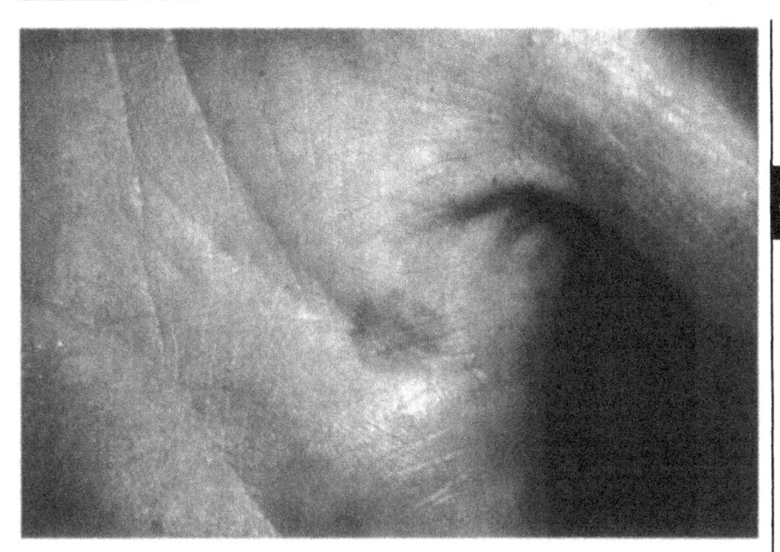

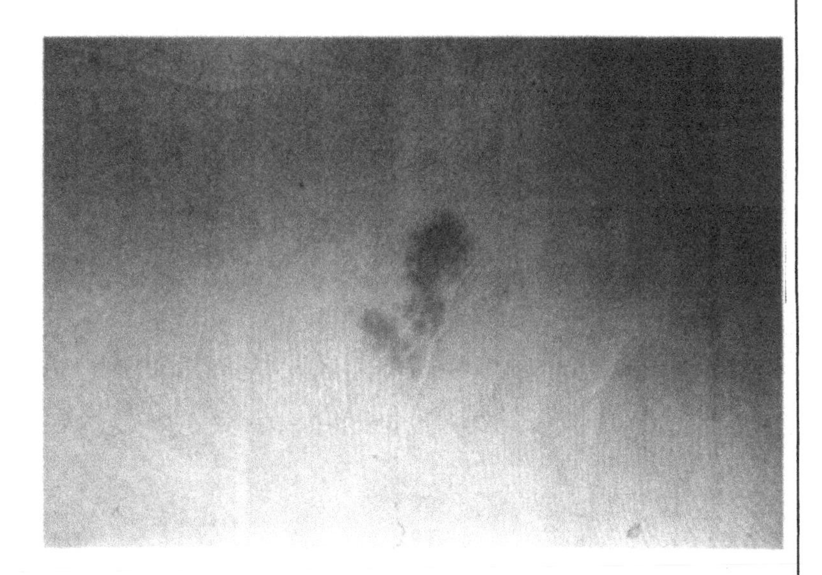

females. There are many cases in children. The most frequent site of involvement is on the palms (tinea nigra palmaris) and on the soles (tinea nigra plantaris), and just a few cases in the arms, legs, neck and trunk have been reported. Distribution is asymmetric and usually unilateral with characteristic hyperpigmented light-brown to dark-brown patches. These spots have rounded edges sometimes covered with fine scales (Fig. 4.1 and 4.2). Its course is usually chronic and asymptomatic; there is no erythema or induration. Minimum pruritus is unusual. Spontaneous healing is common. Pigmented nevi, melanoma, malignant lentigo, Addison's disease, or simply chemical pigmentation by agents like silver nitrate comprise the differential diagnosis.

LABORATORY FINDINGS

The diagnosis is confirmed by the Scotch tape test, and the scales are examined microscopically with potassium hydroxide 20% (KOH). Dark green or brown branched and septated hyphae of approximately 3-5 µm in diameter and sometimes with blastoconidium of similar size are observed. In Sabouraud's agar with antibiotics of 25-30° C, creamy appearing dark colonies develop in 1-2 weeks; they eventually transform into hairy colonies. Biopsy is not helpful.

SELECTED READINGS

1 Interamericana/McGraw-Hill. 1993: 83-86.
2 Herbrecht R, Koenig H, Waller J, Liu KL, Gueho E. Trichosporon Infections: Clinical manifestations and treatment. J Mycol Med 1993; 3:129-136.
3 Miranda MFR, Brito de A, Salgado V, et al. Tricosporonose genitocrural: Estudio de quarenta casos (Genitocrural tricosporonose: The study of forty cases). An bras Dermatol 1994; 69(5): 377-382.

Tricosporonosis

Roberto Arenas and Martin Arce

Yeast from the gender Trichosporon, especially *T. cutaneum*, can cause superficial infections of the hair, skin, or fingernails known as nodosa trichosporia (white rock). This is characterized by soft nodules of the hair as well as by onychomycosis and otomycosis. It occurs in diabetics and in immunocompromised persons (AIDS), as an accompaniment of pneumonia or disseminated infection.

GEOGRAPHIC DATA

Trichosporonosis is found in Europe, the East Asia (particularly in Japan and Russia), Latin America, and less frequently, in the United States. It is most common in young adults. It can be epidemic. Until 1993 there were 128 reported cases of deep infections, 331 superficial, and 82 disseminated.

ETIOLOGY

Six species are involved in human infections: *T. asahii, T. asteroides, T. inkin, T. mucoides, T. ovoides,* and *T. cutaneum (T. beigelii)*. This last is a filamentous yeast with arthrospores that affect cells of the hair, but does not penetrate them. Transmission is by fomites like combs, hair brushes, shampoo containers, and cosmetics. Involvement of skinfolds, feet, and fingernails are observed in diabetics, and systemic forms present in immunocompromised individuals.

CLINICAL PICTURE

The organism infects the scalp, hair, and less often the beard, mustache, pubic, and perineum. Sometimes irregular, adherent fusiform nodules, 1-1.5 mm in diameter, can be noted. They are white-yellowish or reddish-brown, sometimes transparent, and soft (Fig. 5.1). In the genital area, lesions are erythematous and peeling with light itching—similar to candidiasis (*An bras Dermatol* 1994; 69(5):377-382). There is erythema on the feet, peeling and hyperkeratosis. And on the toenails there can be subungueal hyperkeratosis, or unicolisis with discoloration, as with tinea pedis or unguium. Most Trichosporon infections are an

Tropical Dermatology, edited by Roberto Arenas and Roberto Estrada. ©2001 Landes Bioscience.

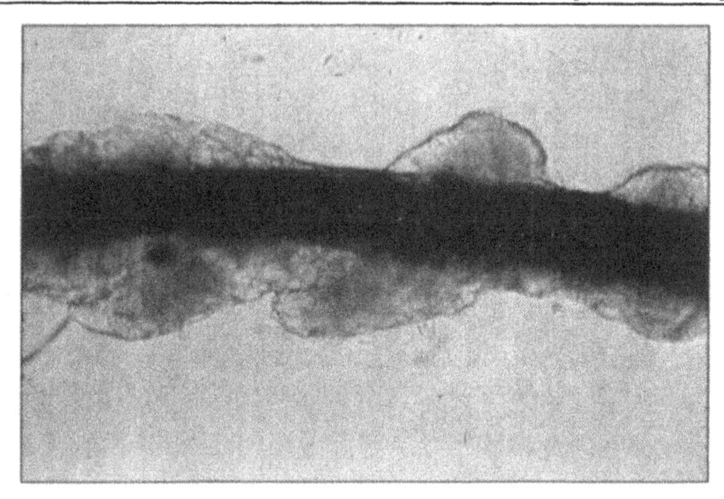

Fig. 5.1. White rock, Microscopic exam (KOH, 10X).

opportunistic mycosis characterized by fever, fungemia, and maculopapular lesions in the skin. It can be fatal in neutropenic patients, in those who receive chemotherapy or corticosteroids in AIDS (*Arch Dermatol* 1993; 129: 1020-23; *Infect Dis Clin North Am* 1996; 10:365-400), and especially in transplant patients.

LABORATORY DATA

There is no fluorescence with a Wood's lamp although lesions can be somewhat greenish. Direct microscopic examination demonstrates, among the cuticle cells, filamentous aggregates, 2-4 mm in diameter, of rectangular, rounded, or oval arthrospores (ectothrix).This is best seen with blue Parker dye (Fig. 5.2) or with histopathologic methods, PAS and Gomori-Grocott. The organism is cultured in Sabouraud's medium. In the disseminated or systemic forms, filaments are observed in the biopsy.

TREATMENT

Hair can be cut from the lesion. Solutions of tincture of iodine 1%, salicylic acid 50%, or any of the imidazole derivatives can be applied. For deep infections in individuals with immune alterations, amphotericin B, and fluconazole and itraconazole are recommended.

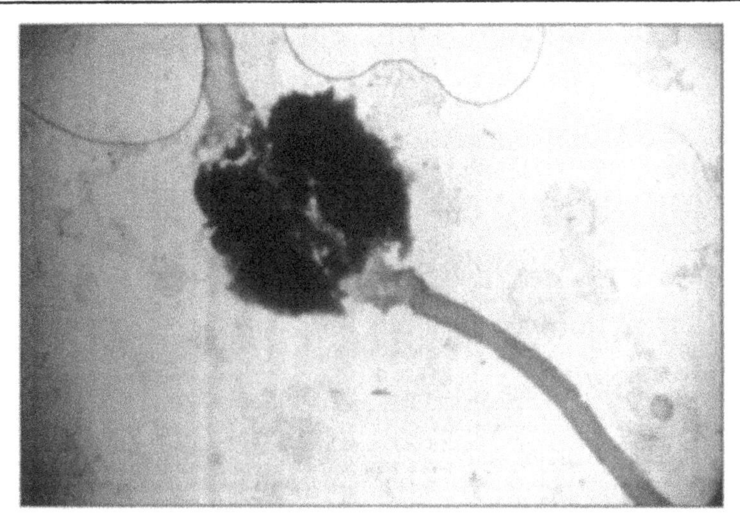

Fig. 5.2 White rock (blue Parker dye 10X).

Selected Readings

1 Arenas R. Micologia Medica Ilustrada (Medical Illustrated Micologia). Interamericana/McGraw-Hill. 1993: 83-86.
2 Herbrecht R, Koenig H, Waller J, Liu KL, Gueho E. Tricosporon infections: Clinical manifestations and treatment. J Mycol Med 1993; 3:129-136.
3 Miranda MFR, Brito de A, Salgado V et al. Tricosporonose genitocrural: Estudio de quarenta casos (Genitocrural tricosporonose: The study of forty cases). An bras Dermatol 1994; 69(5): 377-382.

5

B. Pseudomycosis

Erythrasma

Pitted Keratolysis

Trichomycosis

Prototheocosis

Actinomycosis

Botryomycosis

Erythrasma

Roberto Arenas and Martin Arce

The dermatosis caused by *Corynebacterium minutissimum* affects the stratum corneum. It is localized to large skin folds and interdigital areas of the feet. It has a chronic course characterized by brown patches covered with fine scales.

GEOGRAPHIC DISTRIBUTION

It is widespread but is more frequent in the tropics. It predominates in adult men.

ETIOLOGY

Erythrasma is caused by a filamentous gram positive bacteria, *C. minutissimum*, that is normal flora of the skin and produces a porphyrin. Heat, humidity, diabetes, and poor hygiene favor its growth.

CLINICAL PICTURE

The most frequent sites of involvement are the inguinal folds, axilla, and submammary area. Rarely it spreads to other areas. The lesions are 10 cm, light-brown plaques, with discrete borders, polycyclic, and covered with fine scales (Fig. 6.1). There is mild or no pruritus, and the course is chronic without a tendency to remission. In interdigital spaces and soles, plaques are erythematous with moderate scaling or vesicles (Fig. 6.2). Erythrasma comprises 10% of cases of so-called "swimmer's eczema" (*An Bras Dermatol* 1994; 69(1): 16-20). It is usually associated with candidiasis and dermatophytosis. When nails are involved, they are thickened, have a yellow-orange pigmentation and are striated.

LABORATORY DATA

Lesions fluoresce coral-red or orange with a Wood's lamp, but non-fluorescent lesions have been reported. Microscopy with KOH demonstrates isolated strands

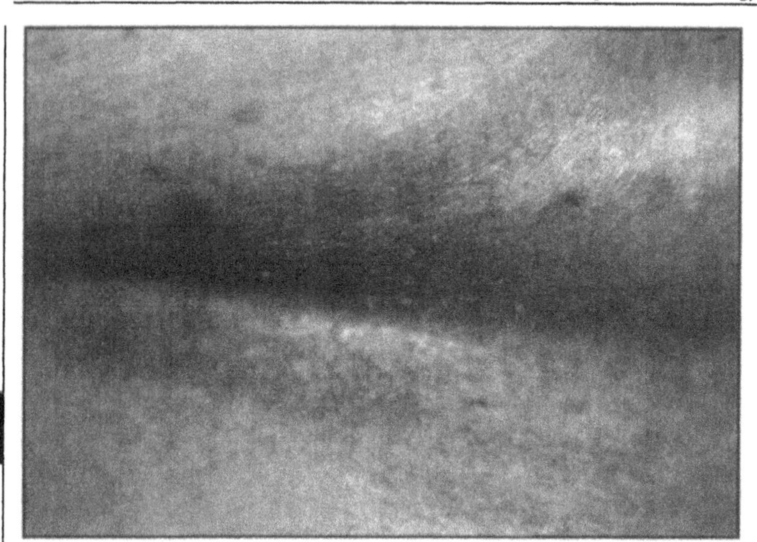

Fig. 6.1. Large skin fold erythrasma.

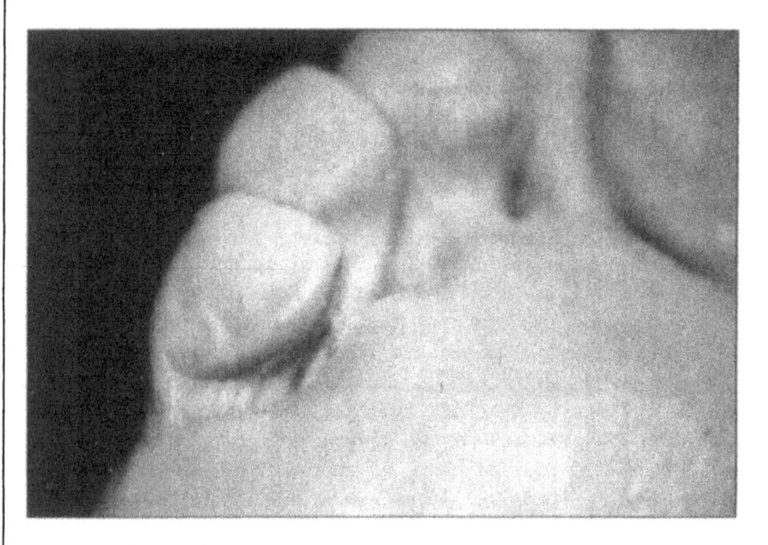

Fig. 6.2. Interdigital erythrasma.

or chains of fine, tortuous 4-7 μm filaments, and coccoidal elements of 1-3 μm. Microscopic examination is better if scales are removed with Scotch tape and stained with methylene blue or Gram or Giemsa stains. Organisms can be visualized better in phase contrast or immersion microscopy. Culture is difficult. From foot lesions, Staphylococcus, Pseudomonas, and Proteus can be isolated. Biopsy is not necessary. Parakeratotic hyperkeratosis is found. Gram, PAS, or Gomori stains demonstrate filamentous, bacillary or coccoidal bacteria. Acanthosis and spongiosis are observed. There is edema, vasodilatation, and lymphocytic infiltration of the dermis.

TREATMENT

Good results are obtained with erythromycin or tetracycline, 1-2 g/day orally for one week minimum or a single-dose of clarithromycin (1 g). Erythromycin is so effective it is used as a therapeutic test. Topical applications are also useful. Sodium hyposulfite 20%, sulfur or keratolytic ointments 3%, Whitfield ointment, or the ointments with imidazolic derivatives, cyclopyroxolamine, fusidic acid, and antibacterial soaps enhance healing. Interdigital forms in the feet are more difficult to treat.

Selected Readings

1 Mattox TF, Rutgers J, Yoshimori RN, et at. Nonfluorescent erythrasma of the vulva. Obst Gynecol 1993; 81(5Pt2):862-64.
2 Steiner D, Cuce LC, Salebian A. Eritrasma interpododigital. An Bras Dermatol 1994; 69(1): 16-20.
3 Harton JR, Wilson PL, Kincannon JM. Erythrasma Treated with Single-Dose Clarithromycin. Arch Dermatol 1998; 134: 671-72.

Pitted Keratolysis

Roberto Arenas

Pitted keratolysis is also known as tropical plantar keratolysis. It is a superficial infection that affects the horny layer. It is localized on the soles, and it is characterized by punctiform depressions and asymptomatic superficial erosions caused by organisms from the genders Corynebacterium, Dermatophilus and Micrococcus. Humidity and macerated skin favor its growth.

GEOGRAPHIC DATA

Distribution is worldwide. It affects equally all races and both sexes. It predominates in the tropics and is more frequent in those who walk barefoot, have their feet exposed to water, or suffer from hyperhidrosis. The incidence is also high in those who use boots or sports shoes such as soldiers and athletes.

ETIOLOGY

Infection is caused by actinomycetes as well as other organisms. It is possible that the casual agent has not been determined and that the proposed microorganisms like *Corynebactirium* sp, *Dermatophilus congolensis* and *Micrococcus sedentarius* act synergistically (*Arch Dermatol* 1972; 105:580-584). Friction, occlusion and maceration favor infection of the horny layer. Pigment changes and the fetid odor are related to the microorganisms involved and depend upon the mixture of thiols, sulfurs, and thioesters. It has been suggested that *M. sedentarius* generates the erosions by the production of two proteinases in the presence of hyperhydration and alkaline pH (*J Appl Bacteriol* 1992; 72(5): 429-34). *D. congolensis* secretes keratinases that can digest the keratin (*Med Microbiol Immunol Berl* 1991; 180(1):45-51).

CLINICAL PICTURE

It is seen on the soles of the feet. It predominates in weight-bearing areas, and it can be unilateral or bilateral (97%). It is rarely seen on the palms. The dermatosis is composed either of punctiform depressions or superficial erosions of 1-3 mm that form rows that coalesce into circular lesions or irregulars lesions of geographic appearance. Their color is gray, greenish, or brownish, and they appear filthy (Fig. 7.1). Rarely there are erythematoses, and painful lesions or tinea pedis. There is a rare hyperkeratotic form (keratoma plantare sulcatum). Its existence is moot. It appears as circinated zones of hyperkeratosis on the plantar arc of the feet, and it can be observed as small depressions. It can be associated with other forms of palmoplantar keratoderma.

LABORATORY DATA

Curetted lesions can be Gram stained. Biopsy by shaving is recommended. A crater-like lesion is observed. It is 0.5-3 mm with well-defined walls that are vertically cut. There are septated filaments of 0.5-1.5 µm and gram positive diphtheroides are seen. They are basophilic with hematoxylin and eosin stain, but are better observed with Gomori-Grocott and PAS (Fig. 7.2). They are not acid-fast. On the surface one finds an opaque material that can be dirt.

TREATMENT

The elimination of predisposing factors is important. Drying powders or powder to control hyperhidrosis can be used. Aluminum chloride 10-20% is useful, as well as Whitfield's ointment (vaseline with benzoic acid 6% and salicylic acid 3%), clioquinol 3%, benzoyl peroxide 5%, or topical antibiotics like clindamycin,

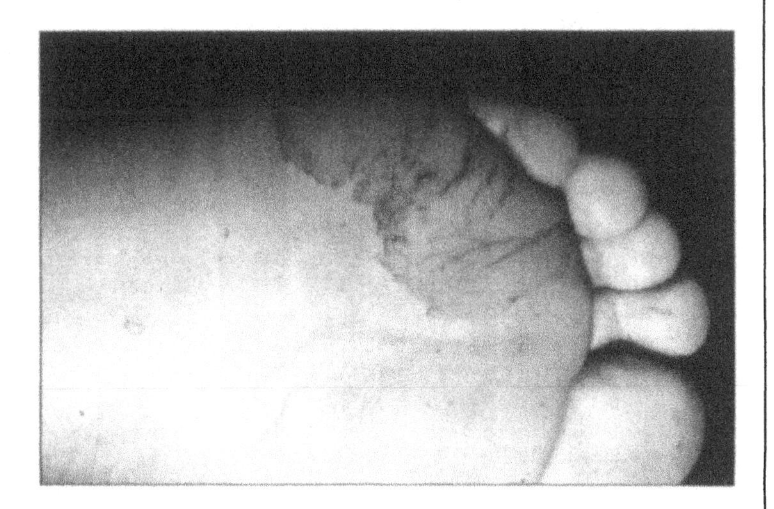

Fig. 7.1. Pitted keratolysis.

Fig. 7.2. Biopsy by shaving, coccoidal forms and filaments (PAS 40X).

erythromycin, mupirocin, or fusidic acid can also be used. Formaldehyde 1% in aqueous solution is very efficient, and good results are observed in one or two weeks.

Selected Readings

1. Arenas R. Jimenez R, Diaz A et al. Queratolisis punteado. Estudio clinico-epidemiologico, histopatologico, y microbiologico en 100 pacientes (Pitted Keratolysis. The clinical-epidemiologic, histopathologic, and microbiologic study in 100 patients) Dermatologya Rev Mex 1992; 36(3):152-58).
2. Shah AS, Kamino H, Prose NS. Painful, plaque-like, pitted keratolysis occurring in childhood. Pediatr Dermatol 1992; 9(3): 251-54).

Trichomycosis

Roberto Arenas

Trichomycosis nodosa is a pseudomycosis caused by *Corynebacterium tenuis* which affects the axillary hair (trichomycosis axilaris) and rarely the genital hair (trichomycosis pubis). It is characterized by a tender, whitish husk which covers the hair. It is chronic and asymptomatic, produces a fetid odor, and is associated with perspiration and poor hygiene.

GEOGRAPHIC DISTRIBUTION

It is widespread although it predominates in hot climates and is more frequent in young men. It is somewhat contagious. It is observed with relative frequency in dermatologic practice although there are only about 20 reports in the medical literature.

ETIOLOGY

C. tenuis, or other Corynebacteria penetrate an erosion of the hair and form a colony that develops distally along the shaft and produces a mucoid lipid material. Some authors believe that dessication of the apocrine sweat glands produces an insoluble cement-like substance and secondary colonization by Corynebacterium.

CLINICAL PICTURE

It affects axillary hair and rarely the genital hair. A sheath, a soft, white and irregular sleeve that does not destroy the follicle, forms around the hair. Sometimes it is nodular (Fig. 8.1). Depending on the coloration, the following varieties can occur: flava, rubra and nigra. There are no symptoms, but the odor (bromhidrosis) and color of sweat (chromhidrosis) can change. The course is chronic and frequently relapsing.

Tropical Dermatology, edited by Roberto Arenas and Roberto Estrada. ©2001 Landes Bioscience.

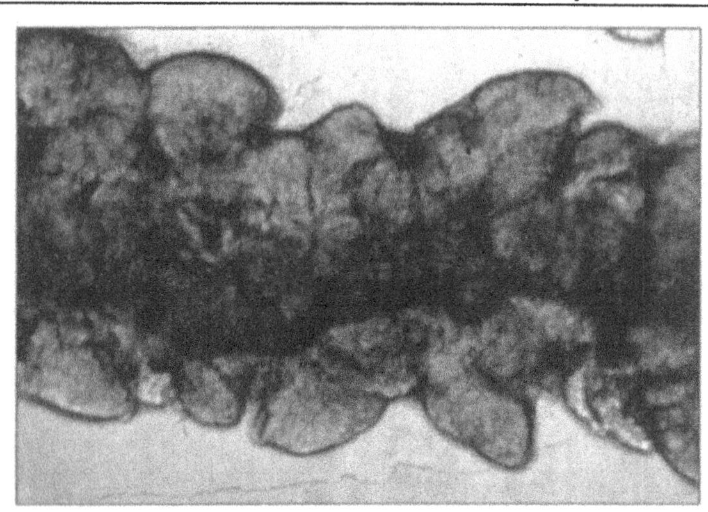

Fig. 8.1. Microscopy of axillaris trichomycosis (Lugol 20X).

LABORATORY DATA

On direct microscopic examination, a granular sheath is seen. This sheath is composed of cocci and gram positive bacterial filaments in a fan-like array or as a homogeneous, mucilaginous dough that does not penetrate the hair (Fig. 8.1). Culture is difficult; sheep's blood medium is required at 37°C.

TREATMENT

Shaving and appropriate hygiene, preferably with antiseptic or antibacterial soaps, is done. The following lotions are used: formaldehyde in alcohol 1%-2%, salicylic acid 3% lotion, iodine tincture 1%, shampoo or cream with selenium disulfide 2.5%, clindamycin and erythromycin topical solutions, and naftifine cream (*Int J Dermatol* 1991; 30(9): 557-69). All of these treatments must be applied twice a day.

SELECTED READINGS

1 Levit F. Trichomycosis axillaris: A different view. J Am Acad Dermatol 1988; 18:788-89.

2 Rosen T, Krawczynska AM, McBride ME, Ellner K. Naftifine treatment of trichomycosis. Pubic Int J Dermatol 1991; 30(9):557-69.

Protothecosis

Victor Fernando Munoz-Estrada and Jorge A. Mayorga-Rodriguez

Protothecosis is a primary or opportunistic chronic infection caused by achlorophyllic algae of the genus Prototheca. These algae enter through minor trauma or surgical incisions. Protothecosis is a uncommon disease with clinical manifestations that vary and with systemic, cutaneous, subcutaneous and bursal involvement. It is diagnosed without difficulty by histopathology and culture. There is no specific treatment. However, triazolic derivatives combined with surgery yield good results.

GEOGRAPHIC DISTRIBUTION

Protothecosis occurs worldwide. It affects all ages and races, and either sex. About 100 cases have been reported in 16 countries. In 1992 Iocovello et al published a review of 60 cases. In the United States there were more than 40 cases. Protothecosis also affects more than 10 species of domestic and wild animals; it manifests as a mastitis (Lerch, 1952) or generalized illness.

ETIOLOGY

The genus Prototheca was described by Kruger in 1894. It includes four algae species (West, 1916). The algae pigment-free unicellular, aerobic or microaerophylic, heterotrophic, and without chlorophyll: *Prototheca wickerhamii* (Tubaki and Soneda, 1959), *P. zopfii* (Kruger, 1894), *P. moriformis* and *P. stagnora* (Cooke, 1968). Only *P. wickerhamii* and *P. zopfii* are human pathogens. Animals are the main source of human infection (Davis, 1964). In 1968 Klintwoth described the first opportunistic infection. Cancer, AIDS, treatment with glucocorticoids and immunosuppressants drugs favor infection.

These microorganisms are found as soil saprophytes, on vegetables and decaying matter and in water. They have also been isolated from skin, feces and sputum, and from animals without symptomatic disease. They are spherical, ovoid, or elliptical; they measure 8-30 µm and have a thick wall. They multiply by binary division and endospores (autospores) of 4-11 µm that form a sporangium containing 2-20 daughter cells (Figs. 9.1 and 9.2).

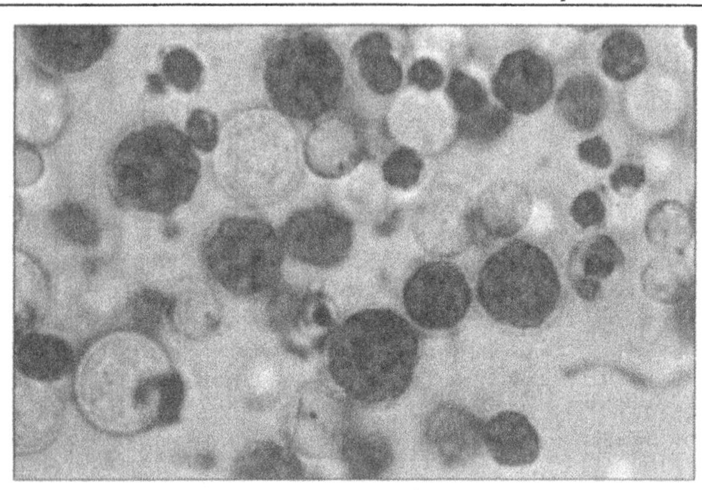

Fig. 9.1. Prothoteca sporangium (PAS, 20X).

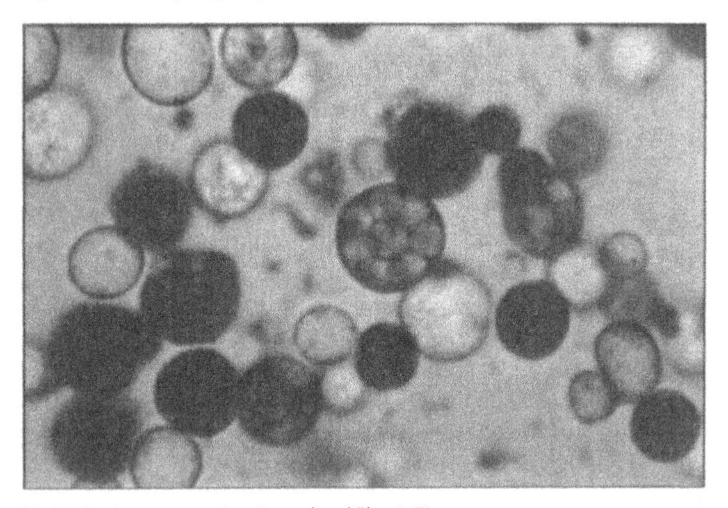

Fig. 9.2. Prothoteca sporangium (Lactophenol Blue, 20X).

CLINICAL PICTURE

Clinical manifestations occur in three varieties: cutaneous and subcutaneous infection, bursitis, and generalized illness. The cutaneous and subcutaneous form are the most frequent. They account for about 50% of cases (*Clin Infect Dis* 1992; 15:959-67; *J Am Acad Dermatol* 1995;32:758-64). Generally infection is limited to exposed skin. It develops slowly as papules, nodules (Fig. 9.3), erythematous, eczematous, verrucous plaques, herpetiform vesicles, and ulcers. Lesions can be

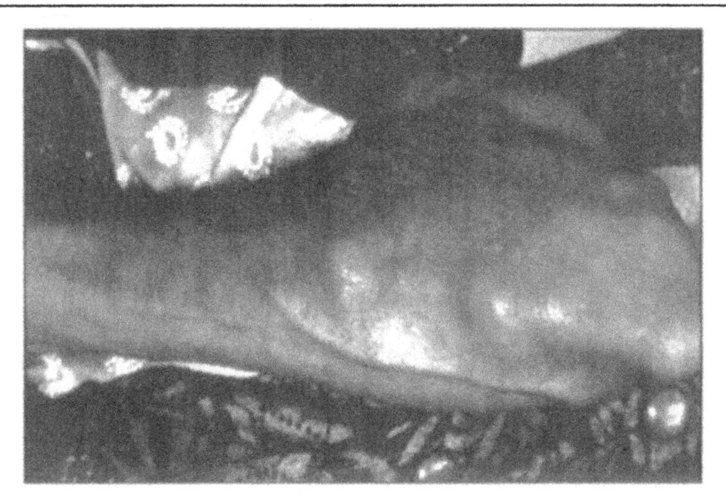

Fig. 9.3. Protothecosis with nodular lesions.

asymptomatic, pruritic, or painful. The articular form is next most common. It results from trauma or surgery, usually affecting the retroolecranon as a bursitis with soft tissue inflammation and moderate to intense pain. The generalized or systemic form is the least common. It is seen as subcutaneous lesions or lesions affecting internal organs. It is associated with diabetes, renal transplantation, systemic lupus, cancer, AIDS, peritoneal dialysis, chemotherapy and radiotherapy. Prototheca has been isolated from blood, peritoneal abscess, kidney, synovial tissue, Hickman catheter, spinal fluid, and nasopharyngeal tissue.

LABORATORY DATA

A skin biopsy of protothecosis shows hyperkeratotic, parakeratotic, and acanthotic epidermis with a mild dermal cellular infiltrate with lymphocytes, neutrophils, histiocytes, and giant cells. The parasite is spherical, 8-26 μm diameter with a thick wall that contains small spores in its interior. They are identified with Gomori and PAS stains (Fig. 9.1).

Growth on Sabouraud glucose agar is rapid. In three days white or beige, opaque, yeast-like colonies are obtained (Fig. 9.4). The optimum temperature for growth is 25-35°C. The theca are 8-26 μm with internal septation; autospores of 9-11 μm (Fig. 9.2) are observed. Prototheca does not grow in the presence of cycloheximide (Actidione). On microscopy, Prototheca cells can be confused with sclerotic cells and *Histoplasma duboisii* yeasts.

Protetheca species can be identified by size, by number of endospores, and by assimilation tests. P. *zopfii* assimilate dextrose, galactose, levulose, and ethanol, but not trehalose. *P. weckerhamii* assimilate trehalose, and its sporangia are smaller.

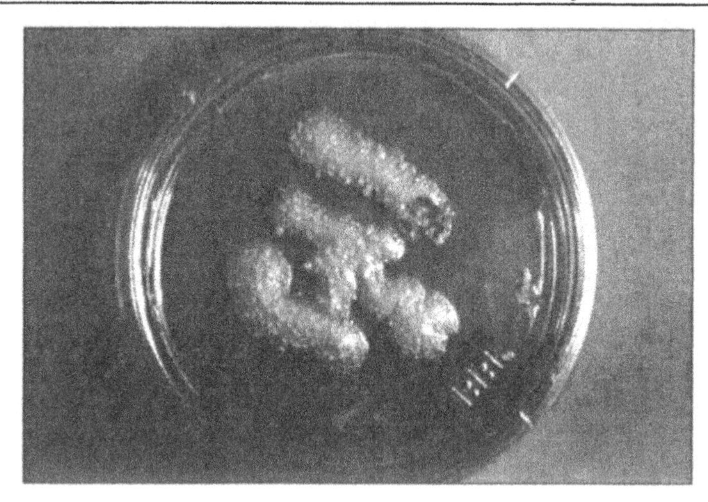

Fig. 9.4. Prototheca culture.

TREATMENT

There is no specific treatment. Several drugs have been tried with a poor response. Among them are: pentamidine, griseofulvin, nystatin, amphotericin B, potassium iodine, 5-fluocytosine, and tetracyclines. Recently, antimycotics such as ketoconazole and itraconazole have been used with better results, especially when combined with surgical excision when this is feasible.

Selected Readings

1. Arenas R. Micologia Medica Illustrada (Illustrated Medical Mycology) Mexico. Interamericana/Mc Graw-Hill. 1993: 328-330.
2. Boyd AS, Langley M, King LE Cutaneous manifestations of prototheca infections. J Am Acad Dermatol 1995; 32: 758-64.
3. Iocovello VR, De Girolami PC, Lucarini J et al. Protothecosis complicating. Prolonged endotracheal intubation: Case report and literature review. Clin Infect Dis 1992; 15:959-967.
4. Rippon JW. Medical Mycology. The pathogenic fungi and pathogenic actinomycetes. Philadelphia: Saunders 1988: 651-680.
5. Woolrich A. Koestenblatt E, Don P et al. Cutaneous protothecasis and AIDS. J Am Acad Dermatol 1994; 31:920-4.

Actinomycosis

Alexandro Bonifaz

Actinomycosis is a chronic pseudomycosis characterized by granulomatous lesions with purulent draining sinuses. It is most commonly caused by *Actinomyces israelii*. It is frequently seen in three areas: cervicofacial, thoracopulmonary, and abdominal.

GEOGRAPHIC DISTRIBUTION

Actinomycosis occurs worldwide but is more common in temperate climates. In underdeveloped countries, it is associated with poor hygiene, especially of the teeth.

ETIOLOGY

It is produced by a filamentous, anaerobic or microaerophilic actinomycetes of the oral mucosa, dental cavities, tonsil crypts, and vagina. Most cases are due to species of the genus Actinomyces (80%, *A. israellii*). Other species, *A. naeslundi*, *A. bovis*, *A. viscosus*, and *A. odontolyticus*, are etiological agents less frequently found. Rarely there are reports where *Arachnia propionica* and *Rothia dentocariosa* have been isolated. Infection is generally opportunistic; it is associated with diabetes or immunosuppression. In the cervicofacial region, infection usually arises from dental caries, dental extraction, oral surgery, or trauma. Pulmonary infection arises from aspiration of the actinomycetes from the oral cavity or from tonsil crypts. In the intestines, infection is due to swallowing microorganisms; usually it happens after abdominal surgery, especially for appendectomy. Cervicovaginal cases have recently increased, probably related to the more common use of intrauterine devices (IUD) which increases the anaerobic flora. In an analysis of 1,520 Pap smears, 11.4% grew out actinomycetes (*J Reprod Med* 1994; 39:385-87).

CLINICAL PICTURE

Actinomycosis occurs at all ages; however it is rare in childhood and in the elderly. Most cases are seen between 30-40 years of age, with a slight increased incidence in females. The incubation period is not known; however, most report

Tropical Dermatology, edited by Roberto Arenas and Roberto Estrada. ©2001 Landes Bioscience.

it is 1-4 weeks. Actinomycosis may be localized or disseminated.

Cervicofacial actinomycosis involves the maxillary—usually lower—region and rarely affects the oral mucosa. It is characterized by an increase in volume, tissue deformity, and purulent draining tracts. This purulent material contains parasitic forms called "sulphur granules". Patients experience intense pain, prurituis, and sometimes trismus (Fig. 10.1.1). In chronic cases, periostitis and osteolysis can occur.

In the pulmonary form of actinomycosis, fever, cough, and expectoration are always present. It may present with skin involvement. In abdominal and pelvic forms, it is usually seen at the ileocecal level, but the pelvis and the perianal region are also affected. The clinical picture is most commonly mistaken for appendicitis. In most chronic cases the skin is involved with abscesses and purulent draining sinuses (Fig. 10.1.2). The most important differential diagnoses of the cervicofacial form are tuberculosis, lymphomas and odontogenic sinuses tracts, while in the thoracic form, tuberculosis and mycetoma and in the abdominal form, appendicitis and neoplasms.

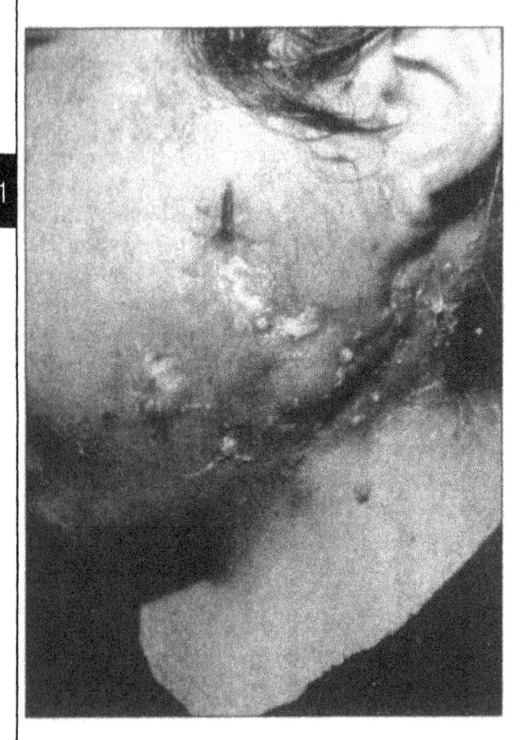

Fig. 10.1.1. Cervicofacial actinomycosis (Courtesy of Amado Saul).

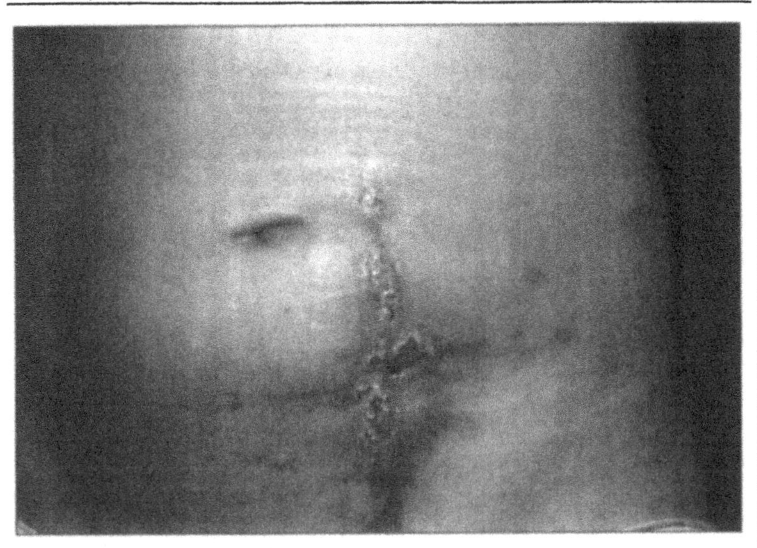

Fig. 10.1.2. Actinomycosis of the abdomen.

LABORATORY DATA

It is easy to establish a diagnosis with direct exam of the pus when the "sulphur granules" are visualized. A mass of yellow-white micromycelia is multilobulated, soft, 50-3,000 µm in diameter. The biopsy should be stained with H&E, Gram, PAS, and Gomori-Grocott as used in all chronic, purulent, and granulomatous processes. In order to distinguish the granules from Nocardia sp, Ziehl-Nielsen stain must be done (as Nocardia is partially acid-fast and Actinomyces granules are negative). Culture confirms the species and must be anaerobic in Brewer's agar or thioglycolate liquid.

TREATMENT

The drug of choice is penicillin, 50-100 million IU. An initial schedule is procaine penicillin, 800,000 IU/day up to 40-50 million IU.followed by benzathine penicillin 1,200,00 IU/week to a total of 100-120 millions of IU, can be administrated. Prolonged treatment with penicillin or other antibiotics is necessary to avoid recurrence. If penicillin is contraindicated, sulfamethoxazole-trimethoprim, tetracyclines, cloramphenicol, clindamycin, minocycline, amoxicillin plus clavulanic acid or cephalosporines can be used.

Selected Readings

1. Bennhoff, DF Actinomycosis: Diagnostic and therapeutics considerations and a review of 32 cases. Laryngoscope 1984; 94:1198-1217.
2. Brown, JR Human Actinomycosis. A study of 181 subjects. Human. Pathol 1973; 4:319-330.
3. Foster, SV., Demmier, GJ, Hawkins, EP., Tilman, JP. Pediatric cervicofacial actinomycosis. South Med J 1993; 86:1147-1150.
4. Warren, NG Actinomycosis, nocardiosis, and actinomycetoma. Dermatol Clin 1996; 14:85-95.

Botryomycosis

Alexandro Bonifaz and Clemente Moreno-Collado

Botryomycosis is a pseudomycosis caused by several nonfilamentous bacteria. It causes chronic granulomatous lesions. Most cases are caused by *Staphylococcus aureus* and *Pseudomonas aeruginosa.* There are two clinical types, cutaneous and visceral.

GEOGRAPHIC DISTRIBUTION

Distribution is worldwide although most cases have been reported in the United States, England, and France.

ETIOLOGY

Staphylococcus aureus (40%) and *Pseudomonas aeruginosa* (20%) (*Int J Dermatol* 1996; 35: 381-8) are the most common infecting agents. Isolated strains have low virulence. Other causative agents reported are: coagulase-negative Staphylococci, *Micrococcus pyogenes*, Streptococcus spp, *Escherichia coli*, and *Proteus* spp. The pathogenesis is not yet very well known. Most authors consider that the responsible bacteria are generally not particularly virulent. In the cutaneous cases, bacteria penetrates as a result of trauma; meanwhile in most of the visceral cases the infectious agents are caused by normal flora like Pseudomonas sp (*Arch Dermatol* 1976; 112:1568-1570), and infection usually occurs after surgery or as a nosocomial infection. Most patients with botryomycosis have cellular immune defects, particularly with regard to the total number of T lymphocytes.

CLINICAL PICTURE

Cutaneous botryomycosis is a chronic, suppurative granulomatous process. It begins after trauma. It predominates in the hands and feet, but it has also been reported on the neck, head, cheeks, trunk, and buttocks. The lesions are similar to those of mycetoma and actinomycosis. Most cases present as nodules, sinuses, abscesses, and/or ulcers. In some cases they present as tumors, cysts, or as verrucous lesions (Fig. 10.2.1). By invasion they affect muscle and bone causing osteomyelitis. Lesions are usually localized, but sometimes they can spread from a

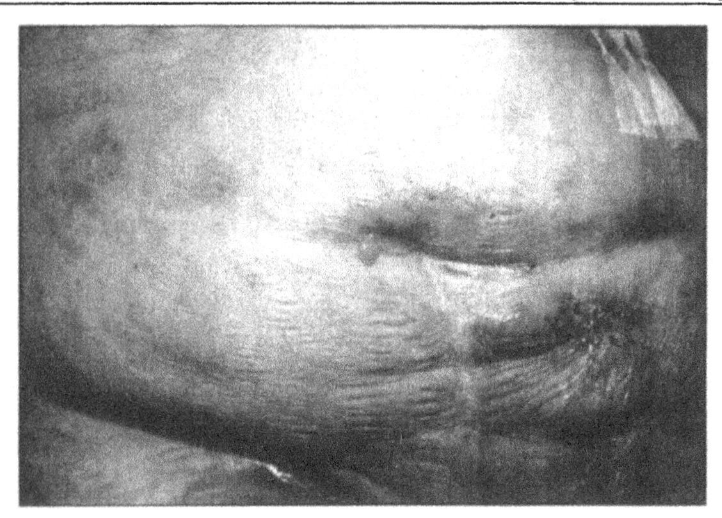

Fig. 10.2.1. Abdominal botryomycosis.

cutaneous focus. Visceral botryomycosis occurs in patients who have undergone recent surgery and who are immunosuppressed. Most cases have been reported in lung, kidney, liver, brain, and the gastrointestinal tract. The differential diagnosis for cutaneous lesions includes mycetoma, actinomycosis, epidermal cyst, tuberculosis, osteomyelitis, and for visceral cases, carcinomas and actinomycosis.

LABORATORY DATA

Round, soft,1-3 mm white-yellow clumps are observed on direct exam with KOH or Lugol's solution. It is important to distinguish the granules from those of *Actinomadurae madure* and *Actinomyces israellii* that are formed by a mass of microfilaments. Culture as well as biochemical tests confirm the species. Anaerobic media must also be employed. Biopsy is important. H&E is proper stain, but PAS and Gomori-Grocott yield better results. Histology is consistent with a chronic, nonspecific, inflammatory process.

TREATMENT

Treatment depends on the causative agent; it is important to determine effective antibacterial agents in vitro. For most reported cases, treatment given for several weeks yields good results with sulfamethoxazole-trimethoprim, minocycline, erythromycin, and cephalosporins. In some cases, combinations of antibiotics are recommended. Occasionally surgical excision is indicated, especially with cystic and tumoral lesions. Carbon dioxide laser treatment has been reported to be effective.

Selected Readings

1 Bonifaz A, Carrasco-Gerard E. Botryomycosis. Int J Dermatol 1996; 35:381-388.
2 Hacker P. Botryomycosis. Int J Dermatol 1983; 22:455-458.
3 Mehregan DA, Su WP, Anhalt JP. Cutaneus botryomycosis. J Am Acad Dermatol 1991; 24:393-396.
4 Moreno-Collado C. Botriomicosis. Reporte de siete casos y revision de la literatura (Botryomycosis: The report of seven cases and revision of the literature) Dermatologia Rev Mex 1995; 39(3):129-136.

C. Subcutaneous Mycosis

Mycetoma (Madura Foot)

Sporotrichosis

Chromoblastomycosis

Lobomycosis (Jorge Lobo's Disease)

Entomophtoromycosis

Rhinosporidiosis

Mycetoma (Madura Foot)

Roberto Arenas and Pedro Lavalle

Mycetoma is an chronic inflammatory process which affects subcutaneous tissue, but sometimes also bones, and rarely viscerae. The most frequent site of involvement is the foot, with tumefaction, deformities, and draining sinus tracts which discharge a seropurulent exudate with parasitic granules. It begins with traumatic abrasion of the skin and the inoculation of a fungus or an actinomycete, and is known as eumycetoma or actinomycetoma, respectively. The highest incidence is in the rural population.

GEOGRAPHIC DISTRIBUTION

Distribution is worldwide, but it is more frequently seen in the tropics. It is also common in subtropical climate. It predominates in India, Sudan, Venezuela and Mexico. In Mexico and Guatemala, Actinomycetes, specifically Nocardia, are the most common pathogens. It also occurs 1500-2000 m above sea level and in very dry zones (*A. madurae*). *S. somaliensis* is the infectious agent in desert areas. It predominates in Somalia and the Sudan where rainfall is only 50-250 cm annually. *A. pelletieri* is frequently the pathogen in West Africa. Eumycetomas occur worldwide. The most common are caused by *Madurella mycetomatis* and *M. grisea* (black granules), and by *Acremonium* sp, *Fusarium* sp, and *P. boydii* (white granules). They represent 2.5% of cases in Mexico.

ETIOLOGY

Actinomycetoma is caused by aerobic actinomycetes: small, white granules caused by *N. brasiliensis*, *N. asteroides*, and *N. caviae*; large; yellow granules caused by Nocardia and *Streptomyces somaliensis*; and red granules caused by *Actinomadura pelletieri*. Eumycetomas caused by the following organisms discharge black granules: *M. mycetomatis*, *M. grisea*, *Pyrenochaeta romeroi*, *Leptosphaeria senegalensis*, *L. tompkinsii*, *Exophiala jeanselmei*, *Phialophora verrucosa*, and *Curvularia lunata*. White granules are discharged by *Pseudoallescheria boydii* (*Scedosporium apiospermum*), *Acremonium falciforme*, *A. recifei*, *Neotestudina rosatii*, *Fusarium* sp. *Aspergillus nidulans*, and *Trichophyton* sp. The causal agents live as saprophytes in the soil. They are introduced into the skin through a minor injury, usually by a thorn, a splinter, rocks, tools, or animal bites. After incubation, the

Tropical Dermatology, edited by Roberto Arenas and Roberto Estrada. ©2001 Landes Bioscience.

microorganisms produce filaments in the tissue. They agglomerate into compact colonies that discharge from draining sinus tracts as "granules". In the tissues the actinomycetic granules are surrounded by suppurative reaction characterized by the presence of polymorphonuclear cells, fibrosis, and neovascularization. The response to eumycetomas is a granulomatous reaction. Infection spreads by contiguity. Small mycetomas (minimycetoma) indicate decreased virulence of the infecting strain and/or a strong host immune response. Mycetomas have been observed in individuals receiving glucocorticoids and other immunosuppressants.

CLINICAL PICTURE

Incubation ranges from weeks to months and even years. Infection predominates in men. More than 60% affects rural men who walk barefoot or with sandals. In most cases, the age ranges from 16-40 years. Prior to 15 years of age, the incidence in both sexes is equal. The onset is chronic and progressive, sometimes lasting 2-3 years, but varies from three months to several decades. Mycetoma increases in size during pregnancy. The lower extremities are involved in 64% of the cases. It predominates on the foot, but it can also be observed in the thigh, knee and lower leg, (Fig. 11.1). Hand, forearm, arm, and shoulder are involved in 14% of cases (Fig. 11.2), and the back in 17-25% of cases. The abdominal wall and chest can also be affected, but it is rare in the head or face.

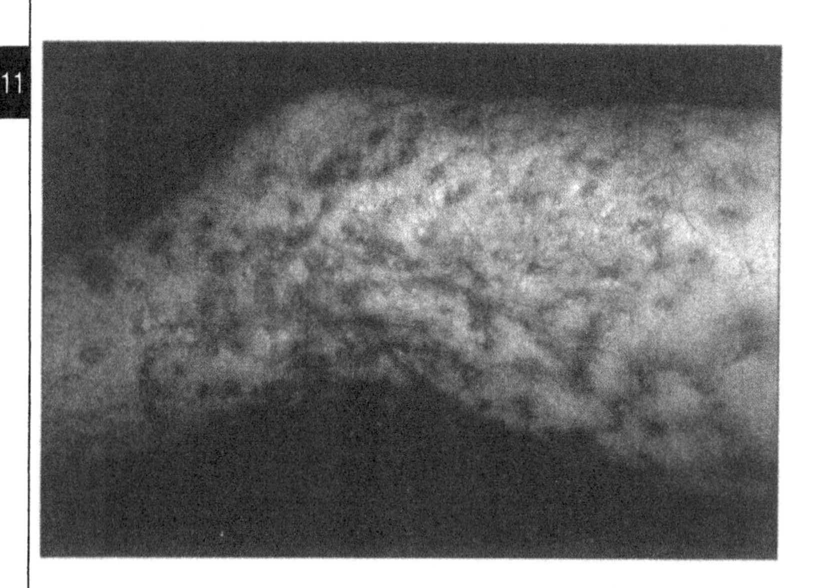

Fig. 11.1. Mycetoma in thighs and leg.

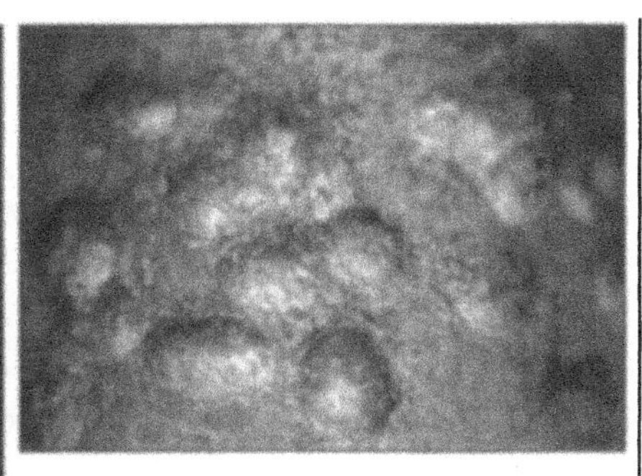

Fig. 11.2, left. Mycetoma causesd by *N. brasiliensis* in the elbow. Fig. 11.3, above. Fistulous lesions with nodular appearance.

Mycetoma is characterized by tumefaction and abundant draining sinus tracts (Fig. 11.3). Sometimes granulation tissue appears nodular (Fig. 11.4). Also there can be ulceration with meliceric crusting and scarring. The course is slow and inexorable without spontaneous regression, except in women after pregnancy. Symptoms are not important. Subcutaneous tissue, muscle, and bone may also be involved. Small bones of the foot and vertebrae may be destroyed. Large bones such as the tibia and femur seem resistant to severe involvement (Fig. 11.5). Mycetomas affecting the vertebrae can involve the spinal cord causing paraplegia. Mycetomas of the thorax can invade pleura and lung. There can be functional incapacity.

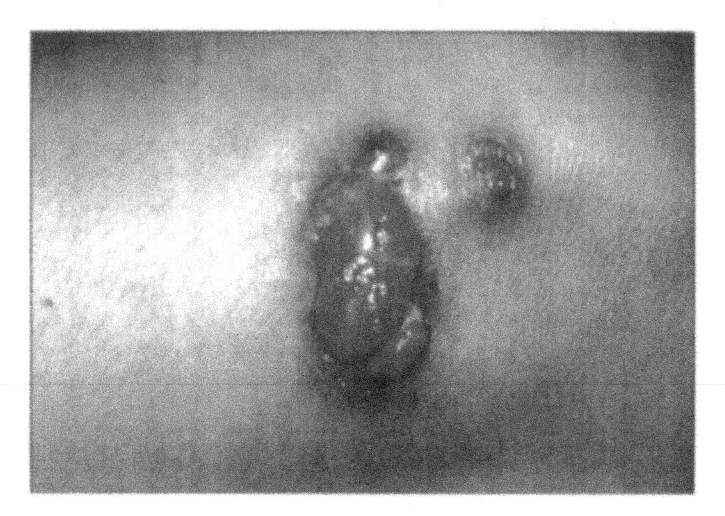

Fig. 11.4. Exophytic lesions caused by Nocardia.

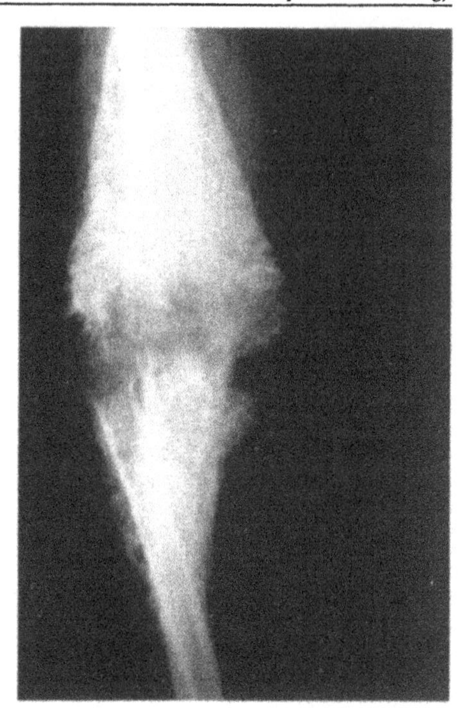

Fig. 11.5. Radiography of osseous lesions.

In the actinomycetomas, often there is a secondary bacterial infection which causes pain and signs and symptoms of general illness, such as weight loss, anemia, and fever. Visceral amyloidosis has been reported. In severe cases it may be lethal.

Mycetomas are described as atypical according to their location, number, size, and other aspects. For example, a single mycetoma on the face, without lesions elsewhere on the body, is exceptional. Lesions on the inguinal region may represent a "metastatic lesion" from a mycetoma of the foot. Mycetoma without fistulae, intraosseus mycetoma, or simply periosteal forms have been reported. Single or multiple, small lesions, so-called minimycetomas, can also occur. These are characterized by the absence of tumefaction and only a few fistulae, and have been reported in children, youngsters, or adults (Fig. 11.6). The infectious agent is *N. brasiliensis*. Deep involvement is uncommon and they usually respond to dapsone. Eumycetomas are usually well-circumscribed whereas actinomycetomas from *N. brasiliensis* and *A. pelletieri* are very inflammatory, have abundant fistulae, and generally involve bones. Those caused by *A. madurae* and *S. somaliensis* cause less inflammation and have smaller sinus tracts (Fig. 11.7).

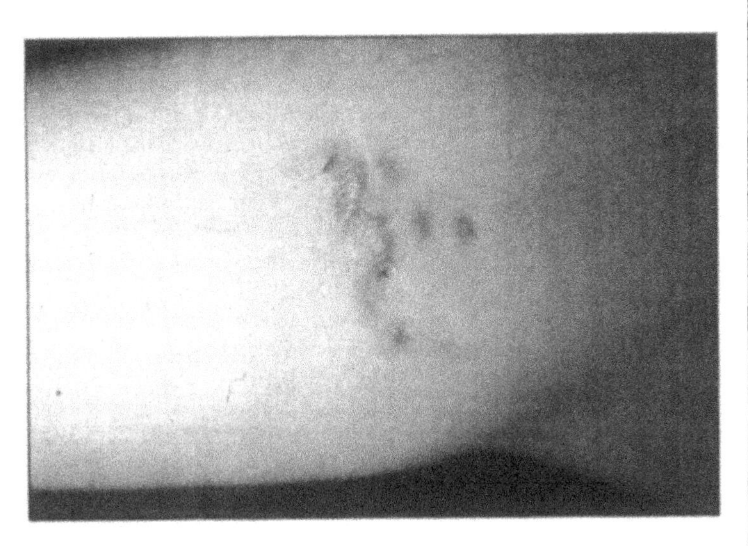

Fig. 11.6. Minimycetoma due to *N. brasiliensis*.

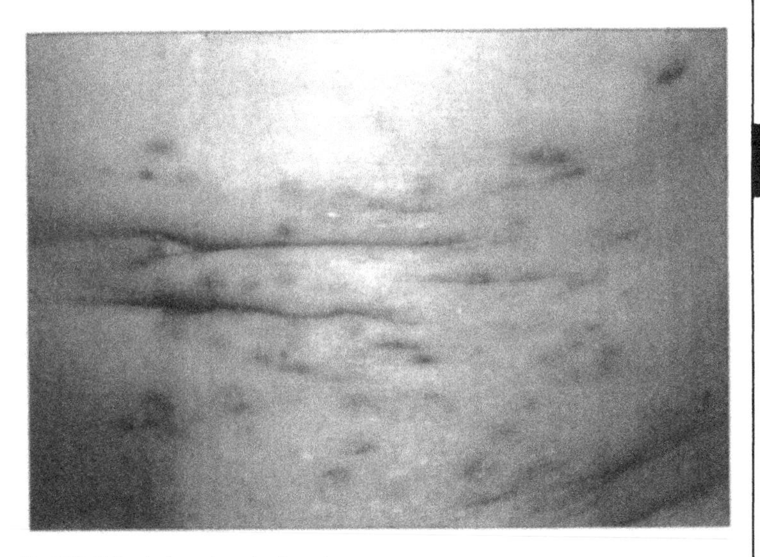

Fig. 11.7. Abdominal mycetoma by *A. madurae*.

LABORATORY DATA

Skin biopsy material stained with hematoxylin and eosin is nonspecific. In the presence of soft granules, there are polymorphonuclear cells, fibrosis and vascular dilatation. With hard granules, a true tuberculoid granuloma may be seen. The characteristics of the granule and the affinity for certain stains are important for diagnosis. Fine needle aspiration cytology has been reported as a good diagnostic tool (*Acta Cytol* 1996; 40(3): 461-464). On direct examination the fungus. granules can be seen easily because their big size. Others can be stained and observed microscopically (Fig. 11.1). In general, Nocardia, *A. pelletieri*, *A. madurae* and *S. somaliensis* (Figs. 11.8 to 11.11) can be diagnosed by direct exam and biopsy (Table 11.1). Causal agents can be cultured on Sabouraud glucose agar at room temperature. Nocardia produces white-yellow colonies that looks like "popcorn". Its identity can be confirmed by casein hydrolysis or xanthine, hypoxanthine and tyrosine tests (*N. brasiliensis* hydrolyzes casein whereas *N. asteroides* does not). *A. madurae* forms beige or pink colonies that grow well on Lowenstein-Jenssen media. Colonies of *A. pelletieri* are red. The true fungi, *M. mycetomatis* and *M. grisea*, grow best at 33°C and 25°C, respectively (Fig. 11.12). Radiographic studies demonstrate soft tissue and osseous involvement (Fig. 11.5). Cavities produced in the bone are called "geodes". Nocardia is very osteophilic; *A. pelletieri* causes microgeodes, and *M. mycetomati* and *A. madurae* cause macrogeodes.

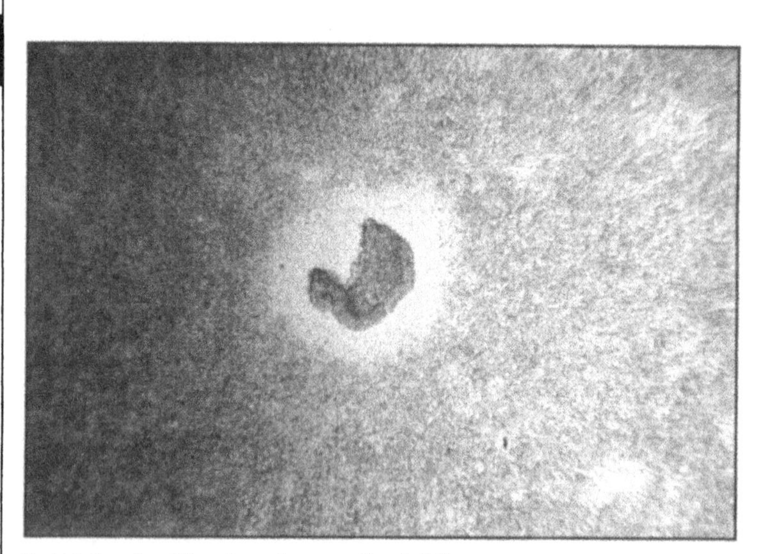

Fig. 11.8. Granules of Nocardia on direct exam (Lugol, 40X).

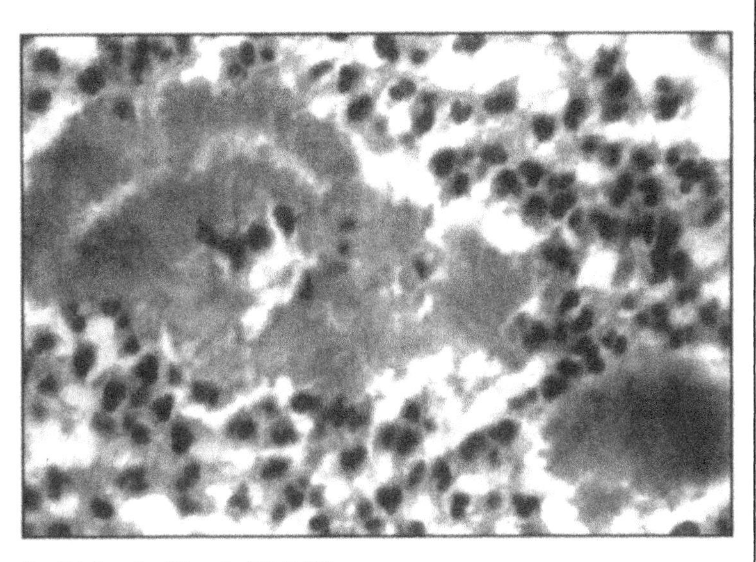

Fig. 11.9. Granule of Nocardia (H & E 40X).

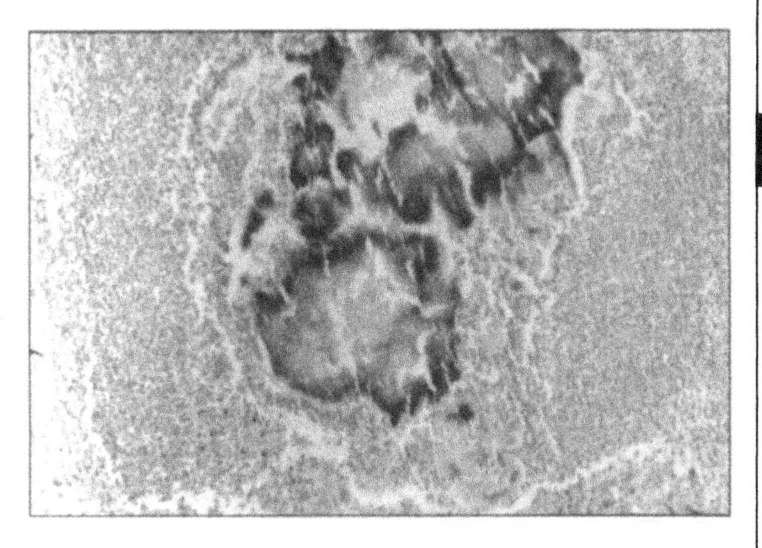

Fig. 11.10. Granule of *A. madurae* (H & E 10X).

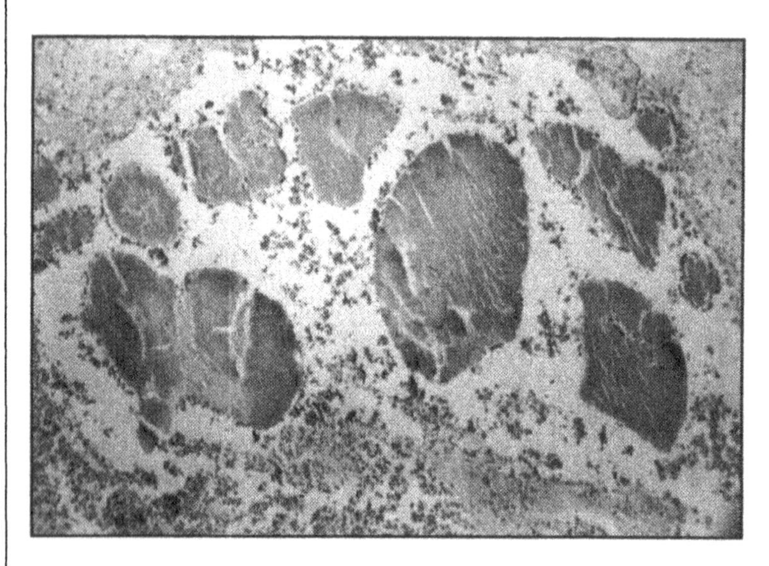

Fig. 11.11. Granules of *S. somaliensis* (H&E 10X).

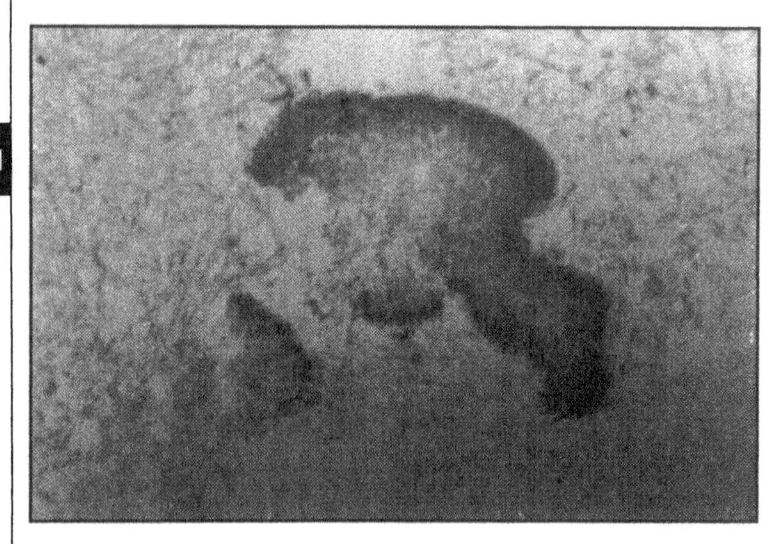

Fig. 11.12. Granule of Madurella sp (H&E 10X).

Table 11.1. Characteristics of mycetoma granules (Arenas, R. Dermotologia. Atlas, diagnostico y tratamiento. Mexico: Interamericana/McGraw-Hill. 1996: 356).

Causal Agent	Consistency	Size	Color	Clubs	Aspect	Hematoxylin and Eosin Stain
Nocardia	Soft	50 to 200 µm	White-yellowish	+	Renal or vermiform	Amphophilic
A. madurae	Soft	1 to 3 mm	White-yellowish	Fringe	Carographic	Purple
A. pelletieri	Firm	200 to 500 um	Red	-	Broken dish	Red
S. somaliensis	Hard	1.5 to 10 mm	White-yellowish	-	Potato slice	Pale, striae
Madurella	Firm	1 mm	Black	-	Compact or vesicular	Black
Fusarium/ Acremonium	Firme	500 µm	White	-	Oval	Eosinophylic

TREATMENT

The treatment of eumycetomas is surgical. Complete excision eliminates the process, and there is no risk of recurrence. Dimethylsulfoxide with amphotericin B has been suggested. Griseofulvin 500-1000 mg/day, ketoconazole 200-400 mg/day, and itraconazole 200-300 mg/day for months to years can be administered. Also, fluconazole has recently been used.

Amputation is not indicated in actinomycetoma because it is associated with lymphangitic or hematogenous dissemination. The treatment for the mycetoma caused by *N. brasiliensis* consists of sulfonamides (diaminodiphenylsulfone [DDS]), 100-200 mg/day. The efficacy of long term (2 to 3 years) treatment must be evaluated. The most important complications of this treatment are methemoglobinemia and hemolytic anemia. Trimethoprim-sulfamethoxazole 80/400-160/800 mg/day for a few months or up to 1-2 years is recommended. The combination of sulfonamides with streptomycin 1 g/day, clofazimine 100 mg/day, rifampicin 300 mg twice/day, tetracycline 1g/day, or isoniazid 300-600 mg/day is used. Amoxicillin 500 mg with clavulanic acid, 125 mg/day for 5 months has also been recommended.

Ofloxacin 200 mg/day or other quinolones have been used, but experience is very limited. In vitro studies indicate that cefataxime and ceftriaxone may be effective. In any case, regardless of the treatment DDS (Dapsone) must be administered for several years or for life to avoid recurrence. In patients with bone or visceral involvement, or in those resistant to standard treatment, Amikacin 15 mg/kg (in adults, 500 mg IM q12h) for three weeks has yielded satisfactory results. This treatment can be repeated once or twice although oto- and nephrotoxicity can ensue. The dose must be adjusted by following the creatinine level (*Int J Dermatol* 1991; 30:387-398).

11

Orthopedic rehabilitation often is required. Actinomycetomas without osseous involvement generally respond well to medical treatment. Without treatment or with resistance to it, progressive osseous involvement is inevitable. With lesions of the foot, osseous involvement with functional impairment is common and dissemination via the inguinal lymphatics occurs. When localized on the back or neck, there is a risk of involvement of the vertebrae, thorax and lungs. Involvement of the abdomen is more benign because the parasite does not penetrate the muscular fascia, although the abdominal cavity can be involved via the inguinal lymphatics. Mycetoma can be fatal. Prophylaxis must be aimed at improving general health and hygiene, and the use of closed shoes in rural areas.

SELECTED READINGS

1. Arenas R. Micologia Medica Ilustrada (Medical Illustrated Mycology) Interamericana/McGraw-Hill. Mexico. 1993:131-144.
2. Buot G, Lavalle P, Mariat F, Suchil P. et al Etude epidemiologique des mycetomes au Mexique. A propos de 502 cas. Bull Soc. Path Exot 1987; 80(3):329-339).
3. El-Hag IA, Fahal AH, Gasim ET et al Fine needle aspiration cytology of mycetoma. Acta Cytol 1996; 40(3): 461-4).

4 Lavalle P et al. Agents of mycetoma: In Dalldorf G. Fungi and Fungous Diseases. New York. Ch C Thomas Springfield. 1962: 50-68.
5 Lavalle P et al. Micetomas por Streptomyces en America. Dermatologia Ibero-Latino-Americana 1972; XIV (3): 379-89.
6 Restrepo A et al. Treatment of tropical mycoses. J Am Acad Dermatol 1994; 31(3 pt 2): S91-102.
7 Welsh O et al. Mycetoma. Current concepts in treatment. Int. J Dermatol 1991; 30:387-98.

11

Sporotrichosis

Roberto Arenas

Sporotrichosis is a subcutaneous, granulomatous, subacute or chronic mycosis usually localized on the face and extemities. It affects the lymphatics and rarely the lungs, bones, or joints. The causative organism is the dimorphic fungus, *Sporothrix scheckii.*

GEOGRAPHIC DISTRIBUTION

Infection is widespread; it is more frequent in the intertropical zone, i.e., in South Africa, Japan, Australia, and in the Americas from Florida to Uruguay. It is common in Mexico. The main sources of infection are green or dry decaying vegetation. Rodents and insects are passive vectors. It affects both sexes and all ages although it is most common in children and young adults 16-30 years old. It is observed frequently in peasants, gardeners, florists, and carpenters, and it can be passed on accidently in laboratories. It is considered an occupational disease. It occurs usually in isolated cases and sometimes as epidemics.

ETIOLOGY

The infectious agent is *S. schenckii*, a dimorphic fungus that exists in a parasitic form like a yeast (seen in tissue in 30% of cases) and in a saprophytic form like a mold. In primary cutaneous sporotrichosis, the fungus penetrates the skin through small wounds or excoriations produced by plants or contaminated material. It can be transmitted by animal bites. This mycosis is also very well known in animals, especially in cats (*Mycoses* 1996; 39(3-4): 125-8). One or several chancres appear two weeks after inoculation; followed by involvement of the cutaneous and regional lymphatics. The lesions persist for months or may resolve spontaneously. Sometimes the initial lesion extends by contiguity and produces verrucous plaques.

In primary pulmonary sporotrichosis, the fungus penetrates the respiratory tract and causes a self-limited, asymptomatic pneumonia that induces a specific hypersensitivity, or it may cause a progressive pneumonia that can be a source of hematogenous dissemination. Dissemination can also originate from cutaneous lesions in patients with diabetes, sarcoidosis, Hodgkin's disease, myeloma, chronic alcoholism, AIDS, or in those who have received long term

Tropical Dermatology, edited by Roberto Arenas and Roberto Estrada. ©2001 Landes Bioscience.

treatment with glucocorticoids. In immunosuppressed individuals, the fungus behaves like an opportunist. It can affect bones, joints, lungs, the central nervous system or other organs. Re-infection is manifested by fixed forms or by non-healing lesions. The immune response in sporotrichosis is not very well known. In experimental studies, abnormal phagocytosis has been observed. It is believed that primary infection depends on exposure to a great number of conidia.

CLASSIFICATION

Lymphocutaneous and fixed forms are the most common. Superficial and mycetomatoid varieties, verrucous, and spontaneous involutional forms, disseminated cutaneous and systemic, and extracutaneous (osseous, articular, and other organs) are also recognized.

CLINCAL PICTURE

The lymphocutaneous form (70-75%) is characterized by an initial chancre, usually on the hands or fingers. In children it generally affects the face. It is constituted by a nodular lesion or an ulcerative gummatous lesion (Fig. 12.1), followed in two weeks by a chain of erythematous, violaceous gumma that are not painful. Regional lymphatics are involved and ulceration may occur (Fig. 12.2). The most frequent localizations are: the upper and lower extremities (53% and 18%) and also the face (21%). The differential diagnosis must include atypical mycobacterial infection and tularemia.

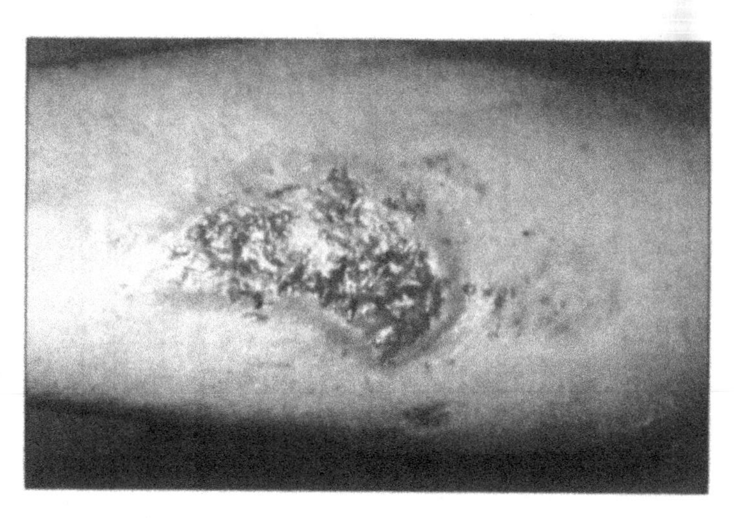

Fig. 12.1. Chancre due to Sporotrichosis.

Fixed cutaneous sporotrichosis is present in 20-30% of cases; it is characterized by a single, infiltrating, verrucous or vegetative crescentic plaque that can be ulcerated and covered with crusts. It is always surrounded by an erythematous, violaceous halo (Fig. 12.3). The superficial or dermoepidermic form is a variant of the fixed cutaneous manifestation. It is characterized by violaceous plaques,

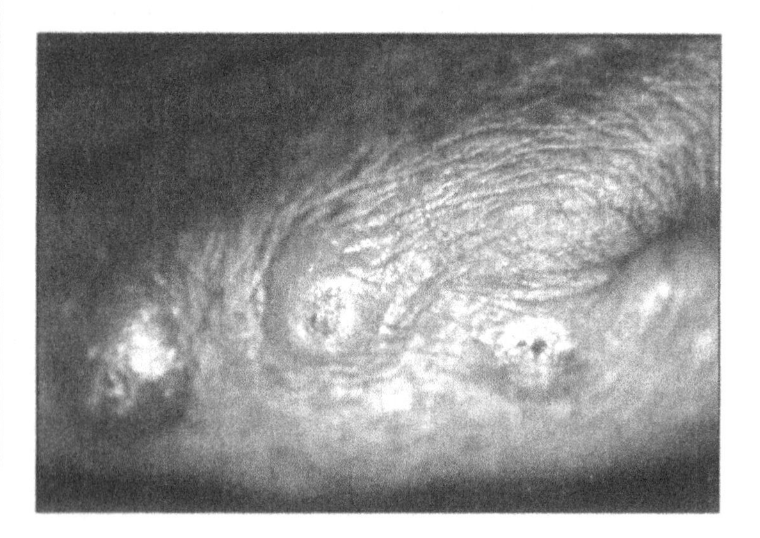

Fig. 12.2. Lymphangitic sporotrichosis.

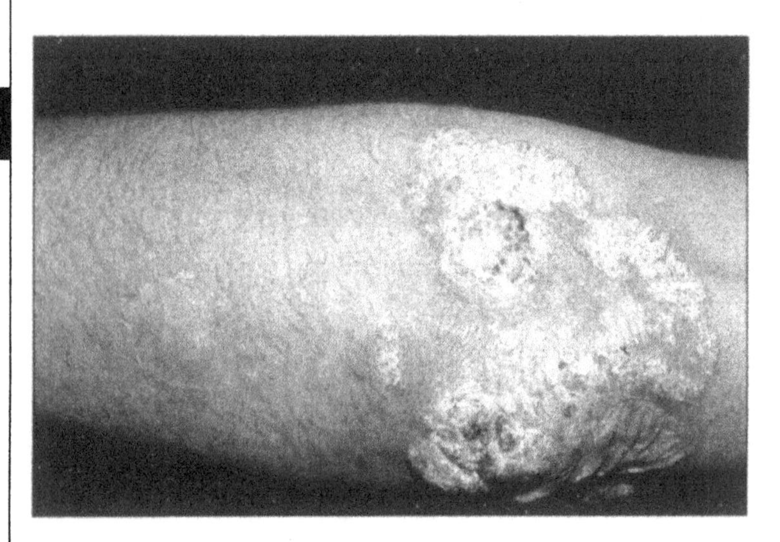

Fig. 12.3. Fixed sporotrichosis.

especially on the face, that follows the superficial lymphatics. Multiple chancres constitute the micetomatoid form which is more frequently seen in the foot. There are chronic forms that spread by contiguity. They can cause intense fibrosis that leads to lymphostasis and elephantiasis. Chronic cases rarely lead to squamous cell carcinoma. The form is known as sporotrichosis recurrens cicatrisans. Disseminated sporotrichosis (5%) presents in two different varieties: disseminated cutaneous and systemic. In the latter there is weight loss, fever, disseminated cutaneous lesions such as ulcerative gumma, verrucous or crusted plaques, and fungemia (Fig. 12.4). The disseminated cutaneous form only affects the tegument, and it responds well to treatment. It is considered the result of multiple inoculations or autoinoculation. The extracutaneous form affects bone and joints. It is the most important mycotic arthropathy; it affects most commonly the knee. There are multiple lytic lesions on the bone with a predilection for the tibia. Internal organs are affected less frequently.

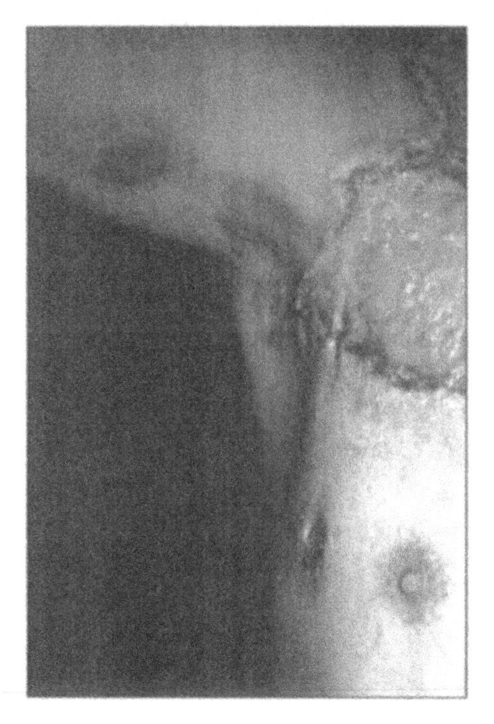

Fig. 12.4. Disseminated sporotrichosis.

LABORATORY DATA

Biopsy is not diagnostic, but it is recommended nonetheless. There is epidermal hyperplasia and sometimes pseudoepitheliomatous hyperplasia. Inflammatory, chronic granulomas with abundant lymphocytes and plasmocytes are found. The typical lesion has a central, suppurative zone; surrounding it is a tuberculoid zone and peripheral to it, a syphiloid zone. Some cigar-shaped or crescent yeast, 3-5 µm and asteroid bodies (yeast surrounded by radiating eosinophilic material) may be observed. All these structures are better visualized with PAS or Gomori-Grocott stain (Fig. 12.5). The intradermal reaction to the sporotrichin skin test after 48 hours is the fastest diagnostic method, although, a positive response doesn't always indicate active sporotrichosis. Yeast stained with PAS or Giemsa are rarely observed on a direct exam The culture confirms the diagnosis (Fig. 12.6). With pulmonary, osseous, or articular involvement, the radiographic findings are non-specific.

In the systemic form, the erythrocyte sedimentation rate (ESR), uric acid, and the alkaline phosphatase may be elevated. The following tests could also be performed: complement fixation, precipitation, latex agglutination, immunodiffusion, immunoelectrophoresis, Western blot, direct immunofluorescence and fluorescent antibodies.

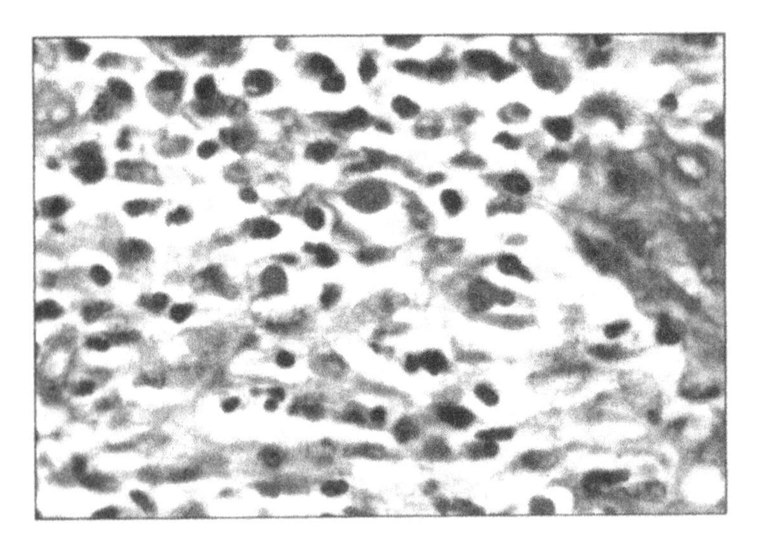

Fig. 12.5. Yeast in skin biopsy (PAS, 40X).

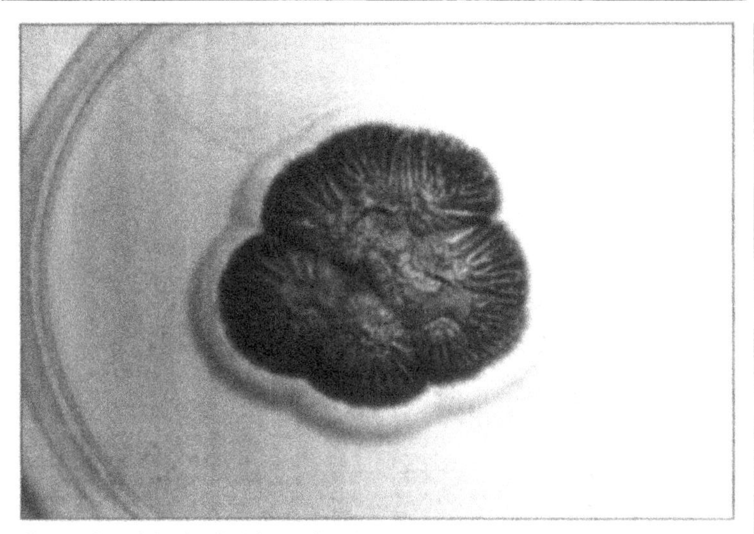

Fig. 12.6. *Sporothrix schenckii* (Sabouraud agar).

TREATMENT

In adults the response to potassium iodide solution (KI) orally 3-6 g/day, in three divided doses for 3-4 months, or for one month after the clinical and mycological cure is achieved, is good. In children the dose is 33-50% of the dose in adults. Side effects include gastritis, sialorrhea, and acneiform eruption. Hypothyroidism is rare, but it can be an important complication in newborns when it is administered to the mother during pregnancy (Cutis 1994; 53(3):128-30).

Amphotericin B, alone or combined with other compounds, is used in extracutaneous forms. Other alternatives are: trimethoprim/sulfamethoxazole 80-200 mg 2/day, griseofulvin 1g/day, ketoconazole 200-400 mg/day, itraconazole 200-300 mg/day, fluconazole 100-400 mg/day, or terbinafin 250 mg/day (*Clin Infec Dis* 1994; 19 (Suppl 1):528-32). All of these medications should be administered until cure is complete; sometimes twice as much time is needed as with potassium iodide. Saperconazole 100-200 mg/day is under investigation.

Selected Readings

1 Arenas R. Micologia Medica Ilustrada (Medical Illustrated Micology) Interamericana McGraw-Hill. 1993: 145-151.

2. Campos P, Arenas R, Coronado H. Epidemic cutaneous sporotrichosis. Int J Dermatol 1994; 33(1): 38-41.

3 Kauffman CA. Newer developments in theraphy for endemic mycosis. Clin Infec Dis 1994; 19(Suppl 1)528-32.

4 Werner AH, Werner BH et al. Sporotrichosis in man and animal. Int J Dermatol 1994; 33(10): 692-700.

Chromoblastomycosis

Alexandro Bonifaz

Chromoblastomycosis is a chronic, subcutaneous mycosis caused by a group of black-pigmented (dematiaceous) fungi, mainly from the genuses Fonsecaea, Phialophora and Cladosporium. It is characterized by warty nodules usually localized to the lower extremities.

GEOGRAPHIC DISTRIBUTION

Chromoblastomycosis occurs throughout the world, but most cases are seen in America and Africa.

ETIOLOGY

The most frequent etiological agents are: *Fonsecaea pedrosoi, F. compacta, Phialophora verrucosa, Cladosporium carrioni* and *Rhinocladiella aquaspersa*. They are dematiaceous hyphomycetes, with slow growth, and heat stability. They tolerate temperatures between 40-42°C and are pathogens of low virulence. They live in the soil and in plants as saprophytes. They have also been isolated from wood. The fungus enters the skin through an abrasion and spreads by contiguity. All the causative agents generate the same parasitic forms, so-called fumagoid cells or Medlar bodies. They are dark, oval thick-walled cells that reproduce by septation or binary fission.

CLINICAL PICTURE

The incubation period is unknown. Chromoblastomycosis occurs mostly in adult males (9:1); it rarely occurs in children. Most patients are peasants. It is most frequent in the lower limbs, especially the foot and leg (85%). It is also observed in the hand, arms, trunk, and head. It is usually unilateral and asymmetric beginning with a papule and slowly forming erythematous, scaly nodules. It spreads by contiguity until it assumes a warty (cauliflower-like) or tumorous appearance (Fig. 13.1). Lesions frequently ulcerate and then are covered with scales. Satellite lesions can be observed. As it resolves, it leaves scars and achromic areas with lymphostasis (Fig. 13.2). In some cases superficial lesions, similar to tinea corporis

Tropical Dermatology, edited by Roberto Arenas and Roberto Estrada. ©2001 Landes Bioscience.

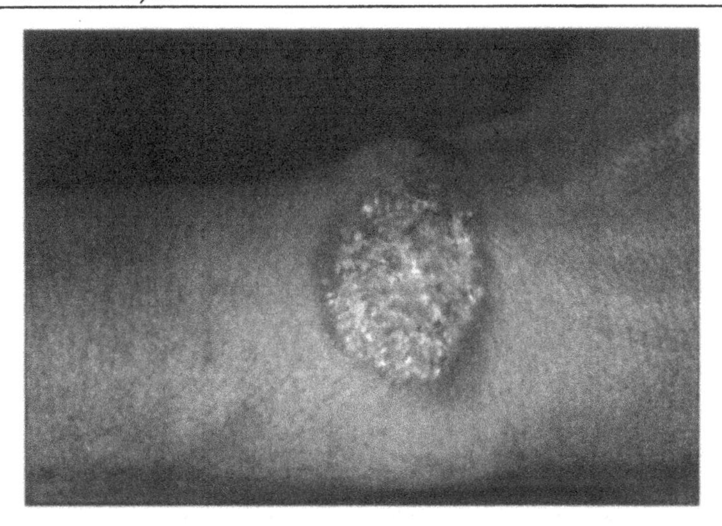

Fig. 13.1. Chromoblastomycosis of the wrist.

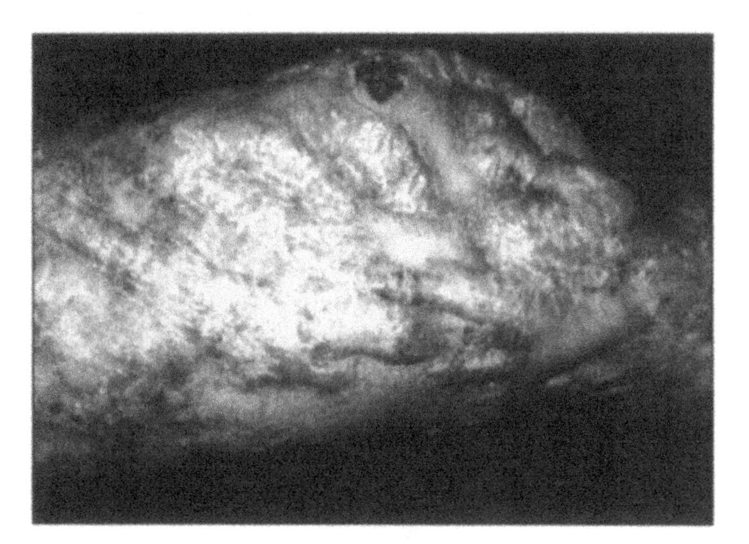

Fig. 13.2. Chromoblastomycosis with a 7 year history.

or psoriasis are seen. Most patients complain of pruritus and tenderness. It is important to mention that chromoblastomycosis is limited to subcutaneous tissue; exceptionally it affects or disseminates via lymphatics. The most important differential diagnoses are tuberculosis verrucous, sporotrichosis, lobomycosis, leishmaniasis, and paracoccidioidomycosis.

LABORATORY DATA

On direct exam with KOH 20-40%, double-walled, brown structures with a diameter of 4-10 µm (fumagoid cells or Medlar bodies) are observed, sometimes with thick and dark hyphae (Fig. 13.3). Culture on Sabouraud agar alone or with antibiotics at 25-28°C confirms the species by micromorphology. The organism grows slowly (25-30 days). Biopsy is useful: hyperplasia or pseudoepitheliomatous hyperplasia is usually seen. In the dermis, a tuberculoid granuloma and the fumagoid cells are observed without special stains.

TREATMENT

There is no single or ideal treatment. In severe cases, therapy is often unsuccessful. The best results have been achieved with 5-fluocytosine at a dose of 100-150 mg/day or with itraconazale 200-300 mg/day (*J Am Acad Dermatol* 1994; 31:S91-92). The duration of treatment varies, but it always is long. For small lesions, complete excision by electrodessication, radiotheraphy, and cryosurgery with liquid nitrogen can be used. At the current time cryosurgery combined with medical treatment is preferred, i.e., itraconazole for 6-8 months as well as a few sessions of cryosurgery. Successful treatment has been reported with KI, calcipherol, intralesional amphotericin B, thiabendazole, ketoconazole, fluconazole, and terbinafin (*Br J Dermatol* 1996; 134 (Suppl. 46):33-36).

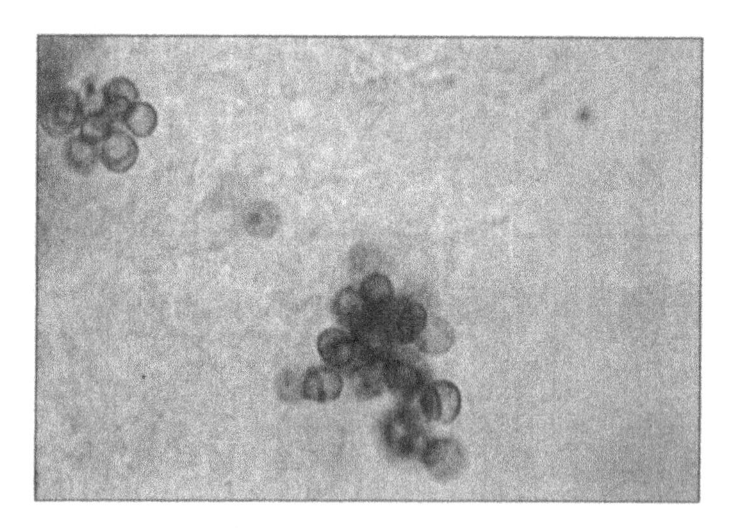

Fig. 13.3. Fumagoid cells (KOH, 40X).

SELECTED READINGS

1 Arenas R. et Chromoblastomycosis. In: Jacobs Ph, Nall L, eds. Antifungal Drug Therapy: A complete guide for the practitioner. New York:Marcel-Dekker, 1990:43-51.
2 Kwon-Chung KJ, Berinett JE et Chromoblastomycosis. In: Medical mycology. Philadelphia: Lea & Febiger, 1992: 337-55.
3 McGinnis MR, Hill Ch et Chromoblastomycosis and phaeohyphomycos. New concepts, diagnosis and mycology. J Am Acad Dermatol 1987; 8:1-16.

Lobomycosis (Jorge Lobo's Disease)

Clarisse Zaitz

Jorge Lobo's disease is a chronic, deep mycosis. The prognosis for survival is good, although it is reserved with regard to regression of the lesions. The disease is also known as Jorge Lobo's blastomycosis, keloid blastomycosis and Lobo's mycosis. It was first described by Jorge Lobo in 1931 in patients from the Amazon Region. In 1950, Trejos and Romero in Costa Rica described the first case outside Brazil. Other cases were then reported in Colombia, French Guiana, Surinam, Venezuela, Peru, Bolivia and Ecuador. In 1978, Zavala-Velásquez and Pérez reported the first case outside South and Central America, in México (*Dermatología Rev Mex* 1978; 1:5-12). The Brazilian experience is mainly concentrated in the Northern Region of the country (Manaus and Belem). The fungus has not yet been cultured. Preliminary studies of the viability of *Paracoccidioides loboi* have been conducted at the Lauro de Sousa Lima Institute (Bauru, São Paulo) in an attempt to obtain a positive culture. Experimental inoculation and therapeutic control of the disease have been undertaken.

GEOGRAPHIC DISTRIBUTION

Jorge Lobo's disease occurs exclusively in the tropical rain forests of Latin America. It is more frequent in males and in individuals aged 21-40 years. There have been 304 reported cases of Jorge Lobo's disease, most of them from the Amazon valley of Brazil involved in agriculture, fishing and in the extraction of nuts, wood or rubber. There is a high incidence of the mycosis among the Caiabi Indians of central Brazil who called the mycosis "Piraip." In 1971 signs and symptoms practically identical to those of Jorge Lobo's disease was described in a dolphin captured along the cost of Florida. The link between the human mycosis and the animal mycosis has not been established. In a case diagnosed in Europe, the patient inoculated himself accidentally with material from a contaminated dolphin captured along the coast of Spain.

ETIOLOGY

Paracoccidioides loboi, also known as *Glenosporella loboi, Blastomyces braziliensis, Glenosporopsis amazonica, Loboa loboi, Lobomyces* is the causal agent. There is no single universally accepted nomenclature. As the fungus has not yet been cultured, a correct classification is still pending. Inoculation of laboratory animals was attempted several times, but an animal model of this disease is not yet available. The skin is considered to be the site of entrance of the fungus. Traumatic injuries from plant fragments or insect bites may permit penetration of the fungus. Inter-human transmission of *P. loboi* has been observed both accidentally and experimentally but is limited to isolated cases. Experimental transmission between animal species has been reported, i.e., from dolphins to mice and from mice to mice of other generations.

CLINICAL PICTURE

The initial lesion is a small wart. The course is slow, with new lesions arising by contiguity or by lymphatic dissemination. The mycosis is limited to the skin and to subcutaneous tissue. In most cases the mycosis is restricted for many years to the site of onset, without involving other areas of the tegument, and with no visceral involvement. The major characteristics of Jorge Lobo's disease are solid, keloid-like lesions of different sizes, which are smooth, pink or dark brown in color, shiny or with small scales and crusts, as well as telangiectasias on the surface (Fig. 14.1). The lesions occur most frequently in the ears (Fig. 14.2) and may be isolated or confluent. One must distinguish them from anergic leishmaniasis and leprosy. The association of dimorphic leprosy and Jorge Lobo's disease has been reported. Lower limbs are the second most common location. The appearance is similar to that observed in the ears, but lesions with a tumoral aspect also may be seen occa-

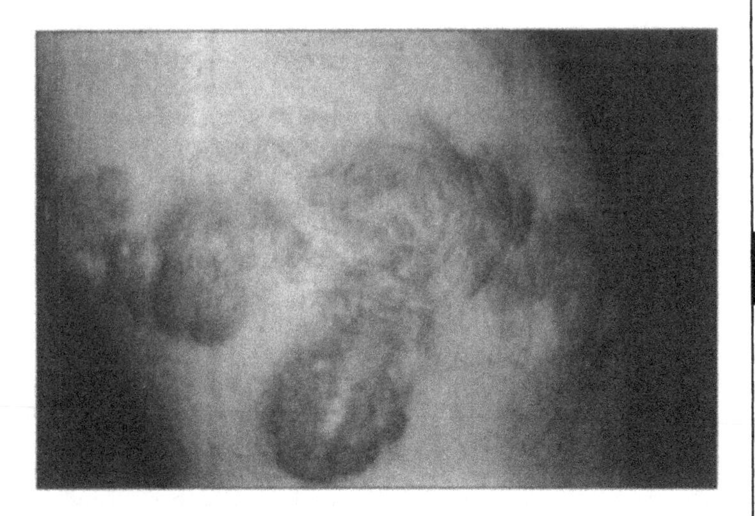

Fig. 14.1. Jorge Lobo's disease. Keloid-like lesions (Courtesy of Prof. Sinésio Talhari, Manaus, Brazil).

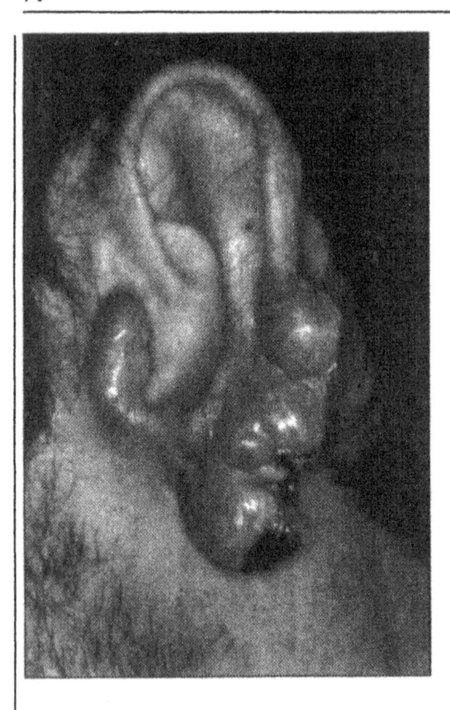

Fig. 14.2. Jorge Lobo's disease. Keloid lesions in the ear (Courtesy of Prof. Sinesio Talhari, Manaus, Brazil).

sionally. On the upper limbs, the third most frequent location, lesions are similar to those observed in other regions although verrucous lesions develop on the palmar edges. In a relatively small number of cases there is cutaneous dissemination. The differential diagnosis of Jorge Lobo's disease includes xanthomatosis, lepromatous (Virchow) leprosy, anergic leishmaniasis and other diseases that have disseminated infiltrations and papulotuberous lesions. Lobo's keloid lesions should always be distinguished from true keloids.

LABORATORY DATA

Isolation of the etiological agent in biopsies or in exudates when present is relatively easy. On direct examination yeast-like cells with a doubly refractive wall, single budding or in chains, are observed in the secretion or in macerated small fragments of the lesion (Fig. 14.3). Culture is not yet obtained. Histopathology reveals epidermal atrophy and rectification of the interpapillary processes; in the dermis, grenz-zone (Unna's band), dilated newly forming vessels, fibrosis and a dense infiltrate with numerous giant cells, especially of the Langherhans type, some eosinophils, and plasmocytes are seen. Large numbers of fungi with a double wall and refractive membrane are observed in this infiltrate. The fungi (Fig. 14.4) are single or budding and are frequently arranged in chains of three, four or even nine elements. The hypodermis also may be involved. Sudan III staining demonstrates lipids in the granuloma and in parasitic structures. Immunological examination demonstrates increased IgG and IgM levels. Complement fixation reveals, cross-reactions between the agents of paracoccidioidomycosis and Jorge Lobo's disease. A depressed cellular immune response is observed only with DNCB.

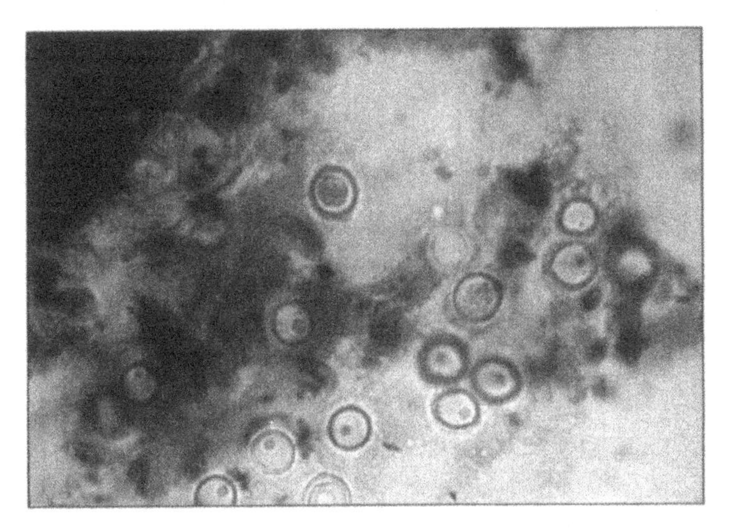

Fig. 14.3. Jorge Lobo's disease. Direct examination (40X).

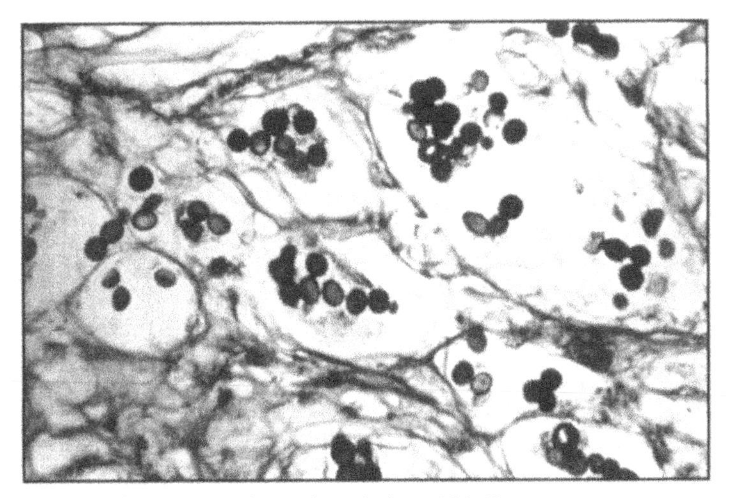

Fig. 14.4. Histopathologic examination on silver-stained material (40X).

TREATMENT

Complete surgical excision or electrofulguration of small lesions is curative. In more extensive lesions, recurrence and progressive mutilation are observed after repeated surgeries. Favorable results have been reported with the administration of clofazimine with initial doses of 300 mg/day during the first month, 200 mg during the second, followed by 100 mg/day for 1-2 years. Ketoconazole, itraconazole and amphotericin B do not yield satisfactory results.

Selected Readings

1 Borelli D. Lobomicosis experimental. Dermat Venez 1961-1962; 3:72-82.
2 Lacaz CS, Baruzzi RG, Rosa MCB. DoenÁa de Jorge Lobo. Editora da Universidade de São Paulo, São Paulo. 1986.
3 Talhari S, Cunha MGS, Schettini APM, Talhari AC. Deep mycosis in Amazon region. Int J Derm 1988; 27:481-4.
4 Talhari, S & Garrido Neves, R. Dermatopatologia Tropical. Editora Médica e Científica Ltda. 1995. Chapter 18-Doença de Jorge Lobo.
5 Trejos A, Romero A. Contribuição ao estudo das blastomicoses em Costa Rica V. Cong. Intern. Microbiol. (Rio de Janeiro). Resumo dos trabalhos, 1950, *apud* Fonseca e Lacaz.

Entomophtoromycosis

Jorge A. Mayorga Rodriguez and Fernando Munoz Estrada

Entomophtoromycosis, or subcutaneous phycomycosis, is a chronic infection caused by the Entomophtorales fungi of the genuses Basidiobolus and Conidiobolus. It is characterized by firm, subcutaneous nodules which rarely ulcerate. In 1925, Van Overeen reported the infection in a horse. In 1956 in Indonesia, Lei-Kian Joe described the first cases in humans; the infectious agent was *Basidiobolus ranarum.* In 1961, Emmons and Bridges defined the etiologic agent *Entomophthora* (*Conidiobolus*) *coronata* in horses, and in 1965 Bras reported a case in a Jamaican native caused by *Conidiobolus coronatus.*

GEOGRAPHIC DISTRIBUTION

It occurs worldwide, but mainly in the tropics and subtropics: (*J Clin Microbiol* 1990; 28(9):1887-90; *J Clin Microbiol* 1990; 28(9): 1887-90). Worldwide there have been about 250 cases reported by *Basidiobolus haptosporus.* All of them affecting trunk and extremities, and usually seen in children. Approximately 160 cases caused by *Conidiobolus coronatus* have been reported. This organism produces nasal and paranasal infection. Only two cases caused by *Conidiobolus incongruus* have been reported. One case in a child with mediastinum involvement, and the second a lethal case in a young female.

ETIOLOGY

The etiologic agents are Entomophtorales fungi which belong to two genuses: Basidiobolus spp. and Conidiobolus spp. (clinical varieties: basidiobolae and coniobolae, respectively). Taxonomically, they belong to the Zygomycetes class. They are saprophytes or parasites of ferns, alga, insects, spiders, horses, intestinal wall and feces of amphibians and reptiles. The infection is transmitted by insect bites or by transepidermal inoculation with contaminated vegetable matter. The entomophtoromycosis is caused by three species: *Basidiobolus haptosporus, Conidiobolus coronatus, Conidiobolus incongruus.* The parasitic forms of the fungi have filaments and few septae and a characteristic eosinophilic halo.

Tropical Dermatology, edited by Roberto Arenas and Roberto Estrada. ©2001 Landes Bioscience.

CLINICAL PICTURE

B. haptosporus (basidiobolomycosis) affects the subcutaneous tissue and the fascia. Lesions are firm and indurated. The skin is hyper- or hypopigmented, but it does not ulcerate. The most frequent sites of involvement are the upper and lower extremities and the lateral sides of the head, trunk (back and shoulders), and buttocks. *C. coronatus* (conidiobolomycosis) produces a chronic nasal inflammation. It affects inferior cornets, sometimes extending to the submucosa, ostia, and paranasales sinuses producing a painless, expanding, bilateral well-fixed erythematous mass, that rarely ulcerates, without fever (Fig. 15.1). This infection can involve the cheeks, forehead, lips, and eye lids causing airway obstruction and deformity. It is more frequent in adults, especially in men.

LABORATORY DATA

The hematoxylin-eosin stain shows a granulomatous reaction with giant, multinucleated cells and a polymorphonuclear inflammatory infiltrate without vessel involvement, and the presence of short hyphae (6-20 μm diameter) with

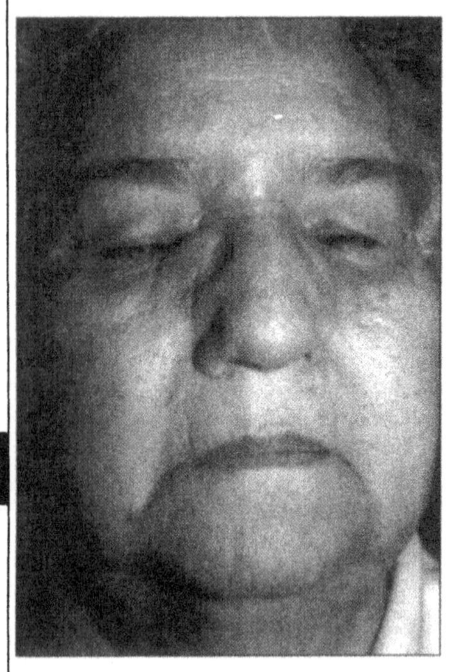

Fig. 15.1. Nasal and paranasal infection by *C. coronatus*.

an eosinophilic halo reaction (Splendore-Hoeppli phenomenon). Gomori-Grocott and PAS stains are also recommended (Fig. 15.2). Culture provides the etiologic agent. *B. haptosporus* colonies appear in 5-7 days at 30°C, although some strains growth best at 37°C. The colonies are flat, wrinkled, folded, oily. They produce sporangiophores and each carries a unicellular sporangium. They form chlamydospores and sporangiola. Secondary spores can be produced by duplication; also cigospores with a diameter of 20-50 µm with a thick, smooth or wavy walls, form. *C. coronatus* produces membranous, folded white-gray colonies at 30-35°C. In old cultures, colonies develop short, white, aeromycelium that turn brown when exposed to light. Sporangiophores are unbranched; spores are 25-40 µm in diameter with a prominent papilla on the wall can form. Secondary spores as well as spores with multiple short appendices that look "fuzzy" give the colonies a crown-like appearance (Fig. 15.3). Colonies of *C. incongruus* are similar to those of *C. coronatus*, although they do not appear to have this fuzzy crown. Immunodiffusion of filtrates of cultures from *C. coronatus* and *B. ranarum* produces bands of specific antigens that can distinguish these two species (*J Clin Microbiol* 1990; 28:1887-90).

TREATMENT

Infections respond to potassium iodide combined with ketoconazole, itraconazole or fluconazole. Other alternatives are trimethoprim-sulfamethoxazole and diaminodiphenylsulfone (Dapsone). For extensive or nonresponding lesions, surgical excision and 5-fluocytosine or amphotericin B are effective.

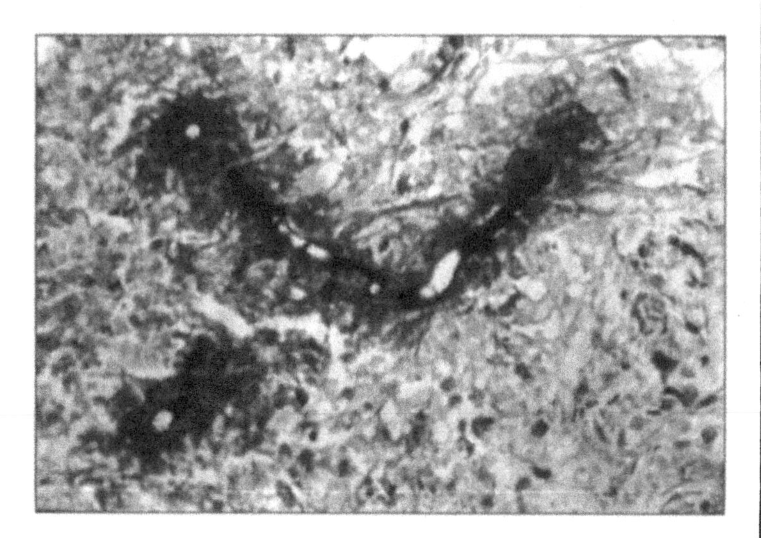

Fig. 15.2. Splendore-Hoeppli phenomenon in conidiobolomycosis (Gomori-Grocott, 20X).

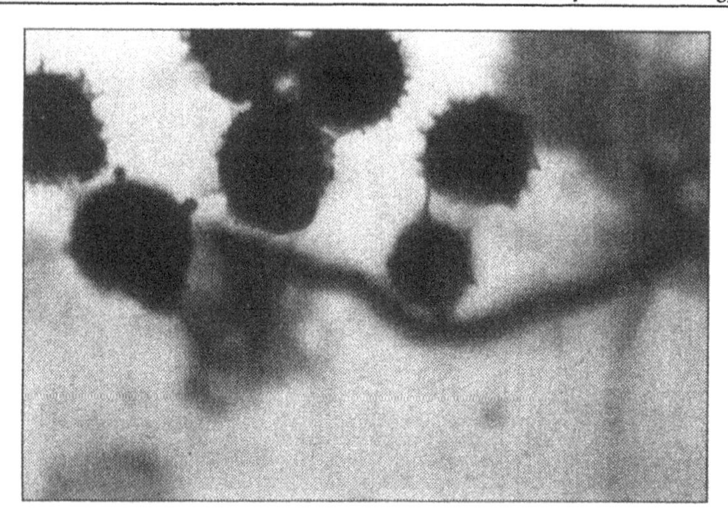

Fig. 15.3. Fuzzy spores of *C. coronatus*.

Selected Readings

1 Bettencourt AL. Entomophtoromycosis. Review. Med Cutan Ibero Lat Am 1988; 16(2): 93-100.

2 Mayorga-Rodriguez JA, Munoz-Estrada VF, Arosemena-Serkisian R. Infeccion nasal y paranasal por C. coronatus: primer caso en Mexico (Nasal and paranasal infection by C. coronatus: First case in Mexico) J Clin Microbiol 1990; 28(9): 1887-90.

3 Marques Da Fonseca AP, Marques Da Fonseca WS, Costa Araujo RD et al. Zigomicose rinofacial. Relato de quatro novos casos. Med Cutan Iber Lat Am 1996; XXIV: 66-72.

4 Rippon JW. Tratado de micologia medica (Medical Mycology) Mexico. Interamerican/McGraw-Hill. 1990: 735-771.

Rhinosporidiosis

Jorge Vega-Nunez

Rhinosporidiosis is a chronic, inflammatory infection produced by *Rhinosporidium seeberi.* It affects the mucosae, especially of the nose. It causes polypoid lesions.

GEOGRAPHIC DISTRIBUTION

It occurs throughout the world, and is endemic in south India and Sri Lanka. In Europe it has been reported in immigrants from India. In 1992, 17 cases were reported in the Balkans, where it was previously unknown, but the source of infection was not determined (*J Trop Med Hyg* 1995; 98 (5): 333-37). It affects all ages especially middle age. In children both sexes are equally affected, while after puberty, it predominates in men. This is probably related to occupational factors such as contact with pond water, mud, and dust.

ETIOLOGY

In 1900, the parasite was identified that was similar to Coccidium (Seeber). After that, it was called Rhinosporidium (O'Kinealy) and *R. seeberi* (Wernicke). It is a Phycomycete; in the division, Mastigomycota; in the suborder, Chrytidineae, and in the families Coccidioidaceae and Olympidiaceae. So it was considered a fungus, but its identity is moot. Even its role as a causal agent is doubtful, and the true nature of its sporangia and endospores—considered "nodular and lysosomal bodies" loaded with cellular residue—is unclear. (*J Submicrosc Cytol Pathol* 1992; 24(1): 109-14). Some studies suggest a phylogenetic relationship with *Loboa loboi*, and it has been classified as an alga. The life cycle has several stages (Fig. 16.1). In its earliest stage, it is a 5-10 µm sphere with a very thin membrane. It begins to grow as it penetrates the tissue (trophic stage). It reaches 50-100 µm, acquires a double membrane, and then it turns into a spherule of 300 µm or larger. It undergoes nuclear division (sporangium or nodular body). Its membrane has an internal cellulose layer, an external layer of chitin and an orifice or pore. As nuclear division continues, it forms up to 20,000 endospores of 5-10 µm. The mature sporangium expels the endospores that disseminate along the tissues and each one

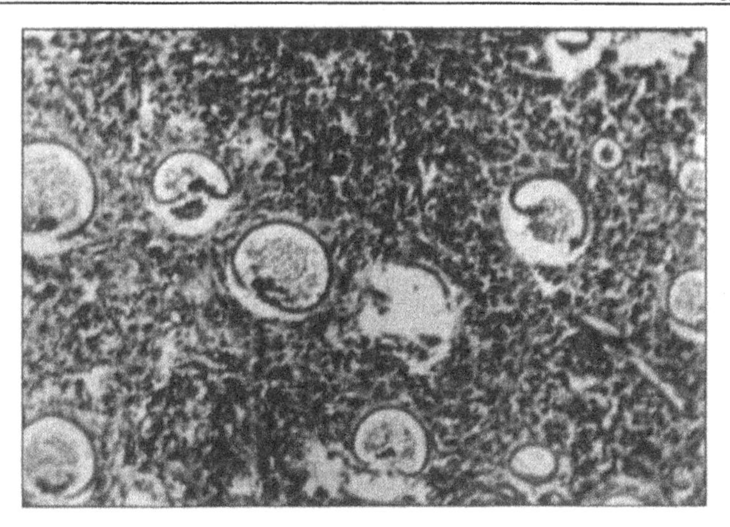

Fig. 16.1. Sporangia of Rhinosporidum (H.E. 20X).

restarts the parasitic cycle. The organism has not been cultured nor has infection been transmitted by inoculation. The mode of transmission is not known. Most cases are related to immersion in water or mud which suggests them as the natural reservoir of the parasite. It has been suggested that fishes and geese are intermediate hosts or that the infection could occur primarily in animals, and that man is just an accidental host. Report of infected brothers suggests the possibility of interhuman transmission (*Dermatologia Rev Mex* 1996; 10(3): 498-508).

CLINICAL PICTURE

Mucosae are affected, especially the nasal and nasopharyngeal. The mucosae of the urethra, vulva, trachea, bronchus, and auditory meatus are less frequently involved. The skin is secondarily involved, although there have been reports of primary infection on the palms and soles. It begins as a small, pink or purple-red tumor (Fig. 16.2). This lesion is sessile at the outset, grows slowly becoming pedunculated and multilobular. The surface is smooth with a mucoid or papillomatous appearance, with multiple fine whitish or yellowish spots that form the mature sporangia located beneath the epithelia or on its surface. These tumors are very friable. In the nostril, they can grow to the pharynx and the palate, or externally through the nasal orifice invading the upper lip. Lesions are associated with mild pruritus, nasal obstruction, dyspnea, or dysphagia. In the eye the conjunctiva is congested and tearing; photophobia, the sensation of a foreign body, and eversion of the eyelid also occur. In the skin the lesions have a warty appearance, they grow slowly, may last for many years or may regress spontaneously.

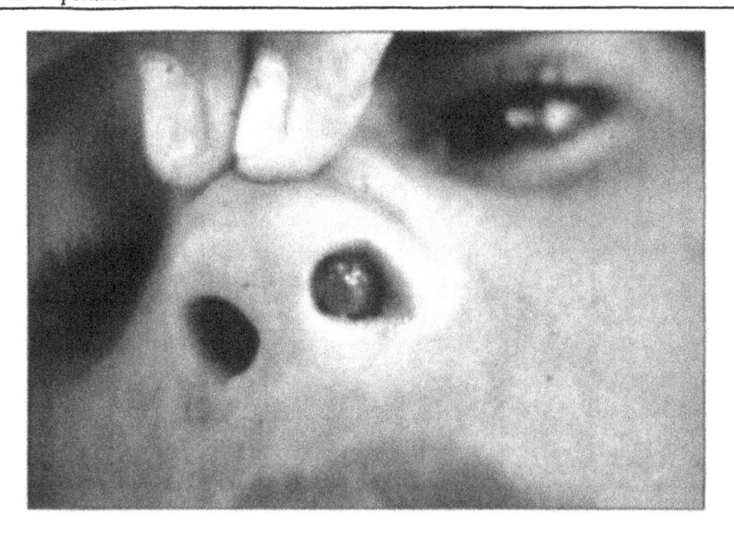

Fig. 16.2. Infantile rhinosporidiosis.

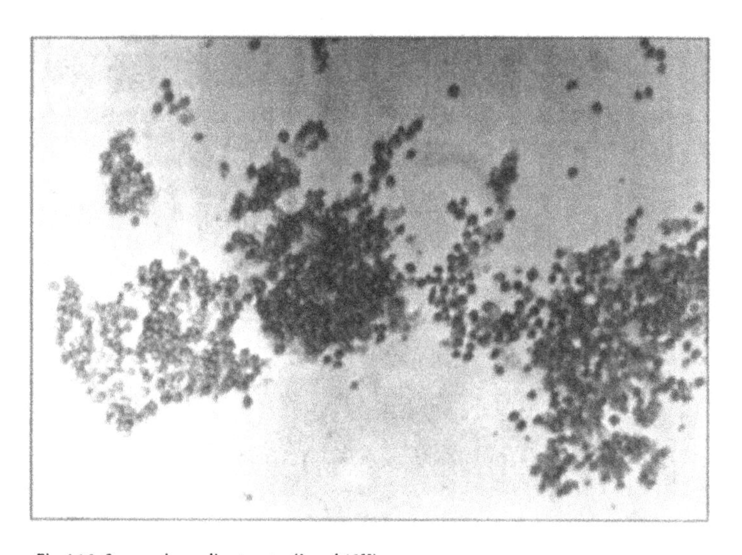

Fig. 16.3. Sporangia on direct exam (Lugol 10X).

Usually not affecting the patient's general health, some have died of laryngeal obstruction and of hematogenous dissemination to bones or viscerae.

LABORATORY DATA

On direct examination of the exudate or fragments of the mucosa, brown spores can be observed (Fig. 16.3). Hematoxylin and eosin stain shows epithelial hyperplasia, numerous vessels, connective tissue edema and a chronic inflammatory reaction with polymorphonuclear cells, lymphocytes, plasmocytes, and some giant cells. Abundant parasites are in evidence when a large vesicle—50-350 µm in diameter with spores in their interior, 7-12 microns— is seen (Fig. 16.1).

TREATMENT

The ideal choice is complete surgical excision with electrodessication of the tumor base to prevent recurrence. Antimony derivatives, such as intravenous neoestibosan, have been used. Also a long course with intralesional amphotericin B and dapsone were reported to be effective (*J Laringol Oto* 1993;107(9): 809-12).

Selected Readings

1 Ahluwalia KB. New interpretation in rhinosporidiosis enigmatic disease of the last nine decades. J Submicrosc Cytol Pathol 1992; 24(1): 109-14.
2 Vega-Nunez J, Herrero A. Dos casos de rinosporidiosis familiar (Two cases of familiar rhinosporidiosis) Dermatologia Rev Mex 1996; 10(3):498-508.
3 Vucovic Z, Bovic-Radovanovic A, Latkovic Z et al. An epidermiological investigation of the first outbreak of rhinosporidiosis in Europe. J Trop Med Hyg 1995; 98(5): 333-37.

D. Systemic Mycosis

Coccidioidomycosis

Paracoccidioidomycosis

Histoplasmosis

Coccidioidomycosis

Oliverio Welsh

This systemic mycosis was first reported in 1892 by Alejandro Posadas and Roberto Wernicke in Buenos Aires. These authors described the infection in Domingo Escurra, a soldier from the Gran Chaco in Argentina. In 1894 in the United States, Rixford and Gilchrist described a similar case in California in a Portuguese worker, observing the organism which corresponded to the etiological agent and thought, as did Posadas, that it was a case of Coccidia and calling the organism *Coccidioides immitis.* Ophls and Moffitt in 1900 defined the etiology by culturing the fungus and describing its characteristics. Blacks, Philippines, and Hispanics are more prone to develop the disseminated systemic disease than people of Caucasian origin and its incidence increases in immunosuppressed patients whether due to another illness such as AIDS, cancer or systemic use of corticosteroids and other immunosuppressive drugs.

GEOGRAPHIC DISTRIBUTION

It occurs predominantly in the western hemisphere, mainly in the southern area of the United States such as California, Arizona, Texas and New Mexico and in the northern states of Mexico such as Baja California, Sonora, Sinaloa, Chihuahua, Coahuila, Nuevo Leon and Tamaulipas. Cases have been also been found in Durango and Colima. It is endemic in some regions of Central and South America such as Guatemala, Venezuela, Colombia, Argentina, Paraguay and Uruguay. Isolated cases have been reported in Africa.

ETIOLOGY

C. immitis is a dimorphic fungus characterized by a saprophytic (infecting) phase in the form of arthrospores and a parasitic phase characterized by spherules containing endospores in their interior. The fungus is found in the soil and in the vegetation in semiarid areas with moderate winters, hot summers, with sandy and alkaline soils and with low annual pluvial precipitation. It occurs predominantly in areas where cacti and a shrub called *Larrea tridentata* grow. Rodents, dogs, cats, horses, oppossums, and other animals can be infected and suffer the disease. Usually the fungus is introduced to the lungs through the airway tract. Infected individuals are asymptomatic in 60% of cases and in the rest (40%) the illness manifests as a nonspecific upper respiratory infection. Approximately 5% of the patients develop a chronic pulmonary infection and/or infection disseminated to other organs. Disemination to other organs is frequently related to a depressed immunocellular response in the host. This has been corroborated in experimental infections in athymic mice and rats which were more susceptible to the infection than normal mice and rats. Activation of Th1 lymphocytes and their lymphokines contributes to the cure of the infection while the predominance of Th2 lymphocytes and their lymphokines promote dissemination (*Annu Rev Immunol* 1989; 7: 145-73).

Precipitating antibodies and complement-fixing antibodies (CF) are produced by the host. The former last 4-6 months; the latter remain positive until complete remission of the illness, whether it happens spontaneously or with treatment.

Other means of transmission, such as transcutaneously caused by the saprophytic phase of the fungus and a person-to-person infection brought on by the parasitic phase, are very rare. The latter mode of infection has been reported only in accidental cases of punctures received while performing autopsies.

CLINICAL FEATURES

Symptomatic patients have fever, chills, asthenia, anorexia, productive cough and occasionally chest pain. These symptoms disappear in about 8 weeks. In about 5% of cases, a chronic infection develops with pulmonary infiltrates, pleural discharge, cavernous lesions and coccidioidomas. Disseminated forms of the disease affect most organs but predominantly the lymphatic nodes (colliquative form, Fig. 17.1), skin, bones, joints, liver, spleen, kidneys, the central nervous system, and occasionally the retina. Symptoms vary depending on the affected site. In the skin it can manifest as abscesses, fistulae, verrucous lesions, ulcers, and keloidal scars. Central nervous system involvement causes headache, confusion, hydrocephaly and diverse neurological syndromes. Differential diagnosis includes cancer, viral infections, Ricketts, mycoplasma, and mycobacteria infection, principally tuberculosis. It must also be distinguished from other mycoses, among them North American blastomycosis, histoplasmosis, aspergillosis, sporotrichosis, and cryptococcosis.

LABORATORY DATA

Imaging studies such as chest x-ray, scans, and nuclear magnetic resonance help the extent of the fungal infection. The diagnosis is obtained by the identification of the fungal spherules by direct KOH examination or by a PAS stain of exudate, pus or the cerebrospinal fluid (Fig. 17.2). Biopsy of affected tissue shows a granuloma with polymorphonuclear infiltrate, plasma cells, macrophages, giant cells, areas of necrosis and the presence of spherules that vary from 10-100

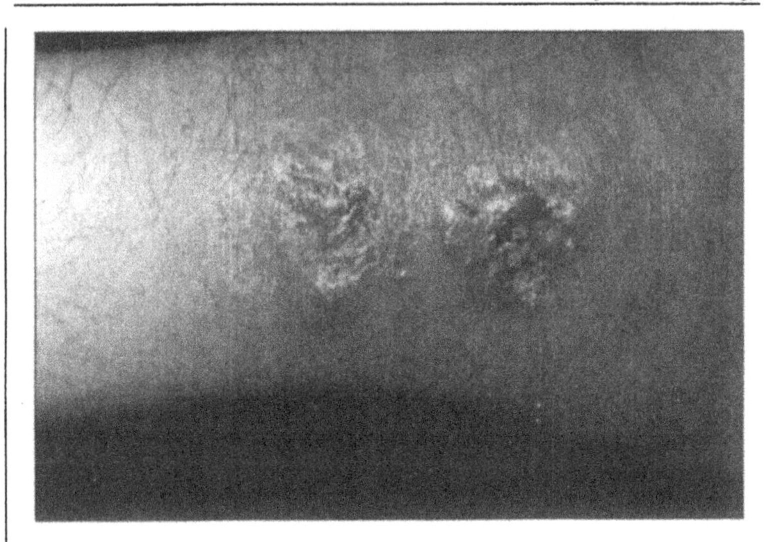

Fig. 17.1. Coccidioidomycosis, sinus tracts on the ankle.

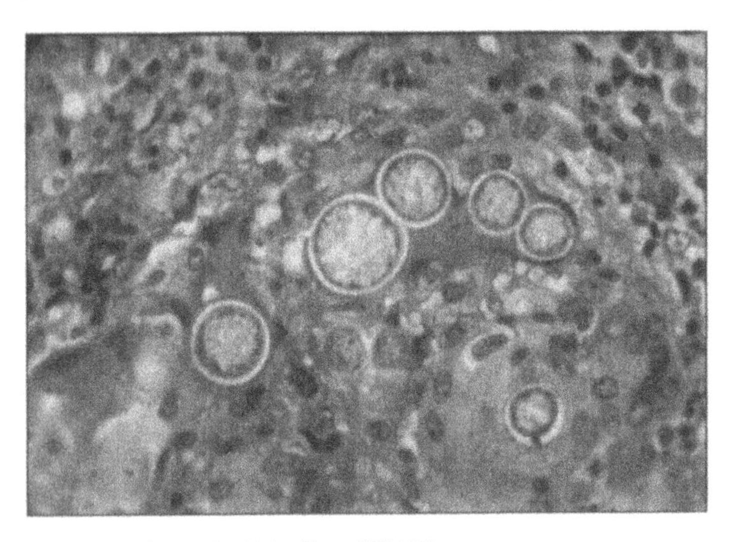

Fig. 17.2. Spherules of *C. immitis*, in a biopsy (H.E. 20X).

microns in diameter. PAS, Gridley and Gomori stains with a green background for contrast facilitate the identification of the fungus especially when the spherules are scarce. Culture requires Sabouraud's medium; it is better if antibiotics (Mycosel) and BHI are added. It is important to perform the culture in closed jars and to process it in a level III security bell since the fungus is extremely dangerous and an accidental infection may occur in laboratory personnel. *Coccidioides immitis* grows for 7-14 days as a white, downy colony but it can present variations in texture and color. When atypical strains occur without arthrospores, the use of techniques with DNA probes and the identification of specific exoantigens permits the precise classification of the fungus. Antigens obtained from the mycelial phase—coccidioidin—and from the spherular phase—spherulin—have been used in cutaneous tests in order to evaluate previous exposure to the disease and response to treatment. Coccidioidin is more readily available and is more commonly used. The three serological tests most commonly used to detect antibodies in the first phase of the infection are: the classic test of precipitin in vitro, the gel diffusion precipitin test, and the test of particles of latex covered with coccidioidin antigens (latex agglutination test). The most reliable is the immunodiffusion test which can be used both in serum and cerebrospinal fluid. These tests can become negative in the first 4-6 months. Complement-fixation serological tests determine the level of IgG antibodies against the antigens of *Coccidioides immitis*. There are two tests: the classic complement-fixation test and the immunodiffusion complement-fixation test. The latter is more commonly used. Other types of serological tests that have been developed have as drawbacks their lack of specificity or reliability.

The immunodiffusion complement-fixation test and/or the complement-fixation test together with the cutaneous test, can be employed to evaluate the results of therapy. A negative response to coccidioidin with high ratios of FC antibodies—from 1:16 to 1:32-indicates a moderate prognosis, whereas clinical improvement accompanied by a strong reaction to coccicioidin and FC antibodies at 1:2 or negative indicates a good prognosis.

TREATMENT

In 1957 Fise introduced amphotericin B for the treatment of coccidioidomycosis. This antifungal was isolated by Squibb Diagnostics, found from *Streptomyces nodosus* in the Tembladero region on the banks of the Orinoco river in Venezuela. Amphotericin B acts on the membrane of the fungus causing defects that allow electrolytes to escape which are essential to the survival of the fungus. The dosage fluctuates between 0.5 to 1.25 mg/kg, administered intravenously in a 5% glucose solution given over a period of 4 to 6 hours. Initially, 1 mg per 30 minutes should be administered to check for anaphylaxis. The adverse side effects are chills, fever, nausea, vomiting and severe headaches. The main adverse medium to long term side effect is nephrotoxicity which may cause renal failure. An IV infusion of 500 ml of saline solution before and after the administration of amphotericin B diminishes the probability of renal toxicity. Other side effects such as hypokalemia, leukopenia, anemia, thrombocytopenia, cardiac arrythmias and phlebitis occur quite frequently. The administration of amphotericin B in meningeal coccidioidomycosis and infection in other sites of the central nervous system is accomplished via an Ommaya reservoir by means of intrathecal and lumbar puncture. Arachnoiditis, bleeding, meningitis and bacterial infection are some of the complications of this treatment. The therapeutic dose administered in this way is 0.025 mg and slowly increased to 1 mg per day.

One important advance has been the incorporation of amphotericin B within lipid complexes (liposomes), which permits the delivery of higher dosages—from 1.25, 2.5, 3 and up to 5 mg/kg three times a week—with less risk of renal damage. Currently there are three commercially available lipid compounds of amphotericin B (1). Abelcet (amphotericin B lipid complex—ABLC—Liposome Co., Princeton, NJ) is composed of bilamellar membranes in the form of a bow which are constituted by the 7:3 molar combination of dimyristoylphospatidylcholine (DMPC) and dimyristoylphosphatidylglycerol (DMPG) with amphotericin B. Amphotericin B in colloid dispersion—ABCD. (2) Amphocil1 (Sequus Pharmaceuticals, Menlo Park, California) is formed in disk-shaped structures of cholesteryl sulfate complexed with amphotericin B. (3) AmBisome2 (Nexstar, San Dimas, California) contains unilamellar vesicles comprising a double layer of lipid membranes of phospatidylcholine and distearoylphosphatidylglycerol derived from hydrogenated soy stabilized with cholesterol in a 2:0.8:1 proportion combined with amphotericin B. These lipid compounds accumulate in organs where the reticuloendothelial system predominates. Studies of experimental models of the infections in animals and severe systemic mycotic infections in humans have increased the cure rate and/or the control of the disease with a lower rate of renal toxicity. This is probably due to the lower affinity that these compounds have for tubular and glomerular cells. They should be administered until the illness is cured or until the side effects no longer permit their administration (*Clin Infect Dis* 1996; 22 (Suppl 2):S133-44). The side effects of amphotericin B in liposomal compounds are similar to those of amphotericin B except that nephrotoxicity is reduced (less than 5%). Reports of its use in the treatment of coccidioidomycosis have involved individual cases and isolated groups of patients. The long-term therapeutic effects and side effects should become clear in the near future.

Ketoconazole, the first orally administered imidazole, it was evaluated at the end of the '70s for the treatment of coccidioidomycosis. It achieved remission of the disease in approximately 40% of patients. This imidazole acts upon the P450 cytochrome system of the fungus, interfering with the synthesis of ergosterol and the oxidative processes of susceptible fungi. The most common side effects are gynecomasty and alterations in hepatic metabolism. Severe and fatal hepatotoxicity have been reported so hepatic function should be checked before administering the drug. The recommended dose is 400-800 mg per day for one year after complete remission of the disease. Relapse can occur if treatment is suspended. Itraconazole, a triazolic, lipophilic compound is orally administered. It concentrates in the tissues and it is more effective, with fewer adverse side effects, than ketoconazole. Yet its pharmacological action is the same as that of ketoconazole. The recommended dosage is 200 mg every 12 hours for one year after remission. It should be administered with food but should not be taken with milk and other products that lower gastric acidity. Itraconazole interacts with coumadins, hydantoin, rifampin, cyclosporine A and other medications that act on the P450 cytochrome system. The most common side effects are gastritis, nausea, vomiting, edema and transitory hypertension. Relapse can occur with the suspension of treatment. Fluconazole, a triazolic compound, is administered both orally and parenterally. It inhibits oxidative enzymatic processes and ergosterol synthesis in susceptible fungi. Oral and parenteral administration attain similar levels in the blood and the cerebrospinal fluid although the simultaneous administration of some drugs can interfere with fluconazole serum levels. The therapeutic dose for coccidioidomycosis is 400-600 mg up to 1,200 mg per day for 6-12 months after remission of the disease. Relapse may occur when suspending treatment.

Response to therapy for coccidioidomycosis is measured by clinical improvement or remission, negative cultures, biopsy and/or the resolution of findings imaging studies. An increase in coccidioidin and/or spherulin tests, accompanied by either negative or decreasing levels of CF antibodies suggests a favorable prognosis. Due to the fact that relapse can occur with all of the currently commercially available fungicides, careful follow-up after remission is important. In the future, the interaction between the host and the parasite will be better defined and the immune response to the infection will be explained. Recently, studies have been conducted on experimental mycotic infections in which the response of the host to the infection has been modified by administering cytokines that enhance the immune response of the host against the infecting agent. Finally, it will be necessary to develop effective fungicidal drugs with fewer side effects in order to achieve the cure for this and other severe systemic mycotic diseases. Furthermore it will be necessary to continue studies in the development of a vaccine that can protect high-risk populations.

Selected Readings

1 Pappagianis D. Epidermiology of coccidioidomycosis. Current topics of Medical Mycology. McGinnis, ed. New York, Springler Verlag 1988; (2):199-238.
2 Mosmann TR. Coffman RL. Th1 and Th2 cells: different patterns of lymphokine secretion lead to different functional properties. Annu Rev Immunol 1989; 7: 145-73.
3 Stevens DA. Coccidioidomycosis. Current Concepts. New Eng J Med 1995; 332 (16):1077-1082.
4 Hiemenz JW, Walsh TJ. Lipid formulations of amphotericin B recent progress and future directions. Clin Infect Dis 1996; 22 (Suppl 2):S133-44.

Paracoccidioidomycosis

Clarisse Zaitz

Paracoccidioidomycosis, Brazilian blastomycosis or South American blastomycosis is a deep and systemic mycosis caused by the fungus *Paracoccidioides brasiliensis*. In 1908, Adolpho Lutz from São Paulo, described for the first time the disease in two patients. In 1912, Alfonso Splendore from São Paulo published four new cases describing the histopathological and mycological findings. In 1930, Floriano Paulo de Almeida from São Paulo established a new genus in the kingdom of fungi, the Paracoccidioides. Before sulfonamides were introduced in 1940 for the treatment of paracoccidioidomycosis, by Domingos Oliveira Ribeiro from São Paulo, this disease—Lutz-Splendore-Almeida—was invariably fatal.This therapy results in improvement or cure in more than 60% of the cases.

What is new in the research on paracoccidioidomycosis?

1) *P. brasiliensis* antigens have been located at the ultrastructure in both yeast and mycelial forms of the fungus. Immunofluorescence and ultrastructural immunolabeling techniques have been performed and the antigen deposits have been observed within the cytoplasm and on the cell wall of the fungus (*J Mycol Med* 1966; 6:1-6).
2) The distribution of *P. brasiliensis* antigens in the skin and mucosa of patients with paracoccidioidomycosis has been studied by double immunolabeling and immunoelectromicroscopic techniques. *P. brasiliensis* antigens were found to be deposited between basal keratinocytes, and granulomatous cells and accumulates within the macrophages.
3) Brazilian researchers are currently investigating gp43 for vaccination in experimental studies.
4) Blood from patients with paracoccidioidomycosis has been investigated by PCR for the presence of *P. brasiliensis.*

GEOGRAPHIC DISTRIBUTION

Although the first cases studied on paracoccidioidomycosis were in Brazil, other cases have been reported from Latin America (Fig 18.1). It is distributed from México (20°N) to Argentina (35°S), but most cases have been reported in Brazil, Colombia, and Venezuela. All patients diagnosed on other continents acquired the mycosis in Latin America where they had previously lived or travelled. Paracoccidioidomycosis is more common in males (15:1), frequently

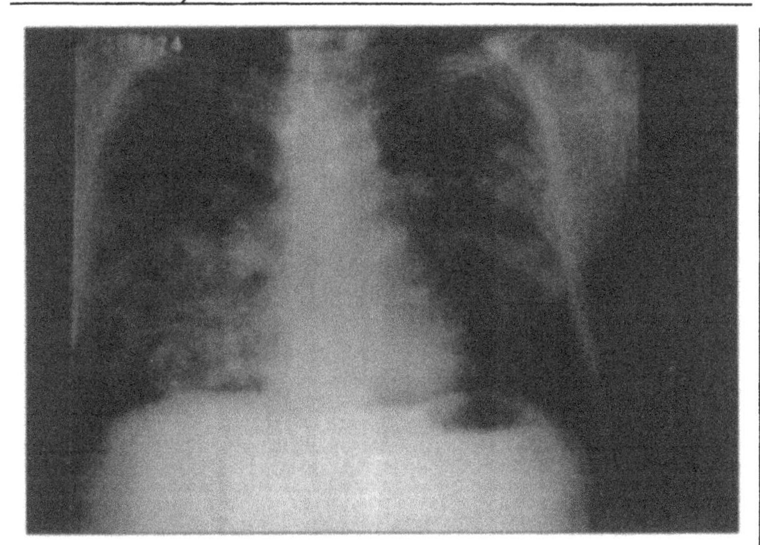

Fig.18.1. Paracoccidioidomycosis. Pulmonary involvement.

seen in rural workers during the most productive period of their lives. However, there is no significant difference between males and females in the frequency of paracoccidioidomycosis infection, when measured by anti-*P. brasiliensis* skin test reactivity. This finding suggests that both sexes are equally exposed to the fungus. An attractive explanation for the finding is that the possible inhibition of specific estradiol receptors in the cytoplasm of the fungus by female hormones prevents the transformation of mycelium into yeast.

ETIOLOGY

Paracoccidioidomiycosis is caused by the thermally dimorphic fungus *Paracoccidioides brasiliensis.* The dimorphic process, characteristic of most of the pathogenic fungi that produce human systemic mycosis, is relevant in the pathogenesis mainly because only one of the forms, the yeast-like phase (36°C), is usually associated with the disease. In nature the fungus is found in its mycelial phase (25°C). The route of entry of *P. brasiliensis* is through the respiratory tract by inhalation. The infection starts in the lungs (Fig. 18.1) and may be disseminated through hematogenous or lymphatic pathways or sometimes by contiguity. Direct inoculation of the fungus is less common (Fig. 18.2).

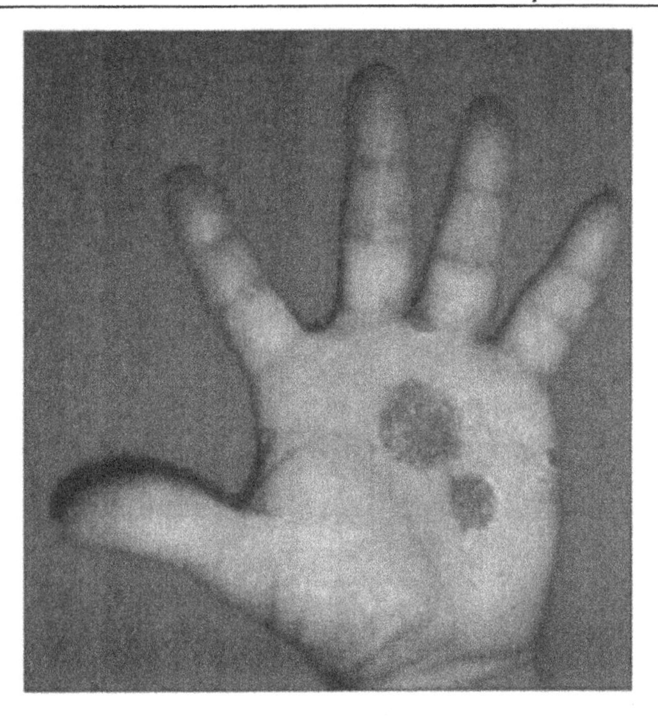

Fig.18.2. Paracoccidioidomycosis. Direct inoculation.

CLINICAL PICTURE

Paracoccidioidomycosis infection can be described by the immune response.

1) chronic disease, in most cases characterized by a Th1 response;
2) acute disease, in most cases characterized by a Th2 response and a depressed cellular immunity; or
3) cicatricial disease (Fig. 18.3).

Most patients affected by paracoccidioidomycosis have a benign form of chronic infection. Chronic pulmonary involvement and regional adenopathy are usually present. Most skin lesions, which occur in 30-50% of cases, are mucosal (muriform stomatitis) (Fig. 18.4) papules, tubercles, vegetations or ulcers. The Th2 response is observed mostly in teenagers. The patients, males or females, have acute disease, a malignant form of disseminated infection (Fig. 18.5). Once established, it progresses rapidly by lymphatic and lymphatic/hematogenous dissemination to the reticuloendothelial system (spleen, liver, lymph nodes and bone marrow) (Fig. 18.6). Skin lesions are rare.

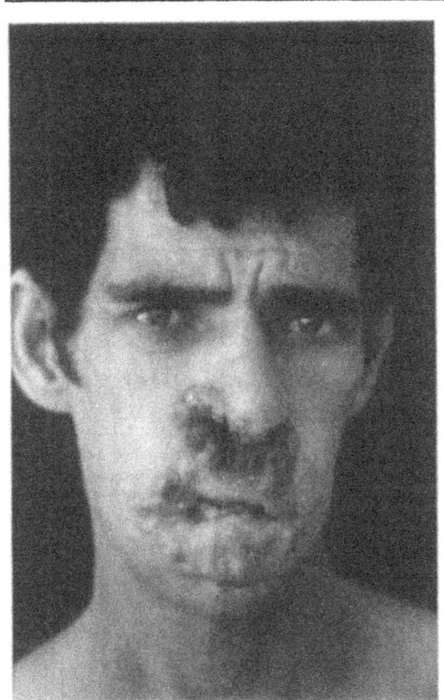

Fig.18.3. Paracoccidioidomycosis. Scarring form (Courtesy of Prof. Silvio Alencar Marques. Botucatu, SP, Brasil).

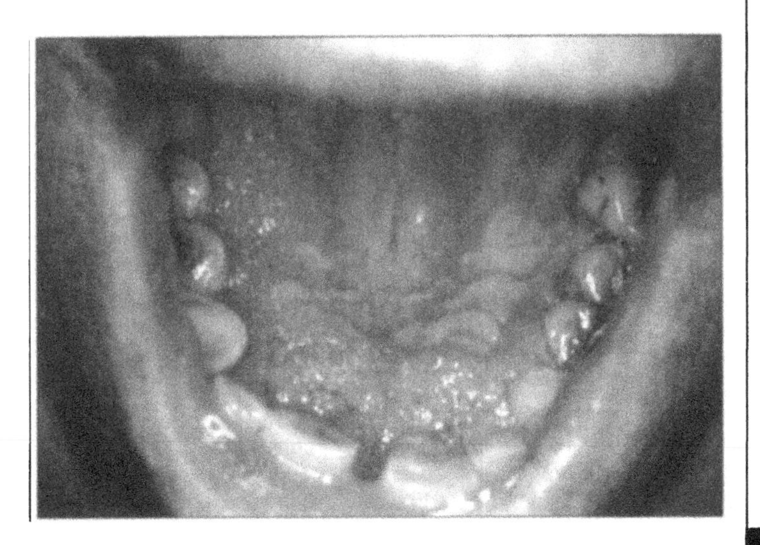

Fig.18.4. Paracoccidoidomycosis. Muriform stomatitis (Courtesy of Prof. Silvio Alencar Marques. Botucatu, SP, Brasil).

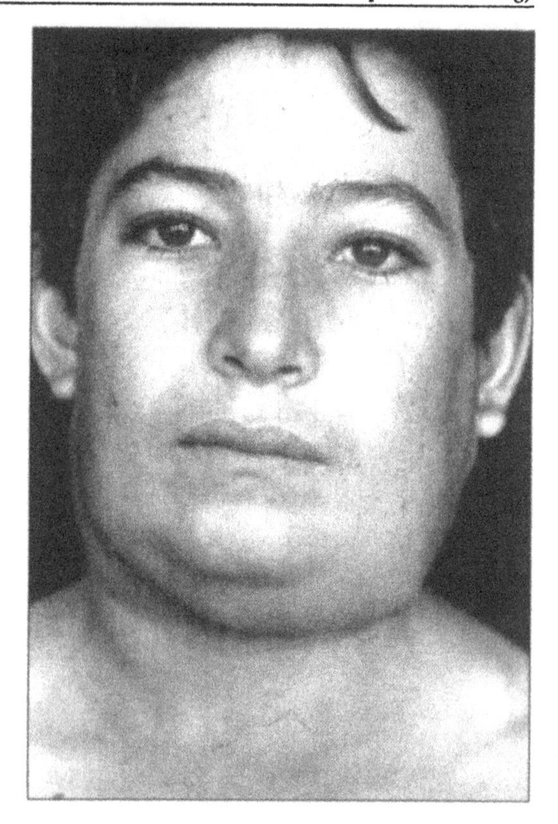

Fig. 18.5. Paracoccidioidomycosis. Juvenile acute form (Courtesy of Prof. Silvio Alencar Marques. Botucatu, SP, Brasil).

LABORATORY DATA

The agent of paracoccidioidomycosis expresses several antigens which may be recognized by antibodies raised in the host. The 43 kDa glycoprotein (gp43) is immunodominant for *P. brasiliensis*. Diagnosis is mainly clinical, mycological—especially by direct examination which shows yeast-like cells with a helm-like arrangement in the yeast phase—and histopathological—compact epithelioid granulomas with few fungi in the chronic cases, and in the acute ones a mixed suppurative and loose granulomatous inflammation with extensive areas of necrosis and large numbers of fungal cells (Fig. 18.7). Serological techniques using the gp43 are also specific for diagnosis and useful to monitor treatment.

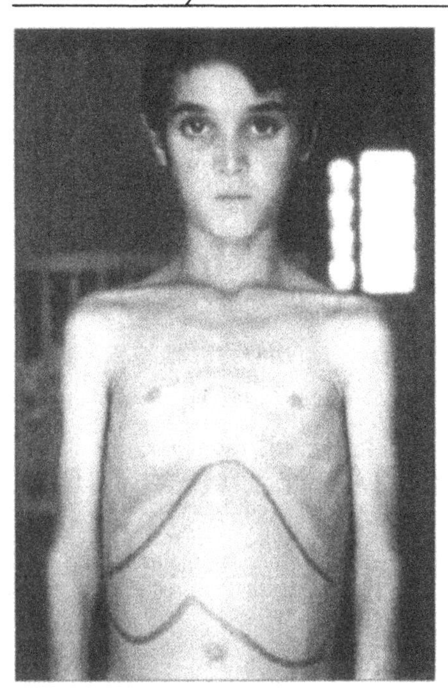

Fig.18.6. Paracoccidioidomycosis. Juvenile acute form (Courtesy of Prof. Silvio Alencar Marques. Botucatu, SP, Brasil).

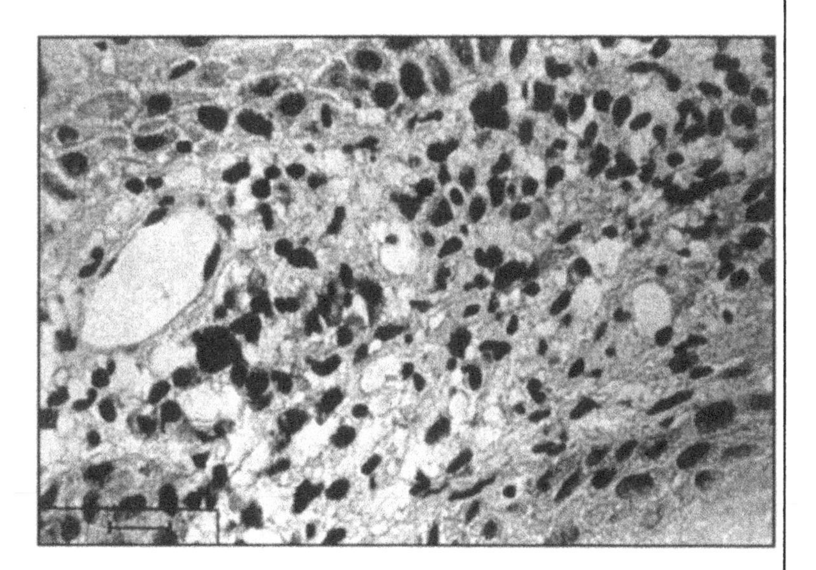

Fig.18.7. Paracoccidioidomycosis. Anatomopathologic examination (HE 40x) - acute form.

TREATMENT

Sulfonamide derivatives such as sulfamethoxipiridazine 1g a day or trimethoprim/sulfamethoxazole, 160/800 mg twice a day until clinical remission and 50% of the dose one more year thereafter as indicated. Also effective are ketoconazole, 400 mg/day until clinical remission and 200 mg/day for at least for three years thereafter; itraconazole 300 mg/day for 12 months and 100 mg/day, one or two more years; fluconazole 200-400 mg for at least 6 months. In severe cases fluconazole is administered by the IV route. Also amphotericin B may be used.

Selected Readings

1 Franco M, Montenegro MR, Mendes RP, Marques SA, Dillon NL, Mota NES. Paracoccidioidomycosis: a recently proposed classification of its clinical forms. Rev Soc Bras Med Trop 1987; 20:129-132.
2 Sandoval MP, del Negro GMB, Mendes-Giannini MJS et al. Distribution of exoantigens and gp 43 in yeast and mycelial forms of P. brasiliensis. J Mycol Med 1996; 6: 1-6.
3 Sandoval, MP et al. Antigen distribution in mucouscutaneous biopsies of human paracoccidioidomycosis. Int J Surg Pathol 1996; 3 :181-88.

Histoplasmosis

Ruben Lopez-Martinez

Two types of histoplasmosis exist in humans: one caused by *Histoplasma capsulatum* var. *capsulatum*, also called American histoplasmosis, and the other caused by *Histoplasma capsulatum* var. *duboisii*, also called African histoplasmosis because it is restricted to that continent. There is a different agent that causes epizootic lymphangitis in horses and mules; the agent is *Histoplasma capsulatum* var. *farciminosum*. Here we will refer only to American histoplasmosis. It is an acute or chronic granulomatous mycosis of the primary pulmonary type. Many infections are asymptomatic, but in those cases with pulmonary manifestations or with systemic dissemination, histoplasmosis becomes very serious and can be lethal in a great number of patients.

GEOGRAPHIC DISTRIBUTION

Histoplasmosis is found in the five continents; nevertheless its prevalence varies widely in accordance with the ecological conditions in the places where the fungus is isolated. Countries with temperate, subtropical and tropical climates are those of higher incidence, such as the United States, Mexico, all of the countries of Central America, Venezuela, Colombia and Argentina (*Rev Inst Med Trop Sau Paulo* 1995; 37: 531-535). Cases have also been reported in Japan, Thailand, India, Malaysia, Indonesia, Singapore and the Philippines. In Europe it has been diagnosed in Italy, Turkey and in the countries of central Africa, South Africa and also in Australia. In the USA the endemic type of histoplasmosis usually predominates. In Latin America tourist groups, miners and speleologists that visit bat caves are often affected by what is considered to be an occupational mycosis. It has recently been described in dogs and cats in a new endemic zone in Texas. In Mexico, the epidemic form of the illness is most often reported, as it usually has a severe course. Skin testing with histoplasmin indicates the prevalence of the infection in endemic zones. The asymptomatic type exists in regions such as the Gulf of Mexico and the southeastern United States where the positive histoplasmin skin test response exceeds 95%. In the United States in general, the histoplamin skin test response is about 80% of those tested. Susceptibility to mycosis is equal in all ages and races and is more common in males. It is greater in patients with AIDS, Hodgkin's disease, lymphoma or leukemia.

Tropical Dermatology, edited by Roberto Arenas and Roberto Estrada. ©2001 Landes Bioscience.

ETIOLOGY

Histoplasma capsulatum var. *capsulatum* is a dimorphic fungus that in the filamentous form produces septated hyphae with echinulate macroconidia and round or pyriform microconidia. This morphology occurs in its natural habitat which contains the guano of bats and birds such as starlings and fowl and also in the media of simple cultures such as Sabouraud's medium, where white or gray-brown cottony colonies develop. In the parasitic form as well as in cultures which have been enriched and incubated at 37°C, it assumes a yeast-like form. The teleomorphic or sexual state corresponds to the ascomycete *Ajellomyces capsulatus* (Kwon-Chung) (McGinnis and Katz, 1979). The isolation of *H. capsulatum* from the soil or from patients and culture yields white (A-strains) or grey-brown (B-strains) colonies. Inhalation of conidia is the most common mechanism of infection. There are few cases which are attributed to cutaneous inoculation. In the majority of healthy individuals, inhalation of conidia causes asymptomatic infection. The mild to moderate symptomatic cases generally resolve spontaneously. *H. capsulatum* is a primary pathogenic fungus that upon penetrating the alveoli causes interstitial inflammation; the conidia are phagocytized by macrophages, lymphocytes and neutrophils. In the primary acute forms, there is a pyogenic response with massive exudates and alveolar tamponade that may cause respiratory failure and death. In the disseminated forms the fungus, transforming itself into its intracellular yeast-like form, is released by the rupture of macrophages to invade new macrophages and migrate to other organs, principally those of the reticuloendothelial system such as reticuloendothelial bone marrow, spleen, liver, Peyer's patches and the superficial and deep lymph nodes. The immunological response, general health and amount of conidia inhaled determine the clinical course.

CLINICAL PICTURE

Histoplasmosis is a systemic infection that can affect multiple organs and tissues; nevertheless, in the great majority of cases, the most affected are the lungs and the tissues of the reticuloendothelial system such as the spleen, liver, bone marrow and the lymph nodes. In accordance with the González-Ochoa's classification (*Rev Inst Salub Enferm Trop* (Mex) 1960; 20:129-145), which described the clinical forms observed in Mexico and other Latin American countries, there are two types of histoplasmosis: the primary pulmonary type—generally not progressive, and the progressive secondary type. Primary pulmonary histoplasmosis is either asymptomatic—the most common form—or symptomatic type. The clinical manifestations are different. The mild form simulates a flu attack. There are no physical or radiological signs. Pulmonary lesions become chronic infiltrates and, with time, calcify. These individuals are histoplasmin-positive. In the moderate form there are radiographically observable lesions that appear as nodular infiltrates collapsed alveoli, and hilar adenopathy (Fig. 19.1). Clinically there is fever and malaise. This form generally resolves spontaneously, but if it is not adequately managed it may be severe. Most severe cases correspond to massive, epidemic infection in individuals who have entered closed spaces such as bat caves or abandoned mine shafts where they have remained for a long period and have inhaled large quantities of conidia of *H. capsulatum*. The febrile pattern dominates, and for this reason, these cases are often mistaken for bacterial or viral infections. The duration is one week to six months. The signs and symptoms are severe; productive cough with hemoptysis, thoracic pain, dyspnea on small effort, asthenia, adynamia, prostration, dysnea and cyanosis are observed. Many of these cases end in respiratory failure and death. The clinical manifestations are accompanied by prominent radiological findings such as nodular infiltrates which simulate tuberculous granulomata that are commonly observed in both pulmonary fields. The abundance of these infiltrates, which sometimes coalesce, is related to the amount of the inoculum. Secondary histoplasmosis, or progressive hematogenous dissemination, is less common. In the disseminated forms, ulcers and

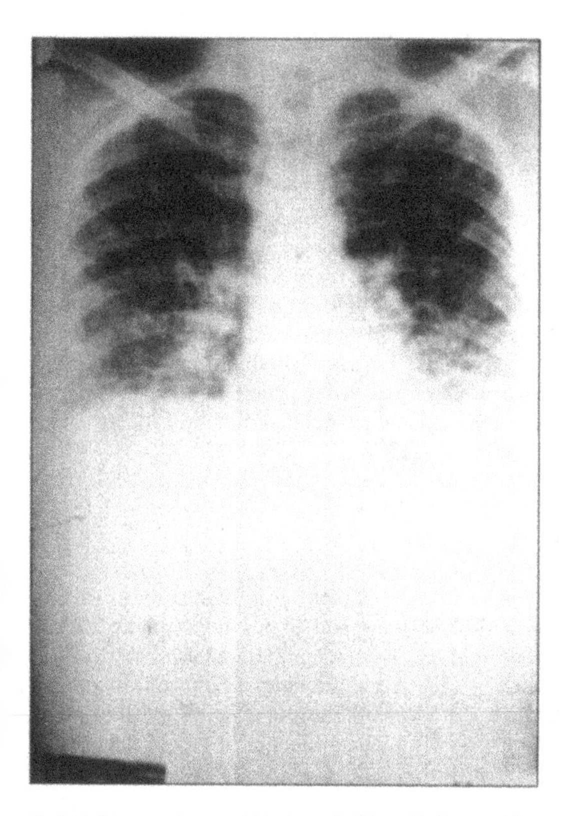

Fig. 19.1. Primary pulmonary histoplasmosis. Bilateral infiltrates of the basal type.

lesions on the liver and spleen, corresponding histopathologically to necrotic and tuberculoid foci, are present. The intracellular yeast-like parasite abounds. This is accompanied by anemia and bone marrow involvement (*Am J Clin Pathol* 1990; 93:367-372). Progressive histoplasmosis occurs as an acute form in children and adults beginning as a primary pulmonary infection that does not decrease in severity. At both extremes of life—childhood and old age—disseminated histoplamosis can be rapidly lethal. In children it manifests as diarrhea due to the invasion of the intestinal lymphatics and is accompanied by hepatomegaly and splenomegaly. In the elderly it presents as a generalized, febrile illness with ulceration of the oral, pharyngeal and laryngeal mucosae. The chronic type is also usually fatal. It appears several years after the primary infection and is mistaken for tuberculous pneumonitis.

Other clinical forms of histoplasmosis are pericarditis and mediastinitis that can be concommitant with the primary pulmonary form or can occur long after the pulmonary symptomatology has settled in. Hilar and mediastinal lymphadenitis are observed. By dissemination, the inflammatory nodular lesions reach the pericardium producing a similar form to that of tuberculous percarditis.

Primary cutaneous histoplasmosis is very rare and has been reported in cases of accidental inoculation in the laboratory. The most common type is mucocutaneous which spreads from disseminated chronic forms in which ulcers and nodules with erythema and exudates are observed in different parts of the skin and mucosae, most frequently in the oral mucosa. Histoplasmosis in AIDS is often diagnosed in areas in which histoplasmosis is endemic. The symptomatology is severe and is accompanied by pneumonitis, hepatosplenomegaly, weight loss and lymphadenopathies (Fig. 19.2). Dissemination to the skin and oral mucosa is also common. Many of these patients die within a short time due to the severity of both diseases (*Clin Infect Dis* 1995; 21(suppl 1):108-110).

Ocular histoplasmosis is a syndrome that has not been proven to be related to *H. capsulatum*. It is observed in very few cases in people who live in endemic zones. It is accompanied by coroiditis and panophthalmia, and patients have a positive skin test to histoplasmin.

LABORATORY DATA

In order to isolate the fungus, a variety of biological materials are processed, such as saliva, cerebrospinal fluid, blood, bone marrow, pus and biopsies. In smears or impressions stained with Wright or Giemsa stains, intracellular yeasts of 3-4 μ in size can be seen. The nucleus stains red-violet and the cytoplasm becomes a very faint, almost colorless blue that looks like a capsule (Fig. 19.3). These products are grown in Sabouraud's medium with antibiotics and are maintained at 25°C for two or three weeks. Two morphological types of colonies have been described: a white colony which corresponds to the A-type (albin) and a brown, which corresponds to the B-type (brown). Both have a smooth, wooly or cotton-like appearance which covers a large part of the culture media. Microscopically, an

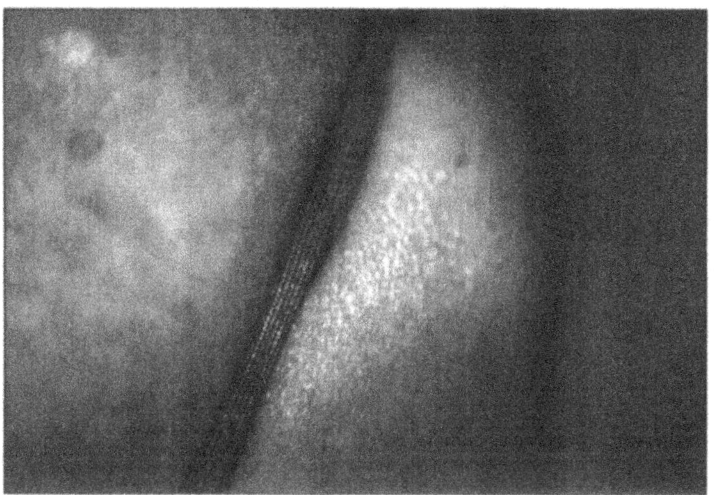

Fig. 19.2. Adenopathy due to histoplasmosis in AIDS (Courtesy of R. Arenas).

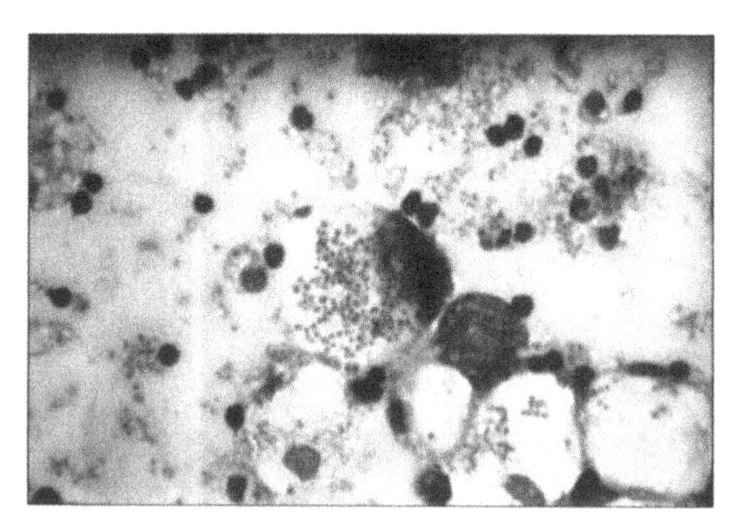

Fig. 19.3. Intracellular yeasts of *Histoplasma capsulatum*.

abundance of tuberculated macroconidia are observed. They are 8-14 μ in diameter, round or oval, and they emerge from short conidiophores at right angles (Fig. 19.4). Also, microconidia can be observed which are 2-4 μ in diameter, spherical or oval, have smooth walls, emerge from short, narrow conidiophores arranged at right angles with respect to the hypha. The B-type produces a greater number of macroconidia while the A-type produces a greater number of microconidia.

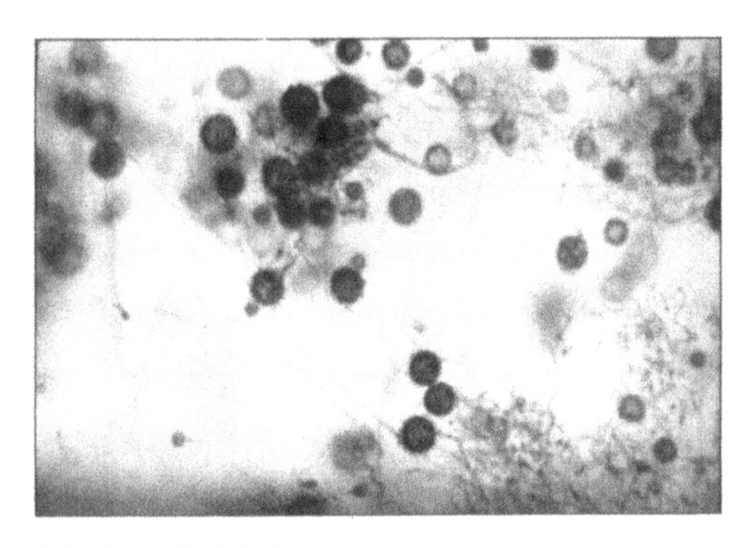

Fig. 19.4. Macroconidia of *Histoplasma capsulatum.*

Also, it is recommended that the products be grown in media of brain-heart infusion, agar-blood or agar-phosphate extract to which antibiotics have been added. To distinguish *H. capsulatum* from other fungi such as *Chrysosporium*, *Sepedonium* and *Renispora*, all of which have similar macroconidia, it is useful to induce the transformation from the filamentous phase to the yeast-like phase for which the colonies are grown in media of brain-heart infusion and incubated at 37°C. Another procedure used to determine the nature of *H. capsulatum* is the test for exoantigens and the hybridization of nucleic acids, which affords a faster identification. Immunological tests are helpful. By precipitation in a capillary tube, precipitins in their early stages can be shown. Nevertheless, they are often neither sufficiently specific nor sensitive, and their positivity disappears after 8-10 weeks. The complement-fixation reaction (CF) is one of the most sensitive and specific tests. Antibodies appear in approximately the second week after the initial infection and they persist all the time while the histoplasmosis is active. A ratio equal to or greater than 1:16 generally indicates an active or progressive illness. This test is positive in approximately 70% of cases. During the course of the illness, the CF test should be periodically employed to monitor the course of the disease and to determine the prognosis. The immunodiffusion test (ID) is carried out by means of Ouchterlony's technique and is very effective when it is used in combination with the CF test. It is positive in the third or fourth week of infection. It is important to distinguish between two different bands. The M-band occurs during the early phase or during recuperation from the illness, while the H-band appears only while the infection is active. The intradermal reaction (IDR) indicates past or current exposure to the etiological agent. In active histoplasmosis, it has prognostic value, along with serological tests, because when it is negative and the antibody ratios are elevated and the prognosis is severe. On the other hand, when it continues to be positive as the antibody ratios decrease, the prognosis is good. The IDR test will remain positive during almost the entire lifetime of the subject. When the ID test is conducted after the IDR test with histoplasmin, an M-band can appear which could lead to an error in the serological diagnosis.

Other tests for detecting circulating soluble antigens are the radioimmunoassay (RIA) and ELISA which are particularly useful in patients suffering severe immunodeficiencies such as AIDS. On biopsy, intracellular yeasts are observed in cross sections stained with PAS, Giemsa and Wright. In the acute form numerous yeasts appear within the histocytes, and there are few neutrophils, plasma cells and lymphocytes. The yeasts should be differentiated from *Leishmania* and from *Toxoplasma.* In the subacute or chronic forms, epithelioid granulomas form that contain plasma cells, lymphocytes, macrophages, neutrophils and giant cells as well as a few yeasts. In moderate to severe primary pulmonary infection, hematologic tests are variable. In fact, leukocytes remain within the normal range, the erythrocyte sedimentation is variable and only in very severe cases it is elevated.

TREATMENT

In mild to moderate primary pulmonary infection, it is not always necessary to administer antimycotic drugs. Patients should be hospitalized for observation and rest, nutrition and hydration. In the majority of cases remission is spontaneous. In moderate and chronic forms, itraconazole and fluconazole are useful in dosages of 200-400 mg/day until the remission of symptoms (*Am J Med* 1995; 98: 336-42). In patients with severe pulmonary histoplasmosis with respiratory insufficiency or in disseminated cases, intravenous amphotericin B is required. Initial dosage is 0.1 mg/Kg, increased by 0.1 mg/Kg with each application until 0.8 or 1 mg/Kg, with a maximum of 50 mg per application is achieved. Every other day, 500 ml of a 5% glucose solution with 1000 units of heparin and 100 mg of hydrocortisone should be administered over 4-6 hours. One-half hour before the infusion, an ampule of antihistamine (alphaminopyridine) is administered intramuscularly. Intolerance is manifested by severe headache, diaphoresis, abdominal pain, nausea and restlessness. In this event, the drip must be discontinued and another ampule of antihistamine administered. When the symptoms of intolerance resolve, amphotericin B can be resumed at a slower rate. Toxicity reactions appear when high doses of the medication accumulate in susceptible patients. Nephrotoxicity is the most important side effect, causing glomular necrosis and tubular degeneration. In some cases nephrotoxicity is reversible if the damage is not too severe. Periodic serum creatinine and BUN are essential.

Selected Readings

1 Bava AJ. Histoplasmosis in the Muñiz Hospital of Buenos Aires. Rev Inst Med Trop Sao Paulo 1995; 37: 531-535.

2 González-Ochoa A, Cervantes-Ochoa A. Histoplasmosis epidémica y su prevención. Rev Inst Salub Enferm Trop (Méx) 1960; 20:129-145.

3 Hajjeh RA. Disseminated histoplasmosis in persons infected with human immunodeficiency virus. Clin Infect Dis 1995; 21(suppl 1):108-110.

4 Kurtin PH, McKinsey DS, Gupta MR, Driks M. Histoplasmosis in patients with acquired immunodeficiency syndrome. Hematologic and bone marrow manifestations. Am J Clin Pathol 1990; 93:367-372.

5 López-Martínez R, Mendez-Tovar LJ, Hernández-Hernández F, Castañón-Olivares, R. Micología Médica. Procedimientos para el diagnóstico de laboratorio. 1a ed. México. Trillas. 1995: 83-98.

7 Wheat J, Hafner R, Korzun AH, et al. Itraconazole treatment of disseminated histoplasmosis in patients with the acquired immunodeficiency syndrome. Am J Med 1995; 98:336-42.

E. Mycosis By Opportunists

Penicilliosis Due to *Penicillium marneffi*

Penicilliosis Due to *Penicillium marneffei*

20

Rataporn Ungpakorn

Penicillium marneffei is a thermally dimorphic fungus that is thought to be endemic to southeast Asia (*Am J Trop Med Hyg* 1984; 33:637-644) and southern China (*Rev Infect Dis* 1988; 10:640-652.). It is capable of causing life-threatening disseminated illness involving the skin, bone marrow, liver, lymph nodes, bones and lungs in immunocompromised and immunocompetent persons. Penicilliosis, previously a rare disease, has now emerged as one of the most common systemic opportunistic infection among AIDS patients who lived in southeast Asia or visited areas in which this fungus is endemic, regardless the time since exposure. Management must be promptly initiated while awaiting confirmation of diagnosis by morphology of culture. Current epidemiological data suggest that disseminated penicilliosis should be included as an AIDS-defining condition. Penicilliosis is a fatal disease unless prompt diagnosis and early treatment with systemic antifungal agents ae undertaken.

GEOGRAPHIC DISTRIBUTION

Penicillium marneffei was first isolated in 1956 and identified in 1959 from a bamboo rat (*Rhizomys sinensis*) found in the highlands of central Vietnam. (*J Med Vet Mycol* 1986; 24:383-389). The first accidental human infection was reported in 1959 by Segretain (*Mycopathol Mycol Appl* 1959; 11:327-353). However, the first natural infection in humans appeared in 1973 in a patient who underwent splenectomy for Hodgkin's disease. *Penicillium marneffei* appears to be endemic to southeast Asia, southern China, and Hong Kong. Recently, there have been reports of *Penicillium marneffei* endemic among AIDS patients from northern Thailand (*Lancet* 1994; 344:110-113). Unpublished data revealed that there may be as many as 1,000 cases diagnosed and treated from 1991 to 1996. Among those cases reported from the U.K., France, Italy, the Netherlands, U.S.A. and Australia, the patients invariably had travelled to the endemic area. There have been more than 20 reported cases of disseminated penicilliosis in children from vertical transmission (*Pedriatr Infect Dis* 1993;12: 1021-25).

Tropical Dermatology, edited by Roberto Arenas and Roberto Estrada. ©2001 Landes Bioscience.

ETIOLOGY

The ecological niche and the mode of transmission is still controversial. The organism has been isolated from the internal organs (lungs, liver, spleen, and mesenteric lymph nodes) of two bamboo rat species (*Rhizomys sinensis* and *R. pruinosus*) and recently, in *R. sumatrensis* and *Cannomys badius* indicating that bamboo rats may be a reservoir for *P. marneffei*. It is possible that man may become infected by eating infected rats or by inhaling or accidentally ingesting fungi that have been dispersed into the environment. *P. marneffei* has been recovered from the soil of bamboo rat burrows whereas some studies showed no significant findings (*J Med Vet Mycol* 1986; 24:383-89). Bamboo rats and humans may acquire the infection from a common source, such as contaminated sugar cane or bamboos which bamboo rats feed on. However, *P. marneffei* has not been isolated from these sources in the field. At present, there is no evidence of the mode of transmission between man and bamboo rats. However, evidence of seasonal variation of penicilliosis in northern Thailand with an increase in incidence during the rainy season suggested possible airborne transmission through inhalation when conditions for growth of the fungus are favorable. Air sample culture from the endemic area have not yielded significant results. It is suggested that perhaps a few conidia may be sufficient to cause the disease in susceptible persons. *P. marneffei* is the only dimorphic *Penicillium spp* (Fig. 20.1). The pathogenic potential may be due to its thermal dimorphism and the ability to proliferate within macrophages, thereby, effectively shielding the organisms from the host's cellular immune response. Experimental infection demonstrated it to be highly pathogenic for hamsters, mice and rats with dissemination to the reticuloendothelial system similar to *Histoplasma capsulatum*.

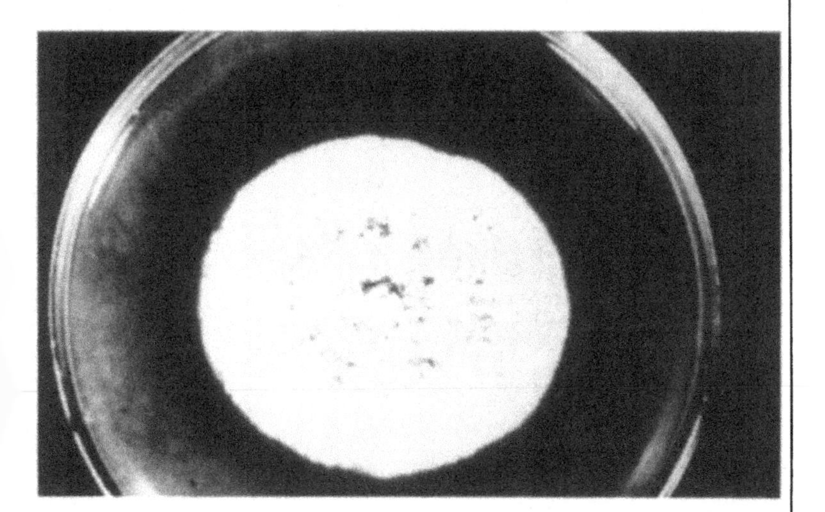

Fig. 20.1. *P. marneffei* culture in sabourau dextrose agar at 25°C showing red pigmenton reverse colony.

CLINICAL

The initial cases were reported in patients with underlying diseases or immunosuppressive therapies. Clinical manifestations were similar among AIDS and non-AIDS patients (J Mycol Med 1993; 4:195-224). The onset of infection may be acute, but more often patients experience several weeks to months of intermittent fever, headache, malaise, and weight loss. Fever was the most frequent sign of disseminated infection present in 95% of the patients with marked weight loss and anemia in more than 75% of cases. Lymphadenopathy, hepatomegaly and pulmonary abnormalities (cough, dyspnea, pleural diffusion, infiltrates on chest roentgenogram) were also common manifestations. Less common presentations included leukopenia, thrombocytopenia, splenomegaly, bone involvement, pericarditis or pericardial effusion, and gastrointestinal symptoms. However, in the AIDS group the clinical presentations were more acute and intense with a higher incidence of septicemia and an increased frequency of oropharyngeal and mucous membrane involvement (*J Mycol Med* 1993; 4:195-224). Absence of osteoarticular lesions was also noted. In endemic areas, the presence of fever with unexplained increase in alkaline phosphatase enzyme level was not an uncommon presentation for penicilliosis in AIDS patients.

Cutaneous manifestations occurred frequently and may be the initial presenting sign in HIV patients. Abscesses were common in non-HIV patients while cutaneous presentations were more diverse among HIV patients. Papules, pustules, nodules, vesicles, acneiform, maculopapular eruption and mucocutaneous lesions have been reported. Skin lesions appeared mainly on the upper half of the body, 70% on the face, scalp, upper extremities and trunk. Umbilicated papules (molluscum contagiosum-like), often with central necrosis, were seen in almost 90% of AIDS patients (*Mycoses* 1991; 34:245-249) (Fig. 20.2). These lesions were

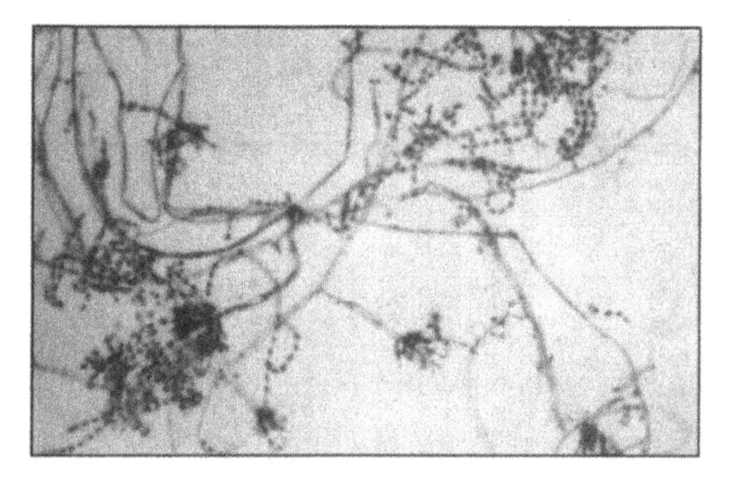

Fig. 20.2. *P. marneffei*, hyphal phase.

characteristic but could also be seen in a few other opportunistic fungal infections, namely histoplasmosis and crypttococcosis. The presence of the skin lesions usually suggested dissemination at the time of diagnosis. All infected children have symptomatic AIDS with the appearance of similar manifestations as seen in adults by the age of 2 years, although the presentation of papular lesions resembling insect bites was less common. Severe anemia, thrombocytopenia and osteomyelitis seemed to be more common in children than in symptomatic adults but whether they were of any importance as diagnostic clues had to be further evaluated. Penicilliosis is a treatable disease but delay in treatment may be fatal with almost a 100% mortality rate in both AIDS and non-AIDS patients. Clinical suspicion and early diagnosis are essential for prompt treatment and prognosis.

LABORATORY DATA

A quick and convenient diagnostic procedure is a direct touch smear and Wright's staining from tissue specimens. Intracellular and extracellular basophilic, spherical, yeast-like organisms with clear central septation are characteristic. Histological examination may be helpful. The most reliable procedure for isolation of *P. marneffei* with 100% sensitivity is culture from bone marrow aspirates or lymph node biopsies. At 25°C in Sabouraud's glucose agar, the mycelial saprophytic phase is characterized by the colony with a flat-down, bluish-gray-green center and white periphery. The reverse colony develops a brick-red color with diffusion of pigments into the agar in as early as 24-48 hours (Fig. 20.3). At 37°C, in vivo and in vitro (blood agar), the morphology of the yeast-like parasitic form is characterized by round, oval or elongated yeasts with planate or central septation. Rarely, short filaments may be visible. Immunohistochemical assays using a monoclonal

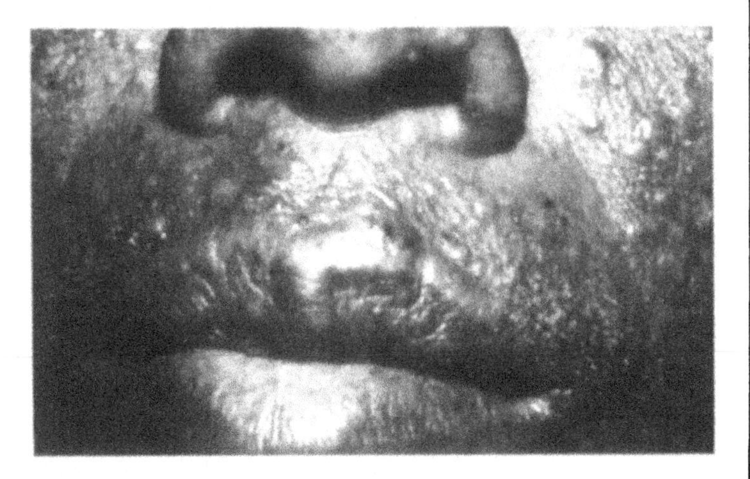

Fig. 20.3. Penicilliosis due to *P. marneffei*, umbilicated pupules.

antibody (EBA-1) directed against an external wall epitope shared by *P. marneffei* and *Aspergillus* species may be an alternative method of tissue diagnosis. Other monoclonal antibodies are being developed for early diagnosis.

TREATMENT

The first successful treatment of focal penicilliosis with 20 million units of oral nystatin (13 times the usual daily dose) for 30 days was reported in 1959. Since then there have been isolated reports of systemic treatment of systemic penicilliosis. Only recently have there been in vivo and in vitro susceptibility studies and treatment reports with various systemic antifungals. Most of the cases were in the AIDS population in the endemic area. Different antifungals have been administered with various results.

Nystatin is a polyene macrolide antibiotic and the earliest drug reported to be effective in an experimental infection in golden hamsters and mice. Amphotericin B (AMB) remains the drug of choice for severe cases of penicilliosis. An initial test dose of 1 mg AMB in dextrose solution (0.5 mg in children less than 30 kg) should be given over 1-2 hours with observation. In adults with normal renal function the usual dose of 0.3-1.0 mg/kg may be given in 4-6 hours. Most patients with systemic fungal infection are treated with 1-2 g of AMB over 6-8 weeks (cumulative dose of 40 mg/kg.) Renal tubular damage is the most serious toxic effect of AMB and it should be used with caution when administering other nephrotoxic and antineoplastic drugs. Seventy-eight percent of the patients treated responded favorably after 2 weeks. Relapses occurred in 17.6% within 6 months after cessation of therapy. Ketoconazole (KTZ) is active for the treatment of penicilliosis with MICs of 0.195 to 0.39 mg/ml. The dose given was 400 mg/d for a period of 4-8 weeks. However, because of its possible effects on the liver and on steroid metabolism, KTZ is not used as the first line of treatment. Transient minor elevation of liver function tests developed in 5-10% of patients on oral KTZ. Liver function tests must be performed before starting treatment and at frequent intervals thereafter, in particular in patients on prolonged treatment or receiving other hepatotoxic drugs. Fluconazole (FCZ) has a variable range of sensitivity as compared to other azole drugs (MIC 0.195-100 (g/ml) with 73% of *P. marneffei* strains resistant. FCZ is not used as an initial treatment but rather as maintenance therapy in a dose of 400 mg/d. However as initial treatment, 36.4% responded to FCZ 400 mg in two divided doses for 8 weeks. The drug is well tolerated but may augment the anticoagulant effect of warfarin and prolong the half-life of antidiabetics. Itraconazole (ITZ), a triazole with a broad spectrum, is used as an initial treatment for mild to moderately severe penicilliosis given at 400 mg for 8-12 weeks. It is advisable not to give the drug to patients with liver disease or to patients who have experienced hepatotoxic reactions with other drugs. Negative blood culture conversion within 24-90 days (mean 57 days) were seen in 75% with 62.5% relapse after 4 months cessation of therapy. Flucytocine (5-FC) is a synthetic fluorinated pyrimidine that inhibits fungal protein and DNA synthesis. There are limited data on the use of 5-FC in the treatment of penicilliosis. The average dose is 50-150 mg/kg given as four divided doses. The most common side effects are diarrhea, nausea and vomiting. Abnormal elevations of liver function tests developed in about 5% of the patients. Thrombocytopenia and leukopenia may occur if excessive serum concentrations are maintained. The effect is reversible if treatment is discontinued.

The drug is best avoided in AIDS patient due to several cases of bone marrow suppression. It is entirely the judgement of the physicians as to the choice of drugs. Selection should depend on clinical severity. The most important prognostic factor in patients with penicilliosis is the severity of the disease at the time of initial drug administration. With early diagnosis and prompt management, the response rate may be as high as 75-80% for AMB or ITZ and 36.4% for FCZ, while 100% mortality is expected if untreated. Because immunity is impaired in HIV infection and because most patients present with disseminated penicilliosis with a high risk of relapse after cessation of antifungal therapy, it is recommended that long-term prophylaxis with either ITZ, 200 mg daily, or KTZ, 200 mg daily, be given for life to prevent recurrence (*Antimicrob Agent Chemother* 1993; 37:2407-2411).

Selected Readings

1. Deng Z, Ribas JL, Gibson DW, Connor DH. Infections caused by *Penicillium marneffei* in China and southeast Asia: Review of eighteen published cases and report of four more cases. Rev Infect Dis 1988; 10:640-652.
2. Supparatpinyo K, Kwamwan C, Baosoung V. Disseminated *Penicillium marneffei* infection in southeast Asia. Lancet 1994; 344:110-113.
3. Sirisanthana V, Sirisanthana T. *Penicillium marneffei* infection in children infected with human immunodeficiency virus. Pedriatr Infect Dis 1993;12:1021-25.
4. Drouhet E. Penicilliosis due to *Penicillium marneffei*: A new emerging systemic mycosis in AIDS patients traveling or living in southeast Asia. J Mycol Med 1993; 4:195-224.
5. Supparatpinyo K, Nelson KE, Merz WG. Response to antifungal therapy by HIV-infection and in vitro susceptibilities of isolates from clinical specimens. Antimicrob Agent Chemother 1993; 37:2407-2411.

F. Mycobacteriosis

Leprosy

Cutaneous Tuberculosis

Atypical Mycobacteriosis

Leprosy

J. Octavio Flores-Alonso

Leprosy is an infectious, contagious and chronic disease caused by *Mycobacterium leprae*. It commonly affects the skin and peripheral nerves, though sometimes the neural damage is not clinically detected. It does not invade the spinal cord or the brain. Although it has always been thought that man is the only reservoir of *M. leprae*, other natural sources of infection have been proposed. Armadillos have been found in the USA and in Mexico with a natural infection indistinguishable from leprosy. And in Africa, it was found in a chimpanzee and in mangabey monkeys. Leprosy has evoked a strong stigma in all cultures in which it has occurred. It is the classic example of a social-medical disease.

GEOGRAPHICAL DISTRIBUTION

Leprosy occurs throughout the world but predominates in tropical and subtropical regions. The problem is greatest in Central Africa and Southeast Asia. Worldwide there are about 5.5 million cases of leprosy, but it is estimated that the total number of cases is actually 11.5 million of which 4 million are in India. In the Americas there are 400,000 cases of which 70% are in Brazil with a prevalence of 1.7 per 1000 inhabitants. The Unites States has more than 7,000 cases, most of them from immigrants. According to WHO the prevalence of leprosy has been modified because the patients that have undergone a course of chemotherapy are not considered active. For the first time the number of reported cases has declined from 5.37 million in 1985 to 3.1 million in 1992. In Mexico, of 16,694 reported cases in 1989 there were only 6,106 active patients, and 7,946 were successfully treated. This gives a prevalence rate of 0.6 per 10,000 inhabitants.

ETIOLOGY

The causative agent is *Mycobacterium leprae* (1873, Armauer Hansen). It is a gram-positive, intracellular, acid-fast bacillus that multiplies within histiocytes of the skin. It has the form of a cane, straight or bent, and measures 3-8 µm. The organisms cluster in packs called globi. When they deteriorate or die, they look granular or fragmented. It has not been possible to culture the organism. When organisms are inoculated in the plantar pad of the mouse, they grow slowly with a generation time of 11-13 days. Inoculates in armadillos of nine bands (*Dasypus*

Tropical Dermatology, edited by Roberto Arenas and Roberto Estrada. ©2001 Landes Bioscience.

novencinctus) reproduce abundantly reaching 100,000 million bacilli per gram of tissue while in man there are 7,000 million bacilli per gram of tissue. Its genes, protein antigens and carbohydrates are very well characterized. Defects in lymphocyte-macrophage interaction are related to impaired cellular immunity in leprosy patients. The populations of helper/suppressor T cells varies in the lesions of different forms of leprosy, and when IL-2 and gamma-interferon appear, they stimulate cellular immunity in lepromatous leprosy (*Dermatol Clin* 1992; 10(1): 73-96).

CLINICAL PICTURE

There are two totally different polar types of leprosy (Rabello Jr, 1938): lepromatous leprosy (LL), and tuberculoid leprosy (TT) as well as two groups of cases: indeterminate (I) and borderline (BL, BB, BT). Lepromatous leprosy is the progressive, systemic type, relatively transmissible, and it does not resolve spontaneously. It affects the skin and mucosae as diffusely infiltrating nodules. It also affects peripheral nerves and all organs and systems except the central nervous system. It is characterized by depressed cellular immunity that manifests as a negative lepromin skin test (Mitsuda's test) at three weeks; the humoral response is normal. Lepromatous leprosy has two clinical forms: nodular and diffuse. The nodular form is characterized by nodules, erythematous infiltrating plaques and hypopigmentation (Fig. 21.1). In the mucosa there is a lepromatous rhinitis that is associated with perforation of the cartilaginous nasal septum causing the so-called "saddle" nose. There is partial alopecia localized in nodules involving the eyebrows, eyelashes, and body. Neural leprosy is bilateral. In the eye it can produce episcleritis, diffuse infiltrative or focal keratitis, iritis and iridocyclitis. It affects all organs of the reticuloendothelium system. Diffuse lepromatous leprosy (Lucio 1852 and Latapi 1946) is characterized by a diffuse, generalized infiltration of the skin, that never becomes nodular, and may present a special type of lepromatus reaction known as the "Lucio's phenomenon" or necrotizing erythema. The diffuse infiltration is almost imperceptible. It gives the skin an edematous, shiny, tense, erythemato-violaceous appearance (Fig. 21.2). In its atrophic phase, the skin becomes thin, wrinkled, dry, and scaly (Fig. 21.3). Rhinitis and total eyebrow, eyelashes and body hair alopecia is found. There is a panneuritis without ocular involvement. Visceral involvement is more severe than in the nodular form. Tuberculoid leprosy affects only the skin and the peripheral nerves. In the skin it presents as one or just a few erythematous, infiltrative and asymmetric plaques. The lesions are red or brown, oval or rounded and raised. Sometimes they are anesthetic and hairless with scaling (Fig. 21.4). They spontaneously involute. Neural involvement is usually asymmetric. The so-called indeterminate cases are characterized by hypopigmentated and anesthetic skin which may also be hairless and anhidrotic (Fig. 21.5). Whereas some cases resolve spontaneously, others progress to lepromatous or tuberculoid leprosy. Borderline cases are unstable. Most of them

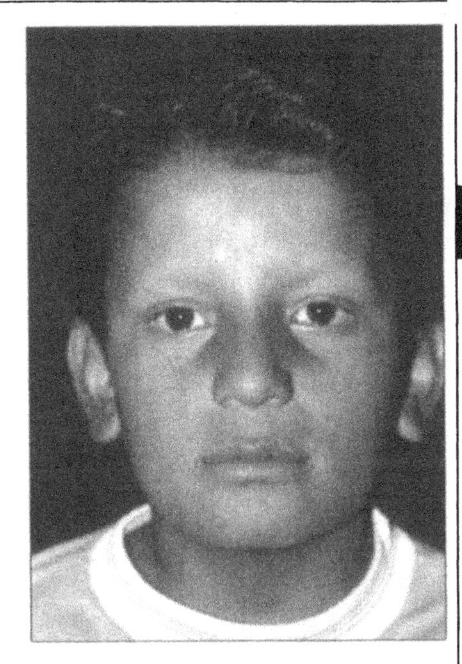
Fig. 21.2. Diffuse leprosy, infiltrative phase.

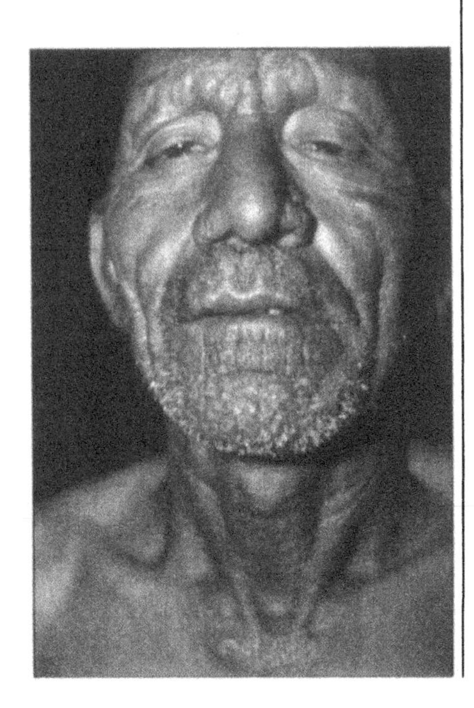
Fig. 21.1. Lepromatous nodular leprosy.

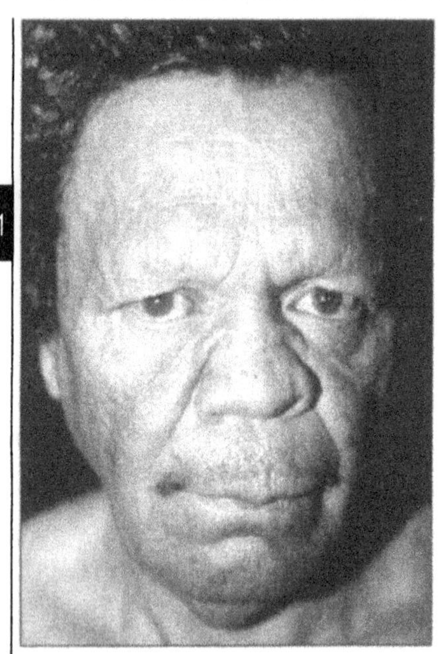

Fig. 21.3. Diffuse leprosy, atrophic phase.

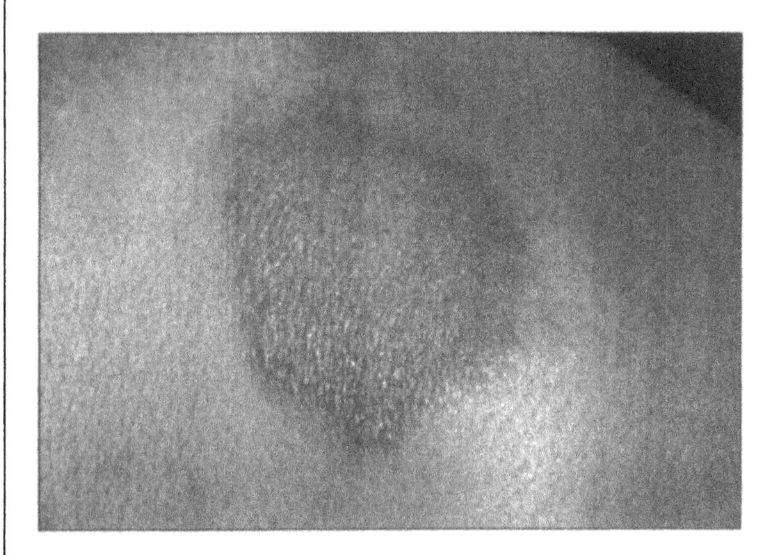

Fig. 21.4. Tuberculoid leprosy.

Fig. 21.5. Indeterminate case of leprosy.

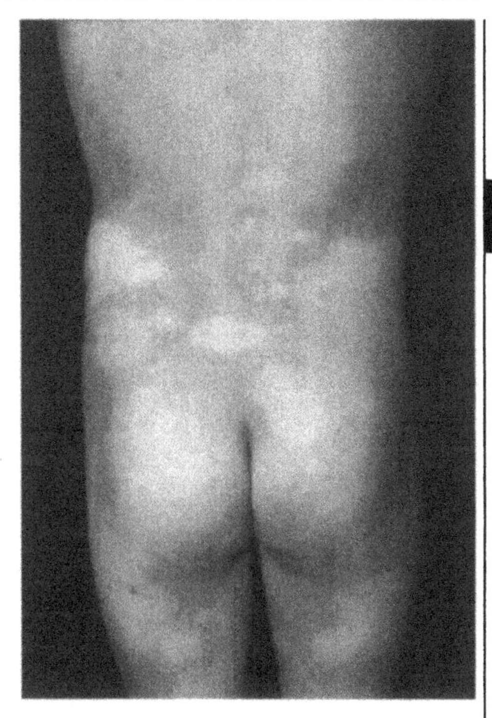

become lepromatous. Ridley and Jopling classify these as borderline tuberculoid (BT), borderline borderline (BB), and borderline lepromatous (BL). They affect the skin and peripheral nerves. Cutaneous lesions are infiltrated, asymmetric, annular, punched-out, with internal and external edges. Some lesions are oval or occur in bands (Fig. 21.6). In BL the lesions are more numerous, shinier, less asymmetric and anesthetic. Neural damage is extensive and frequently results in disability. Bacilli can be numerous or scant. If Mitsuda's test is negative, the patients must be treated as lepromatous.

Sixty per cent of lepromatous cases may present an acute reaction known as leprous reaction or erythema nodosum leprosum (type II). It is characterized by malaise, fever, chills, myalgia, arthralgia, as well as neurologic, visceral and cutaneous symptoms. Erythema nodosum is the main cutaneous manifestation (Fig. 21.6). It is always more extensive and severe than in other diseases (erythema nodosum leprosum). Less frequently multiform erythema may be observed and in diffuse lepromatous leprosy, Lucio's phenomenon (necrotizing erythema) may be present. It is characterized by vasculitic lesions with sharp stellate edges seen more frequently in the legs (Fig. 21.8).

Reversal reaction (type I) occurs in borderline or subpolar cases. It is an immunologic response that generally follows chemotherapy and is characterized

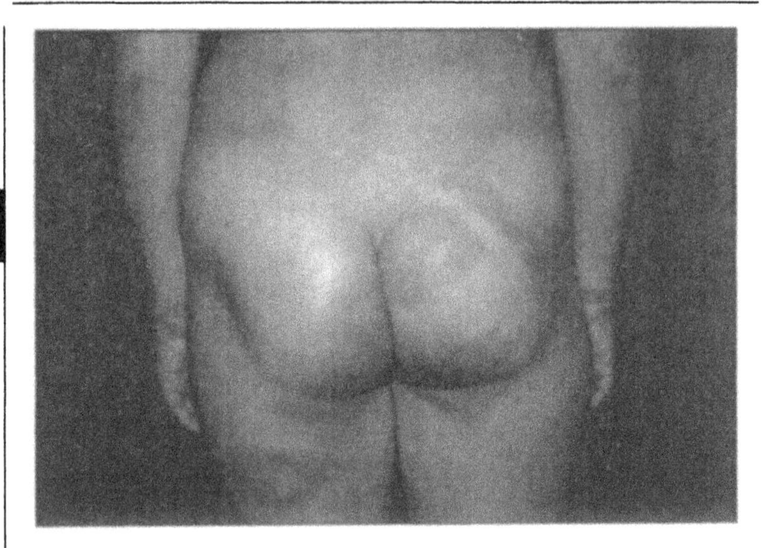

Fig. 21.6. Borderline leprosy.

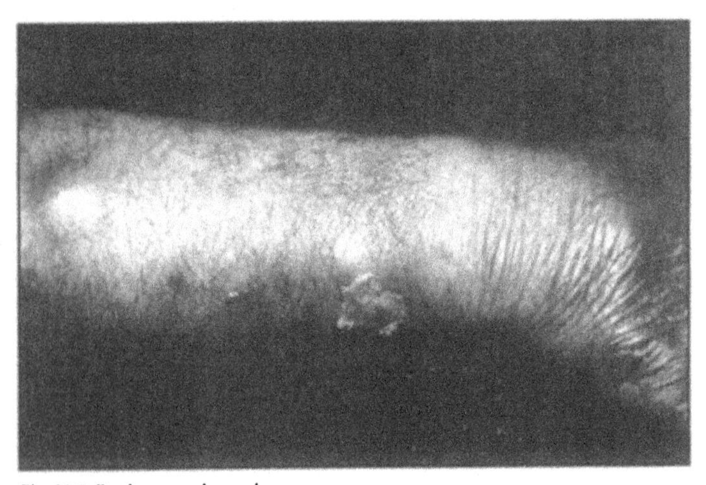

Fig. 21.7. Erythema nodosum leprosum.

by deterioration of the pre-existing lesions and by the development of new edematous lesions and marked exacerbation of neuritic symptoms (Fig. 21.9).

All forms of leprosy can be associated with interstitial and perineural involvement of peripheral nerves. There are sensory and motor changes in advanced cases with dysfunction and deformity (Fig. 21.10).

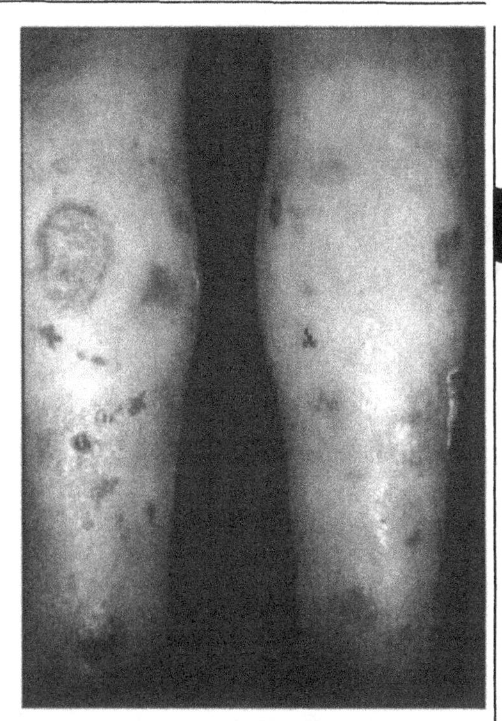
Fig. 21.8. Lucio's phenomenon.

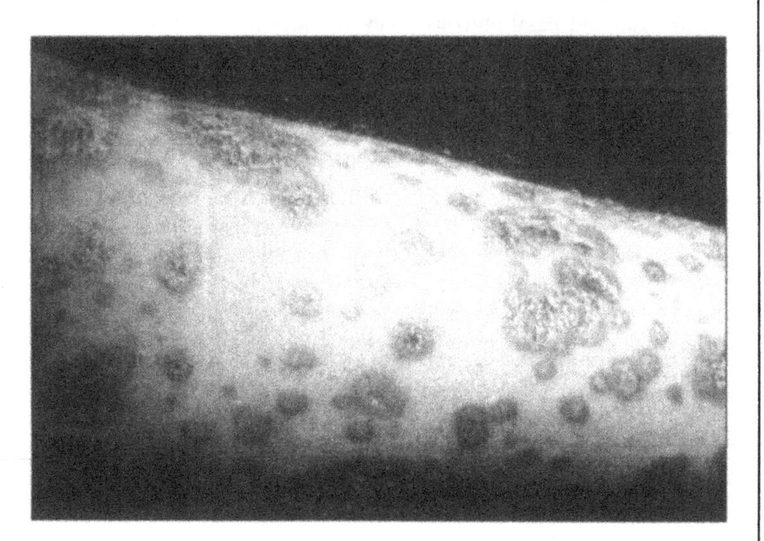
Fig. 21.9. Reversal reaction (type I).

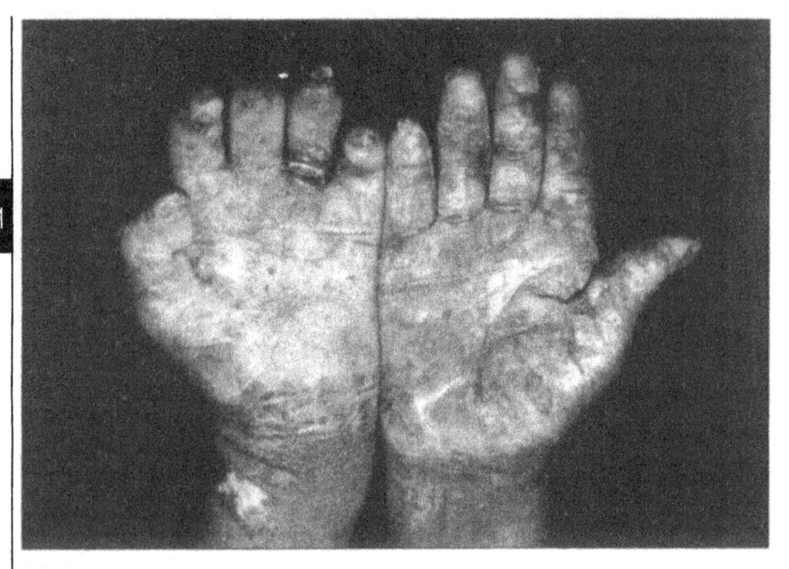

Fig. 21.10. Neuritis in leprosy.

LABORATORY DATA

Acid-fast smears are performed by incision of lesions, knuckle pads, lobule of the ear, and nasal mucous. They are negative in tuberculoid leprosy and in indeterminate cases. They are positive in lepromatous leprosy and sometimes in borderline cases (BL and BB). Biopsy in lepromatous leprosy reveals dermal infiltrates composed of numerous vacuolated histiocytes (Virchow cells). In the nodular form there is a subepidermal band of connective tissue (Unna's band). In the diffuse form of leprosy, the histiocytic infiltrate predominates in the deep dermis and the hypodermis. In both forms bacilli are observed with Ziehl-Neelsen or Fite-Faraco stain. In tuberculoid leprosy a tuberculoid granuloma is observed in the upper dermis which sometimes may damage the epidermis. In borderline cases there is a mixture of vacuolated histiocytes with bacilli and one tuberculoid granuloma. In indeterminate cases there is an nonspecific lymphocytic process. The lepromin skin test is useful for a better classification of cases. It is positive in TT cases, negative in LL cases and variable in borderline cases. Serologic tests are not practical, although they are very useful in theory for early detection of subclinical infection.

TREATMENT

In the pre-sulfa era, a great amount of medicine that did not have any effect was used, such as Chaulmoogra oil. In the sulfa era, derivatives of diaminodiphenylsulfone (dapsone) were used until the optimum dosage of 100 mg/day was found, but this monotherapy was abandoned as resistance developed. High doses are related to methemoglobinemia, hemolytic anemia, and agranulocytosis. Polychemotherapy is the current choice. It began in 1982 and it is considered the most important advance in the management and control of leprosy. The WHO has declared that leprosy will no longer be a public health problem after the year 2000 (less than 1 patient per 10,000 inhabitants). The current treatment regimens are: for multibacillary patients (LL and BL), rifampicin 600 mg plus clofazimine 300 mg and dapsone 100 mg in a supervised monthly dosage and daily in autoadministered form clofazimine 50 mg + dapsone 100 mg for two years; for paucibacillar patients (TT and I), rifampicin 600 mg plus dapsone 100 mg in a monthly supervised dosage and daily in auto-administered form, dapsone 100 mg during 6 months. The treatments must be continuous without interruptions. They are inexpensive, acceptable and associated with few side effects. For the acute lepra reaction (type II), the ideal drug is thalidomide in an adjustable dose. It can be started with 200 mg/day. Another useful drug is clofazimine in a high dose. In the reversal reaction (type I), high doses of corticosteroids must be used to avoid incapacitating neuritis. The following are promising drugs in the near future: clarithromycin 500 mg/day, minocyclin 100 mg/day, ofloxacin 400 mg/day/6months, and other fluoroquinolones like pefloxacin 400 mg twice a day (*Acta Leprol* 1991; 7(4):321-6), and sparfloxacin. Combinations are an alternative for cases of hypersensitivity or resistance. Treatment with polychemotherapy for 2 years and complete eradication of the mycobacteria is the goal of cure, although patients must be examined for relapses for as long as 5 years after having completed the proper treatment. In 1997 WHO proposed to treating multibacillary patients only for 12 months and paucibacillary patients with a single lesion only with a single dose of rifampicin 600 mg, ofloxacin 400 mg and mynociclin 100 mg. Physical disabilities must be treated with rehabilitation. But prevention is still the best approach to the incapacitating sequelae of leprosy.

Selected Readings

1 De Las Aguas JT et Lecciones de Leprologia (Leprology lessons) Fontilles 1973.

2 Ladhani S. Leprosy disabilities: the impact of multidrug therapy (MDT). Int J Dermatol 1997: 36: 561-72.

3 Lepra: Pasado, presente y perspectivas para el futuro (Leprosy: Past Present and perspectives for the future) Publicacion Tecnica del INDRE No 15, Mexico DF, 1992.

4 Manual de Procedimientos Operativos para el control de la lepra (Manual of operative procedures for the control of leprosy) Secretaria de Salud. Mexico, 1996.

5 Saul A.Lecciones de dermatologia, 13 Ed (Dermatology lessons) Mexico. Mendez-Cervantes 1993: 122-28.

6 Thangararaj RH, Yawalkar SJ. La lepra para medicos y personal sanitario, 3 Ed (Leprosy for doctors and sanitary personel) Ciba-Geigy. Basilea, 1988.

7 Waters MF. Chemotherapy of leprosy: Current status and future prospects. Trans R Soc Trop Med Hyg 1993; 87(5):500-3.

21

Cutaneous Tuberculosis

Roberto Arenas

Cutaneous tuberculosis is a chronic infectious disease caused by *Mycobacterium tuberculosis*. Localization depends on the clinical form, and the cutaneous lesions may be nodules, gummas and ulcers, as well as warty plaques and vegetative lesions. Besides the different morphologic aspects, there are atypical features in the immunocompromised patients.

GEOGRAPHIC DISTRIBUTION

It occurs worldwide, especially in impoverished populations. The frequency is low in the developed countries. In non-industrialized countries it constitutes 0.066-0.5% or even as much as 3% of the skin diseases. Unfortunately the incidence is increasing, especially where HIV is common. It predominates in women, with a rate 3:1, and it is more frequent between 11 and 30 years of age; 95% of cases occur in patients under the age of 50 years.

ETIOLOGY

It is caused by *M. tuberculosis*, the Koch's bacillus. In humans 95% of tuberculosis is caused by *hominis* variety, rarely by *bovis*. The bacillus is acid-fast. It measures 2.5-3.5 μm. Infection in humans occurs mainly by inhalation, less often by ingestion or inoculation. Skin involvement depends on virulence of the microorganism, the number of infecting bacteria, the individual's general health, host reactivity, as well as the mechanisms through which the bacteria is introduced into the skin. The relation between immunity and hypersensitivity is controversial. Infection stimulates the production of lymphocytes, lymphokines, interleukins and interferons which promote the accumulation of macrophages and the formation of granulomas. There is generally secondary infection. Reinfection (exogenous infection as in tuberculosis verrucosa cutis) or reactivation (misnamed endogenous reinfection like in scrofuloderma) is common. It is rarely primary.

Tropical Dermatology, edited by Roberto Arenas and Roberto Estrada.

CLASSIFICATION

1) Primary infection: cutaneous tuberculous primary complex.
2) Re-infection
    a) Fixed or localized forms: Tuberculosis colliquativa cutis colicuativa (Scrofuloderma, tuberculous pseudomycetoma, lymphangiitic gummas, hematogenous gummas). Lupus (Lupus vulgaris, lupus-warty tuberculosis, tuberculosis caused by BCG). Verrucous (prosector's wart, tuberculosis verrucosa cutis) Ulcerative, vegetative or ulcerovegetative, miliary.
    b) Hematogenous forms, recurrent or tuberculids: Bazin's indurated erythema and the Hutchinson's form, nodulonecrotic, micronodular, ulcerative tuberculid and tuberculids of the face.

CLINICAL PICTURE

Location and appearance depend on various factors. The cutaneous tuberculous primary complex or tuberculous cutaneous chancre is a rare form of exogenous infection which is more frequent in children and adolescents, and it predominates on the face and extremities. The initial lesion is a painful nodule that ulcerates rapidly. At 3-8 weeks, regional adenopathy appears but resolves spontaneously in 2-5 months (Fig. 22.1). It is usually acquired by managing animal products or taking care for persons who have had active tuberculosis. Tuberculosis colliquativa cutis or scrofuloderma is the cutaneous form of tuberculosis most frequent in Mexico (27-51%). It mainly affects children and undernourished adolescents. It occurs by extension of a tuberculous focus in lymph nodes, bones or joints. It is observed mainly in the neck, axillae and groin. Lesions are painless nodules and gummas that can open and discharge a thick, yellowish pus; sinus tracts and ulceration are observed. These nodules leave behind scars or keloids (Fig. 22.2). The course is slow; there can be fever and weight loss. Tuberculous pseudomycetoma usually results from osseous or articular tuberculosis and manifests as multiple fistulous lesions. The gummatous lymphangitic tuberculosis is characterized by multiple gummas that may be associated with prosectors wart, warty tuberculosis or spina ventosa. They are more frequently observed in the extremities, and adenopathy is not a common feature. Gummatous hematogenous tuberculosis or the abscedants metataic form develops from an occult tuberculous focus and is more frequent in children with immune abnormalities. The gummas are multiple on the trunk and extremities, and they usually ulcerate. Lupus tuberculosis or lupus vulgaris (11-42%) is frequently secondary to reactivation of an endogenous infection. It may be present in the colliquative form. The most frequently affected areas are the central face and ears, although it also appears on the trunk and extremities. There are erythematous and squamous wart-like plaques with centrifugal growth and central ulceration, and atrophic or keloidal scars. The primary lesion is a small, yellow, granular nodule with a tapioca-like appearance

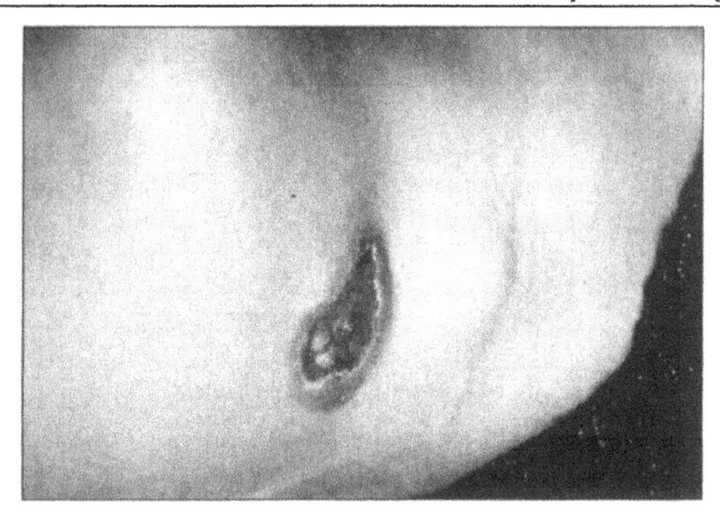

Fig. 22.1. Tuberculous chancre and adenopathy in the neck.

that is called lupoma and is observed at vitropression (Fig. 22.3). The clinical distinction among sarcoidosis, chronic lupoid leishmaniasis, and the so-called lupus miliaris disseminatus faciei is very difficult.

Tuberculous veravcous lupus is a variety of lupus vulgaris with active and exophytic lesions. The tuberculosis that is caused by BCG (bovine bacillus or bacillus of Calmette-Guerin) occurs at the site of BCG vaccination. It is a form of lupus in children and adolescents and also can cause a colliquative lesion in the neck. Tuberculosis verrucosa cutis or warty tuberculosis (8-31% of cases) is due to exogenous reinfection. It occurs in individuals who are in contact with material contaminated by the bacillus. It is localized in the extremities, frequently on the hands, feet, and sometimes the buttocks (Fig. 22.4). The lesions are nodules and warts of different sizes and shapes that are grouped in plaques that tend to grow peripherally with central scarring. If several plaques coalesce, they may affect large areas; in chronic cases in the extremities, lymphostasis and elephantiasis can occur. The necrotic wart or dissector's wart is a variety or early form of warty tuberculosis. When due to accidental or professional inoculation, it is usually localized on the hands or fingers. The initial nodule is inflamed or painful.

Ulcerative or orificial tuberculosis is somewhat common; it originates by auto-inoculation. It appears on the face (periorificial) or extremities and is characterized by ulcers of different sizes and forms. Active extracutaneous tuberculosis must be sought. The vegetative or ulcerovegetative form is exceptional, fast growing and very infiltrative.

Miliary tuberculosis or tuberculosis cutis disseminata is a rare acute form of hematogenous dissemination with a poor prognosis. It affects children who are immune-compromised. It is secondary to visceral or chronic cutaneous tuberculosis. Nodules, pustules, ulcers and hemorrhagic lesions are present.

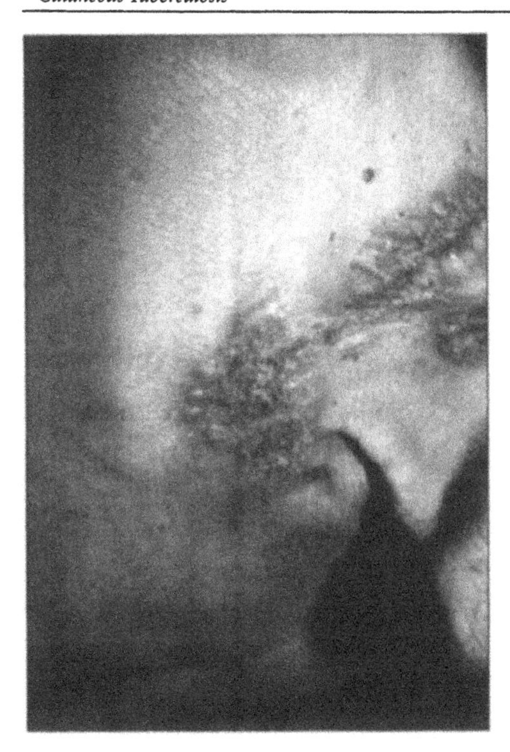

Fig. 22.2. Scrofuloderma and secondary lupus.

Bazin's indurated erythema or deep nodular tuberculid (15-79.5% of cases) predominates in young women. It affects the calf, and it may involve thighs. It is usually bilateral. There are deep, erythematous and painful nodules (Fig. 22.5). It occurs as outbreaks, more frequently in cold weather. When necrosis occurs, it becomes what is known as Hutchinson's indurated erythema in which, in addition, this nodules ulcerate and are covered by purulent exudate that leaves atrophic areas after remission. Recently it has been found through PCR that Bazin's indurated erythema can be a true tuberculosis, not a tuberculid (*Lancet* 1993: 342: 747-8; *Am J Dermatopathol* 1995: 17(4): 350-6; *J Am Acad Dermatol* 1997; 36(1): 99-101).

Nodulonecrotic or papulonecrotic tuberculid is more common in children and adolescents. Frequently it is associated with pulmonary tuberculosis. It has a preference for elbows, knees, thighs, buttocks and trunk. Nodules are less than 0.5 cm in diameter. They are erythematous and violaceous with a black, central necrotic zone that, when it comes off, leaves a varioliform scar (Fig. 22.6).

Micronodular (micropapular) tuberculid, lichenoid or lichen scrofulosorum appears on the trunk and extremities, mainly in children and adolescents. It is characterized by skin-colored or hypochromic 1-2 mm nodules. They grow in various sizes, usually as oval plaques. It is asymptomatic, occurs in outbreaks and

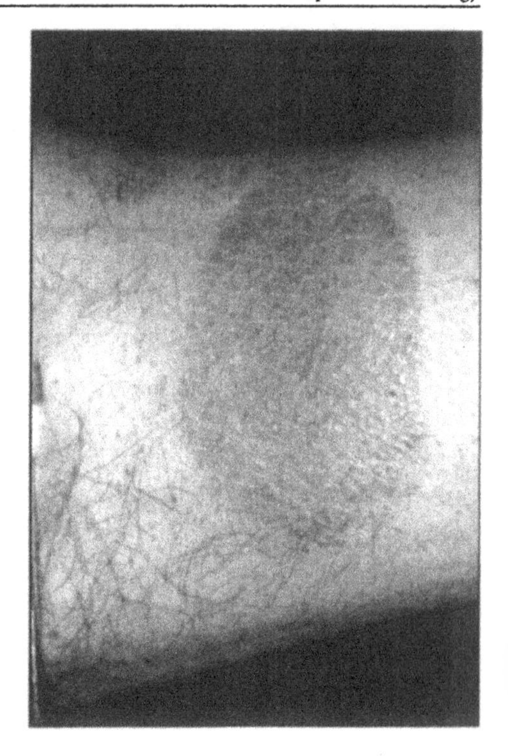

Fig. 22.3 Lupus tuberculosis.

is similar to a follicular keratosis. Sometimes it coexists with other forms of tuberculosis.

Ulcerative tuberculid is a rare form that predominates in women. It affects legs and thighs. It is characterized by ulcerated arciform lesions that grow slowly and are confused with lupus vulgaris and erythema induratum. Many authors do not accept the existence of facial tuberculids, but they are seen in adults as erythematous, small and solid nodules which occur in outbreaks and are localized to the central part of the face and perioral region.

Miliary disseminated lupus (lupus miliaris disseminatus faciei) is considered a pseudotuberculid. The cause is unknown; it appears in adolescents and adults. It is characterized by a chronic eruptive outbreak, usually centrofacial, of red, smooth and firm papules of 1-3 mm. It may also be observed in extremities. On biopsy, a tuberculoid granuloma is found, sometimes with necrosis. The entity may be confused with sarcoidosis or rosacea. It does not resolve with antituberculous drugs, but it does clear with tetracycline and diaminodiphenylsulfone. It may also resolve spontaneously within two years. For some authors only the micronodular and the nodulonecrotic forms are genuinely tuberculids and diagnosis is based on a positive response to PPD.

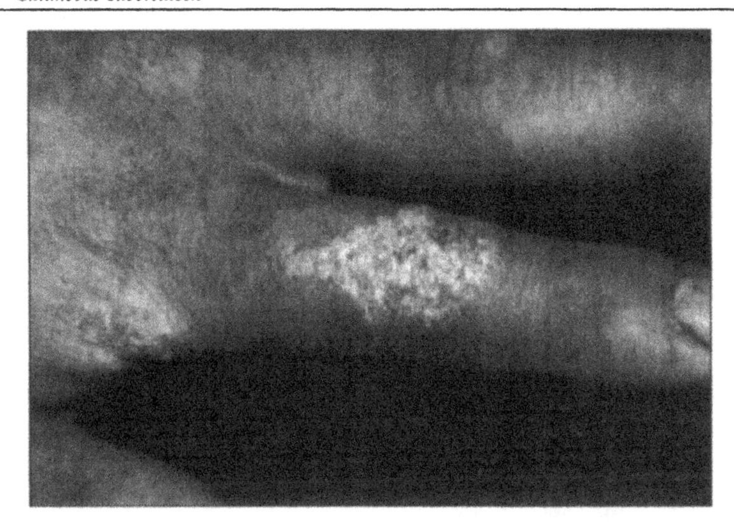

Fig. 22.4. Warty tuberculosis.

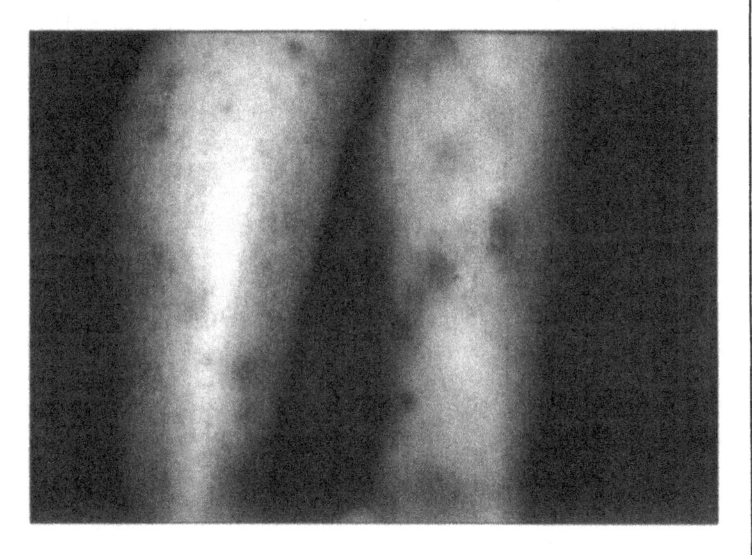

Fig. 22.5. Bazin's induratum erythema.

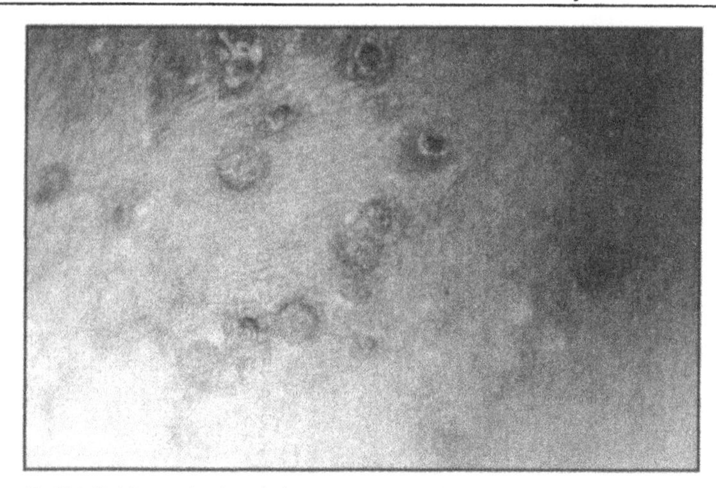

Fig. 22.6. Nodulonecrotic tuberculosis.

Osseous tuberculosis is osteoarticular in 84% of cases. It frequently causes secondary cutaneous lesions. When it affects the short bones of the hands and feet, it is called tuberculous dactylitis. It appears above all in children and adolescents. It is characterized by a painful tumefaction and is associated with ulceration and sinus tracts (pseudomycetoma). It is possible to distinguish by X-ray an expansive form (spina ventosa) or a lytic destructive form.

LABORATORY DATA

On biopsy a typical tuberculoma is found, composed of lymphocytes, epithelioid cells, and multinucleated giant cells of Langerhans type. Various grades of caseous necrosis and vascular lesions are seen. Bacilli are rarely found, and culture is usually negative, although a positive reaction to the PPD skin test (50-70%) has been reported. The intradermal reaction with the purified protein derivative (PPD) does not reflect the gravity of infection. X-rays are useful to diagnose tuberculosis in other regions like lung, bones, and joints. Vaccination with BCG can excite a positive reaction to PPD, but for unknown reasons in 5% of patients with tuberculosis there is a negative response to this antigen. ELISA has been used. PCR can identify DNA of *M. tuberculosis* in sections of paraffin by detecting a specific 123 bp fragment of DNA. Above all, it is useful in paucibacillary forms (*Arch Dermatol* 1996; 132: 71-75; *Int J Dermatol* 1996; 35(3): 185-8). The absolute criteria for diagnosis are demonstration of acid-fast bacilli on Ziehl-Neelsen's staining, recovery of *M. tuberculosis* in Lowenstein-Jensen medium in 3-4 weeks and positive inoculation in 6-7 weeks in guinea pigs. A tuberculoid granuloma on biopsy is suggestive of the diagnosis, as is positive tuberculin or a PPD skin test. The presence of tuberculids has always been questioned for there is not a clinical or a specific histological image, and with the immunologic techniques and the molecular biopsy the concept is changing. Based on PCR in Bazin's indurated erythema and nodulonecrotic tuberculid, the strong relationship between these afflictions and *M. tuberculosis* is confirmed.

TREATMENT

A 2-3 month trial of multidrug therapy is recommended. When the diagnosis is confirmed, the treatment duration is 4-6 months or up to a year, depending on the clinical response. The following is administered: streptomycin 1 g IM every other day (up to 60 g) or rifampicin, 10 mg/kg (600 mg/day), in addition to isoniazid 5 mg/kg (300 mg/day) and ethambutol 20 mg/kg (1200 mg/day), in adults. To avoid secondary neuropathy, pyridoxine 50 mg/day should be added. Rifampicin and isoniazid are intra- and extracellular bactericides in both active and dormant bacilli. The pyrazinamide is an intracellular bactericide. It is administered with rifampicin and isoniazid in the brief regimens. Other compounds are ethionamide, thioacetazone, cycloserine, copreomycin, viomycin, and para-aminosalicylic acid.

For cutaneous tuberculosis the efficacy of short courses of treatment is not completely proved, but multidrug regimens seem to work. For hematogenous forms it is advisable that prednisone 20 mg/day be administered and tapered progressively. For deep nodular tuberculosis, KI can be added. In paucibacillary cases, vitamin D_2 (calciferol) can be added, 600,000 U a week orally for 2-4 months. This vitamin does not act on the bacilli, but helps healing the granuloma. One form of protection for individuals exposed to infection is vaccination with BCG.

SELECTED READINGS

1. Chong LY, Lo KK. Cutaneous tuberculosis in Hong Kong: a 10-year retrospective study. Int J Dermatol 1995; 34(1): 26-9.
2. Jordaan HF, Schneider JW, Schaaf HS et al. Papulonecrotic tuberculid in children. A report of eight cases. Am J Dermatopathol 1996; 18(2):172-85.
3. Sehgal V, Bhattacharya SN, Jain S, Logani K. Cutaneous tuberculosis: The evolving scenario. Int J Dermatol 1994; 33(2):97-104.
4. Sehgal V, Jain MK, Srivastava G. Changing pattern of cutaneous tuberculosis. Int J Dermatol 1989; 28(4):231-236.

Atypical Mycobacteriosis

Roberto Arenas

Acid-fast bacilli other than *M. tuberculosis* cause pulmonary, cutaneous or systemic disease.

GEOGRAPHIC DISTRIBUTION

Atypical mycobacteriosis is considered rare but has increased with the AIDS epidemic. In the United States the prevalence is 1.8 per 100,000 inhabitants. Organisms involved are *M. avium* complex in 61-79% of cases, *M. fortuitum* in 19%, *M. kansasii* in 10-25% and others in 5%.

ETIOLOGY

Infection originates from opportunistic mycobacteria—saprophytes from the ground, natural or stored water, and solutions, that Runyon classified in four groups:

Atypical mycobacteria
- Slow growers:
  1) Photochromogens (*M. kansasii, M. marinum*)
  2) Scotochromogens (*M. scrofulaceum* and *M. xenopi*)
  3) Non-chromogens (*M. avium intracellulare, M. ulcerans*)
- Rapid growers:
  4) (*M. chelonae, M. fortuitum*)

Entry is by inoculation, spreading from a deep focus or by hematogenous dissemination. Illness occurs in individuals with intact immune mechanisms, but above all in individuals immunosuppressed after transplantation or with AIDS.

CLASSIFICATION

Pulmonary illness, lymphadenitis, cutaneous ulcerative mycobacteriosis, mycobacterial abscesses, fish tank granuloma, postsurgical mycobacteriosis, and AIDS.

Tropical Dermatology, edited by Roberto Arenas and Roberto Estrada. ©2001 Landes Bioscience.

CLINICAL PICTURE

The pulmonary illness is indistinguishable from tuberculosis. It can be primary or secondary. In the secondary form there is an increase in gamma globulin. Cutaneous manifestations are diverse: nodules, gummas, abscesses, warty plaques, pustules, and sinus tracts among others (Fig. 23.1). Lympahadenitis is known as nontuberculous scrofuloderma. It affects mainly children. In 99% there is a painless cervical nodule; the best treatment is excision. Cutaneous ulcerous mycobacteriosis or Buruli's ulcer is more frequent in the tropics, and it is characterized by rapidly growing, deeply and undermined skin ulcer with wide zones of necrosis and elevated edges. Mycobacterial abscesses appear 1-3 months after an intramuscular injection. They are commonly localized on the buttocks; they are chronic, cold abscesses and fistulas. Swimming pool, fish tank or aquarium granuloma is acquired by traumatic inoculation and contact with water. It predominates in elbows, knees, and hands. Nodules are single, multiple or sporotrichoid (Fig. 23.2). Postsurgical mycobacteriosis appears after major surgery, mainly in immunosuppressed patients. Nosocomial epidemics involving contaminated solutions and

instruments have been reported. Mycobacteriosis associated with AIDS (2.9%) usually presents with fever, weight loss, anorexia, and lyphadenitis. It is related to low numbers of $CD4^+$ T cells. The infection in these patients is usually lethal; survival is 7-8 months. In the skin there are disseminated papular lesions, pustular abscesses of the ecthyma or cellulite type (Fig. 23.3).

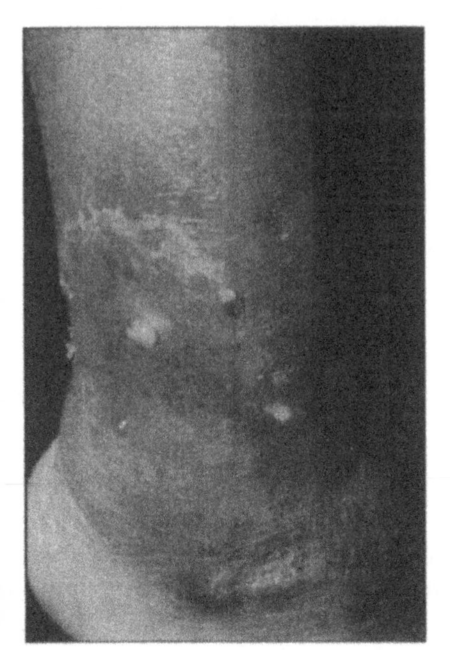

Fig. 23.1. Sinus tracts in cutaneous mycobacteriosis.

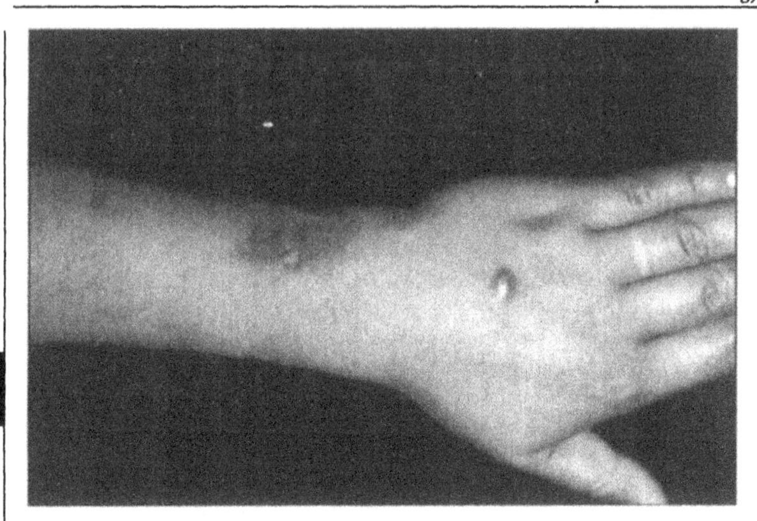

Fig. 23.2. Fish tank granuloma.

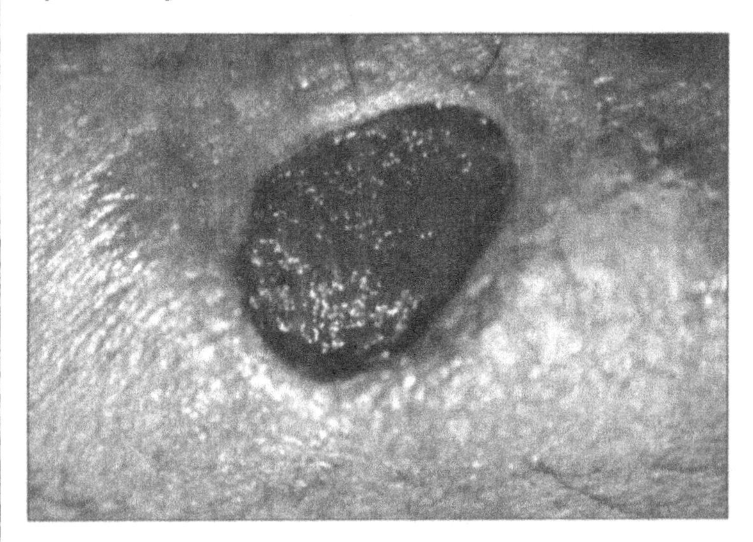

Fig. 23.3. Ecthyma-like mycobacteriosis in AIDS.

LABORATORY DATA

The exact diagnosis is very difficult to establish. It is confirmed with a Ziehl-Neelsen stain of exudate, and by culture, recovering niacin negative bacilli. On biopsy there are lymphocytes, neutrophils and necrosis. Often there are epithelioid and Langerhans giant cells containing acid-fast bacilli. The mycobacteria can be identified by gas chromatography and PCR.

TREATMENT

There are neither controlled studies nor a highly effective drug of choice for treatment of atypical mycobacterial infection (*Pneumologie* 1994; 48(9):711-7). Antituberculous drugs are administered, but usually there is resistance. Several drugs should be administered simultaneously for a prolonged course—at least three medications in usual doses such as amikacin, trimethoprim-sulfamethoxazole, cefoxitin, cefotaxime, procodazol, doxicycline, sulfamides, erythromicin, and new analogs such as clarithromycin, as well as minocyclin, clofazimin, diaminodiphenylsulfone, and quinolones like ciprofloxacin, ofloxacin and norfloxacin. For lymphadenitis and Buruli's ulcer, surgical treatment is important.

Selected Readings

1 Arenas R, Vega-Memije ME, Hojyo MT et al. Micobacteriosis atipicas: Aspectos clinico-epidemiologicos de 44 casos (Atypic mycobacteriosis: Clinical-epidemiologic aspects of 44 cases). Dermatologia Rev Mex 1993; 37(5): 305-315.
2 Butknecht DR. Treatment of disseminated mycobacterium chelonae infection with ciprofloxacin. J Am Acad Dermatol 1990; 23(6):1179-1180.
3 Street ML, Umbert-Millet IJ, Roberts GD, Su WP. Nontuberculous mycobacterial infections of the skin. Report of fourteen cases and review of the literature. J Am Acad Dermatol 1991; 24(2):208-15.

G. Pyodermas

Impetigo

Folliculitis/Furnunculosis

Hidradenititis

Ecthyma/Erysipelas

Necrotizing Faxciitis
(Streptococcal Gangrene)

Staphylococcal Scalded Skin Syndrome
(Ritter's Disease)

Anthrax

Tularemia

Rhinoscleroma

Lyme Borreliosis

Impetigo

Roberto Cortes-Franco

Impetigo is an infectious disease of the superficial layers of the skin. It is the most common pyodermitis in children. Although this disease occurs anywhere in the world, the poor hygienic conditions, poverty and overcrowding in tropical countries make impetigo more common in these regions.

GEOGRAPHIC DISTRIBUTION

It has been reported that 10% of the children in the United States and up to 80% of the children living in tropical countries suffer from impetigo at some time during infancy. It affects both sexes and is much more frequent in children than in adults.

ETIOLOGY

The majority of cases of impetigo (up to 96%) are caused by *Staphylococcus aureus* although in some cases *Streptococcus pyogenes* (beta-hemolytic streptococcus, group A) is the responsible pathogen. Bullous impetigo is caused exclusively by *S. aureus* (group II phage) which produces an exfoliative toxin that induces the formation of intraepidermic bullae at the subgranular level. Initially *S. aureus* colonizes the nasal epithelium; it is from here that it involves the skin. *S. pyogenes* colonizes the skin directly, affecting fibronectin exposed by trauma. Even though *S. aureus* is not considered normal cutaneous flora, approximately 20% of healthy people are asymptomatic carriers in the nares or perineum, and patients who suffer recurring episodes of impetigo are often nasal carriers.

CLINICAL FEATURES

Impetigo is characterized by bullous lesions that are so superficial that they are easily broken leaving erosions and yellowish scabs situated over erythematous skin (Fig. 24.1). Although lesions can appear on any part of the body, they are common around the nostrils, mouth, the outer ear or perineum. In some cases only a few lesions appear whereas in others, lesions appear on all parts of the body. Bullous impetigo is more common in newborns, and infants tend to have many lesions,

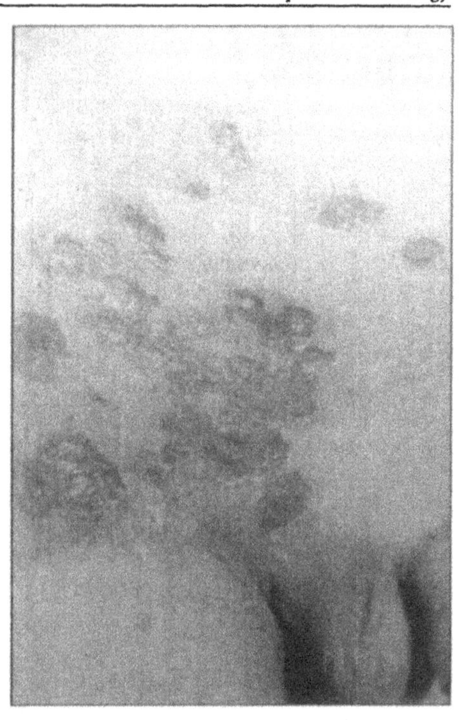

Fig. 24.1. Bullous impetigo.

most of which are bullae with serous and purulent content. It is a very contagious disease, particularly among children and, above all, in tropical countries where hygienic conditions and overcrowding favor its spread. Impetigo is considered primary when it occurs on healthy skin and secondary when it occurs as a complication of another dermatosis, e.g. scabies, varicella, pediculosis, insect bites and atopic dermatitis (Fig. 24.2). Impetigo caused by *S. pyogenes*, although clinically similar to that caused by *S. aureus*, is of particular importance because of the association with post-streptococcal glomerulonephritis which occurs in as many as 5% of untreated cases. In some tropical regions, the majority of cases of acute glomerulonephritis in children are the consequence of an untreated streptococcal pyodermititis.

LABORATORY DATA

Laboratory data is not necessary. If desired, a culture should be grown from the content of the vesicles or even from the skin beneath the scabs.

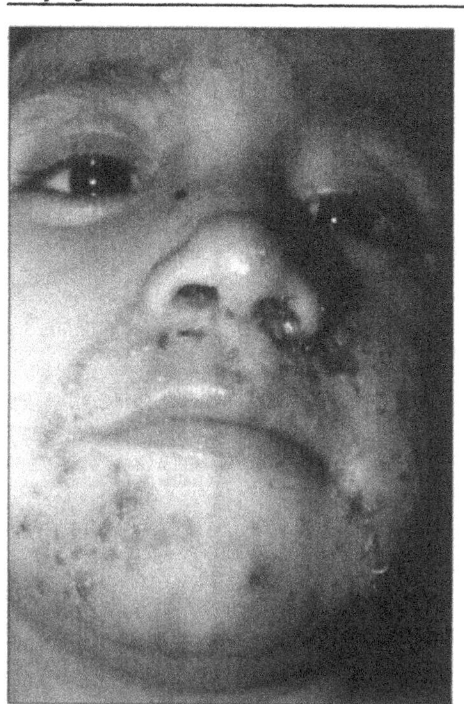

Fig. 24.2. Varicella with secondary impetigo.

TREATMENT

The ideal treatment is determined after culture. Nevertheless, the marked predominance of *S. aureus* and the fact that antibiotics used against it are also useful in treating cases caused by *S. pyogenes*, make culture unnecessary in the majority of cases. Localized and relatively limited cases of impetigo can be treated with topical antibiotics alone. The topical application of a mupirocin ointment b.i.d. is effective in more than 90% of cases. Topical rifampin and fusidic acid are also effective. Extensive infection requires management with systemic antibiotics which are resistant to beta-lactamase since practically all strains of *S. aureus* produce this enzyme and therefore attack the beta-lactamic ring of penicillin and ampicillin. The most widely used are dicloxacillin and oxacyline at a dosage of 500 mg q.i.d. or 20-50 mg/kg/day divided into four doses. Erythromycin is useful in those cases in which the etiologic agent is *S. pyogenes*, but strains of *S. aureus* resistant to this antibiotic are increasingly common. The same can be said of other macrolides such as azithromycin and clarithromycin that on the other hand are more easily tolerated and have a simpler posological regimen than erythromycin. Ampicillin and amoxicillin are destroyed by beta-lactamase producing strains, but combination with clavulanate or sulbactam makes them resistant and a good alternative for the treatment of impetigo whether it is caused by *S. aureus* or *S. pyogenes*. The normal dosage is 500 mg t.i.d. for 7-10 days or 40 mg/kg/day divided in three doses.

In view of the fact that the crusts, which form when the vesicles of impetigo break, are an excellent medium for the bacterial culture, compresses of saline solution, aluminum acetate (Burow's solution), copper/zinc sulfate (Dalibour) or potassium permanganate ($KMnO_4$), 2-3 times a day, are a valuable therapeutic aid.

Selected Readings

1 Chapel KL, Rasmussen JE. Pediatric dermatology: Advances in therapy. J Am Acad Dermatol 1997; 36:513-26.
2 Darmstadt GL, Lane AT. Impetigo: An overview. Pediatr Dermatol 1994; 11:293-303.
3 Shriner-DL, Schwartz-RA, Janniger CK. Impetigo. Cutis 1995; 56:30-2.

Folliculitis/Furunculosis

Roberto Cortes-Franco

Folliculitis and furunculosis are infectious diseases of the skin originating in the hair follicle and almost always caused by staphylococci. They occur most often in adults. They can be observed on any part of the body but are more common on the scalp, the beard and the thighs.

GEOGRAPHIC DISTRIBUTION

They are observed throughout the world but are more common in tropical climates.

ETIOLOGY

Folliculitis and furunculosis are infections of the skin, originating in the hair follicle. Whereas histologically in folliculitis only one acute superficial inflammatory infiltrate exists around the follicle, in furunculosis an additional deep abscess with infiltrates of lymphocytes, neutrophils, plasma cells and multinuclear giant cells is found. *Staphylococcus aureus* is the causal agent of superficial pustular folliculitis (Bockhart's impetigo), of common sycosis (*sycosis vulgaris*), of keloidal folliculitis of the nape of the neck and of furnuculosis. Often those who suffer the disease are nasal carriers of *S. aureus*.

CLINICAL FEATURES

Superficial pustular folliculitis is manifested by follicular papules that quickly become pustules. They disappear spontaneously in 7-10 days without leaving a scar. They are generally asymptomatic although in some cases there can be mild itching and, very exceptionally, pain. It occurs in outbreaks, and it is not uncommon that while some of the lesions are disappearing, new lesions are beginning to appear. The most commonly affected locations are the scalp, the proximal extremities, the bearded area (*sycosis barbae*), the axillae, pubis and the gluteal region (Fig. 25.1). Keloidal folliculitis of the nape of the neck is characterized by the appearance of follicular lesions that when they heal, leave a keloidal scar. As new lesions of folliculitis appear, the keloid grows in size producing a cicatricial alopecia and deformity of the region. (Fig. 25.2). There are two rare varieties of

Tropical Dermatology, edited by Roberto Arenas and Roberto Estrada. ©2001 Landes Bioscience.

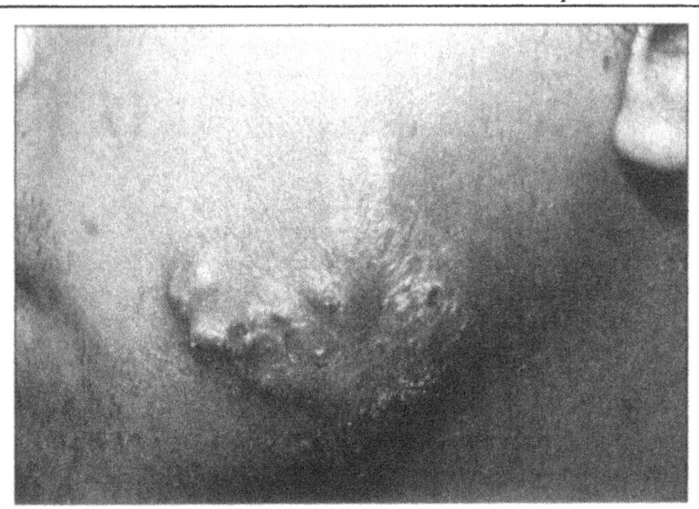

Fig. 25.1. Sycosis barbae.

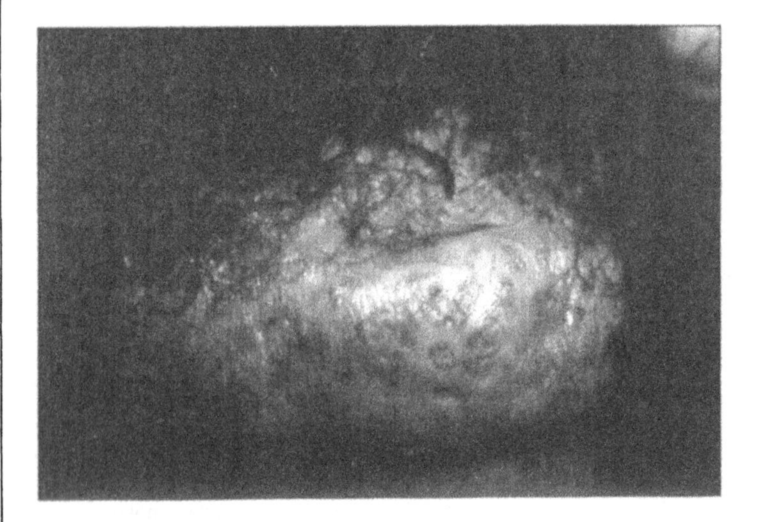

Fig. 25.2. Keloidal folliculitis of the nape of the neck.

folliculitis. Folliculitis caused by gram-negatives (Enterobacteriaceae, Klebsiella, Escherichia, Serratia or Proteus) that is observed in the perioral region in patients with common acne and very seborrheic skin who have been treated over a long period with systemic tetracycline and/or topical antibiotics. It is treated with oral isotretinoin. Folliculitis caused by *Pseudomonas aeruginosa* that occurs in small epidemics related to contaminated bathtubs, jacuzzis and swimming pools is clinically characterized by pruriginous follicular pustules that appear 1-2 days after immersion in the contaminated water (*Semin Dermatol* 1993; 12:336-41). The lesions, although they can appear anywhere on the body, are more abundant on the back, axillae, buttocks and the proximal extremities; even without treatment, they disappear spontaneously in 1-2 weeks. Furunculosis is characterized by the appearance of painful lesions which are initially papular but which turn into abscesses and once they have run their natural course can open to the surface and leave a scar. Generally they occur as single lesions or in scant numbers which most commonly appear in the axillae, groin, buttocks or thighs. When two or more furuncles, with their respective drainage orifices, coalesce, they are called a carbuncle (Fig. 25.3). With some frequency, friction applied to a folliculur lesion can precipitate the appearance of a furuncle. Both folliculitis and furunculosis tend to be more common in obese, diabetic and immunosuppressed patients. The following are predisposing factors for the development of folliculitis and furunculosis: humidity, poor personal hygiene, friction (e.g. between the thighs of obese patients), shaving (legs, pubis), sitting for prolonged periods on seats covered in plastic (folliculitis on the back and buttocks) and, above all, hot and tropical climates.

TREATMENT

If it is a matter of only a very few localized lesions, the use of topical antibiotics such as mupirocin, fusidic acid, or even benzoyl peroxide, can be effective. In extensive or recurrent cases, use of tetracycline 500 - 1500 mg/day, minocycline 50-150

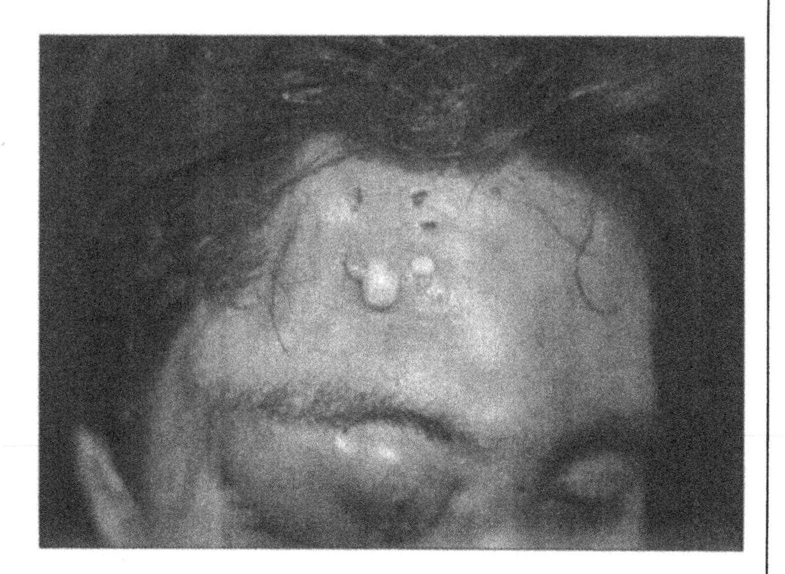

Fig. 25.3. Staphylococcal carbuncle.

mg/day, or dicloxacillin 1-2 g/day is recommended. Keloidal folliculitis is very difficult to manage since it generally does not respond to antimicrobial treatment. Isotretinoin 1 mg/kg/day is reportedly effective, but, in contradistinction to acne cases, when treatment is suspended the lesions of folliculitis reappear. The keloidal area can be surgically excised and allowed to granulate by secondary intention. In cases of nasal carriers of *S. aureus* with frequent recurrences, treatment with bacitracin or mupirocin b.i.d. in the nasal fossae, one week a month, or oral rifampin 600 mg for 10 days can erradicate *S. aureus* and diminish the frequency of recurrences. Bathing with soaps that contain chlorhexidine is also recommended. Furuncles should be treated initially with warm compresses in order to promote liquefaction of their contents and if possible should be drained by means of a small incision. Although treatment based on topical antimicrobial therapy can be sufficient, and prompts the risk of scarring is less if treatment with systemic antibiotics is given early.

Selected Readings

1. Feingold DS. Staphylococcal and streptococcal pyodermas. Semin Dermatol 1993;12:331-5.
2. Noble WC. Gram-negative bacterial skin infections. Semin Dermatol 1993; 12:336-41.
3. Trueb RM, Gloor M, Wuthrich B. Recurrent Pseudomonas folliculitis. Pediatr Dermatol 1994; 11:35-8.

Hidradenitis

Roberto Cortes-Franco

Hidradenitis is a chronic, frequently suppurative, cicatricial inflammatory disease of those areas of the body with a high density of apocrine glands, especially the axillae and the groin. Occasionally it is associated with severe and inflammatory forms of acne such as acne conglobata.

GEOGRAPHIC DISTRIBUTION

It occurs throughout the world and affects both sexes although it is more prevalent in women.

ETIOLOGY

For many years hidradenitis was considered to be a disease of the apocrine glands. Recent studies have shown that the primarily affected adnexa are the hair follicles which become occluded with shed material; folliculitis and cystic formations ensue. The apocrine glands are only affected secondarily, as are also the eccrine glands (*Br J Dermatol* 1995; 133:254-8; *J Am Acad Dermatol* 1996; 35:191-4). In some areas exist (groin, perineum, perianal, buttocks, mammae) in which involvement of the apocrine glands is an infrequent finding. Hyperhidrosis, maceration, use of tight clothing, and the application of chemical substances are considered risk factors for the development of hidradenitis; nevertheless, they are not themselves sufficient to cause infection and a genetic predisposition may also be a factor. Hormones are also important given the fact that the condition is not observed before puberty and, in general, it improves in the climacteric period. However, hormonal levels of these patients are unremarkable.

CLINICAL FEATURES

The initial lesion is an erythematous papule that grows rapidly until it reaches 0.5-2 cm in diameter, affecting several follicles. Within a few days an abscess forms and opens to the surface of the skin and drains purulent material. With time, fistulae and cicatricial areas form that can even limit the movement of the extremity (Fig. 26.1). While some lesions heal, new lesions appear. In addition,

Tropical Dermatology, edited by Roberto Arenas and Roberto Estrada. ©2001 Landes Bioscience.

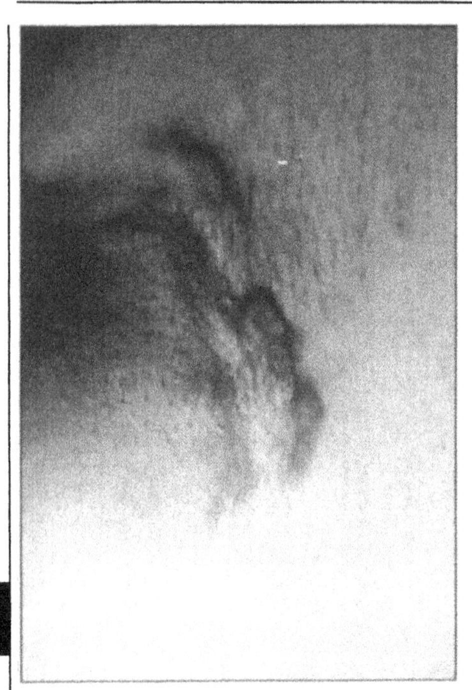

Fig. 26.1 Hidradenitis in the axillae.

chronic forms are accompanied by open comedones. The active disease is often very painful. The differential diagnosis includes folliculitis, furunculosis, epidermoid infammatory cysts, acute bartholinitis, or any other ailment that produces papular lesions and painful abscesses. However, the presence of lesions only in the sites with the highest density of apocrine glands, a history of chronicity and the presence of fistulae and cicatricial areas support the diagnosis of hidradenitis.

LABORATORY DATA

Laboratory examinations are limited to the culture of the purulent secretions in which will usually grow, staphylococci, streptococci or *E. coli*, and in rare cases anaerobes.

TREATMENT

In cases in which there are inflammatory lesions but which are without scars or fistulae, the treatment can be based on systemic antibiotics such as minocycline 100-200 mg/day, dicloxacillin 500 mg q.i.d. or clindamycin 300 mg b.i.d. Antiseptics such as chlorhexidine gluconate or topical antibiotics are only useful as prophylactics. Some very painful papular or abscessed lesions respond well to infiltration with triamcinolone acetonide, 5-10 mg/ml. Chronic cases, recurrent abscesses, fistulae and broad cicatricial areas should be treated with systemic antibiotics and complete excision of the axilla with wound closure by secondary intention (4-6 weeks) or grafting. Non-steroidal anti-inflammatories offer little improvement and systemic steroids have little use given the chronicity of the disease. Isotretinoin 1 mg/kg/day has also been used but with much less effect than in acne cases.

Selected Readings

1. Jemec GB, Hansen U. Histology of hidradenitis suppurativa. J Am Acad Dermatol 1996; 34:994-9.
2. Attanoos RL, Appleton MA, Douglas-Jones AG. The pathogenesis of hidradenitis suppurativa: a closer look at apocrine and apoeccrine glands. Br J Dermatol 1995; 133:254-8.
3. Jemec GB, Heidenheim M, Nielsen NH. The prevalence of hidradenitis suppurativa and its potential precursor lesions. J Am Acad Dermatol 1996; 35:191-4.

Ecthyma/Erysipelas

Roberto Cortes-Franco

Ecthyma and erysipelas are each a pyodermatitis caused in the majority of cases by *Streptococcus pyogenes* (group A beta-hemolytic streptococcus). Both affect the skin from the epidermis to the deeper dermis and are often accompanied by fever and other systemic manifestations. As with other cutaneous infections caused by *S. pyogenes*, treatment with penicillin is often very effective.

GEOGRAPHIC DISTRIBUTION

Ecthyma and erysipelas are both universally distributed, and occur frequently, especially ecthyma, in tropical and subtropical zones. They can be observed in people of any age or sex and they occur predominantly in the malnourished.

ETIOLOGY

Ecthyma is an infectious skin disease caused by *Streptococcus pyogenes* which penetrates by inapparent breaks in the skin. Initially it infects the epidermis and afterwards invades the deep layers of the dermis. If, indeed, it can occur in healthy individuals, it tends to occur more frequently in subjects whose hygiene is poor, who live in overcrowded conditions, are malnourished and alcoholic, or in persons with diabetes mellitus or AIDS or on immunosuppressant treatment (e.g. chemotherapy). Erysipelas is also a dermatosis produced by *S. pyogenes* which penetrates through a break in the skin, infecting the deep dermis and its lymphatics, and only secondarily the epidermis (*N Eng J Med* 1996; 334:240). Cellulitis is a very similar illness but one in which the infection is predominantly in the subcutaneous cellular tissue and in the deep dermis. In a few cases the causal agent is *Staphylococcus aureus* or *Haemophilus influenzae* and only rarely gram-negative bacilli, vibrio, clostridia or other anaerobes.

CLINICAL PICTURE

Ecthyma begins as a pustule that within a few hours opens up becoming an ulcer 0.5 to 2 cm in diameter, with well-defined borders, whose base extends to the deep dermis. The ulcer is surrounded by an erythematous halo (Fig. 27.1). On

Tropical Dermatology, edited by Roberto Arenas and Roberto Estrada. ©2001 Landes Bioscience.

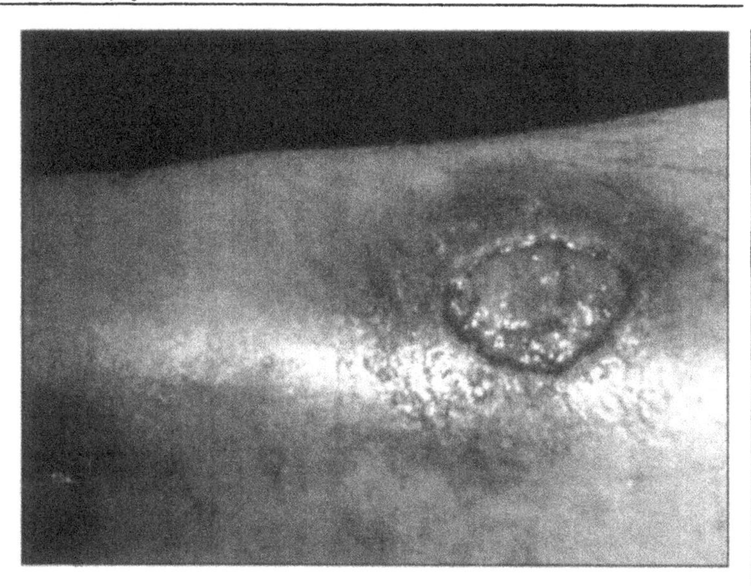

Fig. 27.1. Ecthyma.

some occasions, when the vesicle breaks, a scab or area of necrosis forms and until these lesions disappear, the ulcer remains invisible. Ecthyma occurs more frequently on the lower limbs. Erysipelas is characterized by erythematous plaques, smooth or shiny skin or by skin with an appearance similar to the skin of an orange (peau d'orange). Lesions are generally well-defined (Fig. 27.2). In some cases vesicles or bullae can be observed. Although erysipelas can occur in any location, it most often appears in the lower limbs. Perhaps the most common breaks on the skin for *S. pyogenes* in erysipelas of the legs is by means of interdigital *tinea pedis*, although it can begin in areas of trauma, surgical wounds or other types of dermatosis (e.g., psoriasis or eczema). Patients have severe pain, regional adenopathy, fever of 38°C or higher and malaise. Alterations in lymphatic drainage are among the most important risk factors for the development of erysipelas and cellulitis and given that these infections primarily affect the lymphatics, a vicious cycle is created in which one erysipelas infection leaves lymphedema as a sequela and this condition promotes a new erysipelas infection. The repeated episodes of erysipelas in one extremity can generate *elephantiasis nostras verrucosa*, a severe and untreatable condition of hard lymphedema with verrucose tumefactions. Cellulitis occurs with a clinical pattern that is often indistinguishable from that of erysipelas; nevertheless, cellulitis tends to be peripherally less well-defined and is more often accompanied by bullae or even by superficial abscesses (Fig. 27.3).

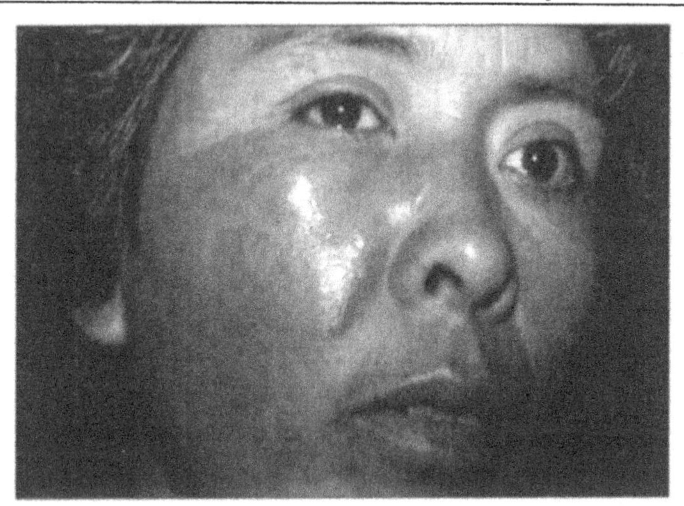

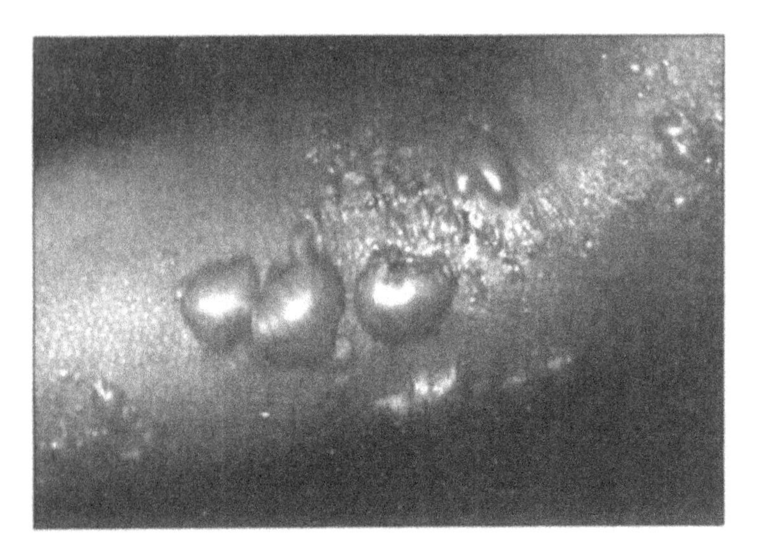

Fig. 27.3. Cellulitis.

LABORATORY DATA

Attempts to culture *S. pyogenes* from lesions tend to be fruitless: injecting sterile water into the active edge of the lesion and then aspirating yields positive culture at the most in 5-10% of cases. CBC shows leukocytosis with neutrophilia.

TREATMENT

Even without treatment, the majority of episodes resolve within 15-20 days. Due to the fact that cutaneous infections of *S. pyogenes* can cause glomerulonephritis, all cases of ecthyma, erysipelas and cellulitis should be diagnosed and treated promptly. Currently, penicillin is still the best treatment. Penicillin V can be used, 250 mg, p.o. q.i.d. for 10 days or procaine penicillin G, 800,000 U b.i.d. for 10 days. In severe cases the addition of clindamycin, 300 mg p.o. b.i.d., is recommended. Erythromycin is a good alternative for those allergic to penicillin or in cases in which oral administration is preferred over parenteral.

Selected Readings

1. Bisno AL, Stevens DL. Streptococcal infections of skin and soft tissues. N Engl J Med 1996; 334:240.
2. Chapel KL and Rasmussen JE. Pediatric dermatology: advances in therapy. J Am Acad Dermatol 1997; 36:513-26.
3. Stevens DL. Group A streptococcus: from basic science to clinical disease. West J Med 1996; 164:25-7.

Necrotizing Fasciitis (Streptococcal Gangrene)

Roberto Arenas

Necrotizing fasciitis is a streptococcal or mixed infection of the soft tissues that ranges from a cellulitis to a serious myositis with necrosis. It occurs at the site of a laceration or surgical wound. It is accompanied by fever and other systemic symptoms.

GEOGRAPHIC DISTRIBUTION

The disease occurs throughout the world and predominates in adults and the elderly. Invasive streptococcal infections have an incidence of 1.5 cases per 100,000 inhabitants and necrotizing fasciitis is present in 6% of them.

ETIOLOGY

It is generally caused by a toxin produced by virulent bacteria such as Group A (rarely Group C or G) streptococci, *S. pyogenes*, Peptostreptococcus spp, *Bacteroides fragilis* group, *Clostridium perfringens*, *Escherichia coli*, *Vibrio vulnificus* and Prevotella spp, among others. When it is monomicrobial it is attributed to cutaneous flora, and if it is polymicrobial it is attributed to enteropathogens. In children it is generally aerobic-anaerobic polymicrobial. It is associated with diabetes, alcoholism, immunosuppression, peripheral vascular disease and intravenous drug abuse.

CLINICAL PICTURE

It is manifested by redness, edema and pain,usually involving the extremities. Later a bluish or purple color appears along with vesicles and bullae that produce an area of necrosis that resembles a third-degree burn (Fig. 28.1). Secondary thrombophlebitis may occur as well as claudication and shock. With scrotal involvement, it is known as Fournier's gangrene; on the abdomen and perineum it is very severe because of visceral involvement. Bacteremia can cause disseminated lesions similar to those of purpura fulminans. There may be pain, fever, hypotension,

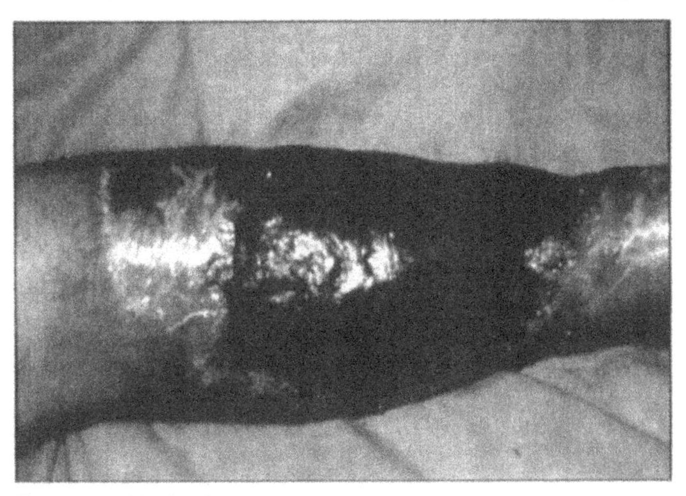

Fig. 28.1. Necrotizing fasciitis.

respiratory distress, coagulopathy and acute renal failure. The mortality rate is 25%, but with adequate treatment it is less than 10%.

LABORATORY DATA

Gram stain and culture of the liquid from vesicles and blood may be effected by aspiration with a fine needle. Culture of the material from the area of the necrosis generally yields nonspecific findings. Some bacteria may be identified by molecular procedures or by serotyping. Histopathology reveals cutaneous necrosis along with intense angiitis with fibrinoid necrosis in the arteries and veins that pass through the involved fascia. Fibrillar thrombi also occur. Polymorphonuclear cell and monocyte infiltration along with the presence of gram-positive cocci can be observed. Radiographic studies can be useful to detect the presence of gas.

TREATMENT

Extensive and repeated debridement is important. Fasciotomy and grafting may be necessary. The initial antibiotic treatment should be of a sufficiently wide spectrum to cover aerobes and anaerobes, especially streptococci, e.g., crystalline penicillin every four hours or clindamycin 0.6 - 1.2 g every six hours. Hyperbaric oxygenation is also recommended. Due to the risk of multiple organ failure, intensive care may be necessary.

Selected Readings

1 Brook I. Aerobic and anaerobic microbiology of necrotizing fasciitis in children. Pediatr Dermatol 1996; 13(4):281-4.
2 Davies HD, McGreer A, Schwartz B et al. Invasive group A streptococcal infections in Ontario, Canada. Ontario Group A Streptococcal Study Group. N Engl J Med 1996; 335(8):547-54.
3 Green RJ, Dafoe DC, Raffin TA. Necrotizing fascititis. Chest 1996; 110(1):219-29.
4 Lille S, Sato TT, Engrav LH et al. Necrotizing soft tissue infections: Obstacles in diagnosis. J Am Coll Surg 1996; 182(1):7-11.
5 Stevens DL. Invasive group A streptococcal disease. Infect Agents Dis 1996; 5(3):157-66.

Staphylococcal Scalded Skin Syndrome (Ritter's Disease)

Roberto Estrada

One of the manifestations of staphylococcal skin infection is the scalded skin syndrome, which is one of its more severe forms. Together with bullous impetigo and scarlatiniform eruption, it has been related directly to epidermolytic toxins produced by *Staphylococcus aureus*, a group II phage which includes types 57, 71, 3A, 3B and 3C (*Semin Dermatol* 1982; 1:101). Ritter von Rittershain called it dermatitis exfoliativa neonatorum, and it is also known as pemphigus neonatorum because it more frequently affects newborn infants and because of its bullous appearance. Because of the extended involvement and its severity, it was confused with and described as a staphylococcal form of Lyell's syndrome. Thanks to studies done in newborn mice (*N Eng J Med* 1970; 43:1114), it was possible to clearly establish the distinct origins of the two entities as well as the production of an epidermolytic enzyme or an epidermotoxin.

GEOGRAPHIC DISTRIBUTION

It occurs worldwide, without racial distinction. It is a rare disease that affects newborns in the first three months of life and children up to the age of five. However, older children and adults with kidney failure or immunodeficiency can be affected (*J Am Acad Dermatol* 1994; 30(2):319-24).

ETIOLOGY

The bacterial focus which generates the condition is not localized in the skin, but rather in other organs such as the pharynx, tonsils, kidneys or conjunctiva. Two types of epidermolytic toxins exist—A and B—which are produced by staphylococci and are responsible for destructive lesions in the skin and affect the granular layer causing cellular separation. The toxins elicit specific antibodies that have been found in up to 75% of persons older than 10, which would explain the absence of the lesions in adults. On the other hand, mature kidneys are able to eliminate the epidermotoxins more efficiently and for this reason, kidney failure can be an important risk factor in adult patients.

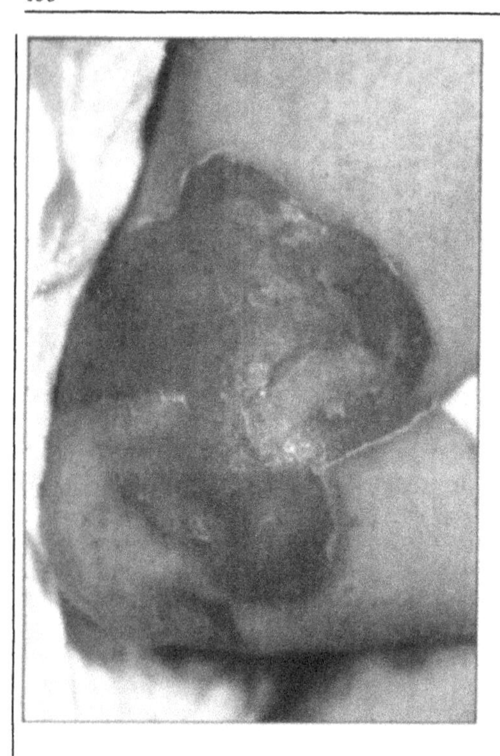

Fig. 29.1. Staphylococcal scalded skin syndrome.

CLINICAL PICTURE

It begins with a minimal infection in an extracutaneous foci, which is shortly followed by skin lesions. At the outset, only a diffuse erythema and increased sensitivity are noticeable. It looks similar to scarlatina, except there is neither pain nor bullae. One or two days later bullous lesions appear which have a very thin wall which is easily broken allowing a clear serous or sometimes purulent fluid to escape. There are high fever and malaise. Occasionally gastroenteritis and kidney damage with proteinuria ensue. Nikolsky's sign is positive. After the bullae have broken, the necrotized skin separates and dries up. There is a noticeable desquamation halo and the surface of the plaque is highly sensitive. The intensity of the manifestations is inversely related to the age of the patient. Although the majority of cases are cured, a mortality rate of 2-3% has been reported. The prognosis is worse in immune compromised individuals. It should be distinguished from staphylococcal impetigo in which the bullae are more localized, meliceric crusts tend to form, Nikolsky's sign is negative, systemic manifestations are absent and the bulla is subcorneal. Lyell's syndrome occurs, above all, in adult patients who have recently taken systemic drugs. It should not be confused with poorly treated burns on children where there is a history of trauma nor with bullous epidermolysis, incontinentia pigmenti or even with erysipelas.

LABORATORY DATA

In general, the diagnosis should be considered when there is intense erythema, spreading bullae, peripheral desquamation, isolation of *Staphylococcus aureus* group II phage, producer of epidermotoxins, and the histological picture is characterized by epidermal fissures or bullae at the level of the granular layer (*Cutis* 1983; 31:431-4). These criteria are reinforced if the patient is a newborn or is less than 5 years old, since it is rare after this age.

TREATMENT

If skin loss has been extensive, the patient is treated as if he had been burned. Electrolyte replacement and prevention of secondary infection must be considered.

Antibiotics specific for streptococci are indicated, such as dicloxacillin or fluoxacillin as well as cephalosporins. Wounds must be kept clean. The lesions do not usually leave scars. It should not be forgotten that the original focus resides in an extracutaneous bacterial location, which must be identified and adequately eliminated. Equal attention must be paid to the detection and treatment of healthy carriers at home, in nurseries or relatives outside the house who are in contact with the child. Corticosteroids are discouraged because of their immunosuppressive effect.

Selected Readings

1 Cribier B, Piemont Y, Grosshans. Staphyloccocal scalded skin syndrome in adults. J Am Acad Dermatol 1994; 30(2):319-24.
2 Melish ME. Staphyloccocci, streptococci and the skin. Semin Dermatol 1982; 1:101.
3 Melish ME, Glasgow LA. The staphylococcal, scalded skin syndrome: Development of an experimental mouse model. N Engl J Med 1970; 282:1114.
4 Falk DK. Criteria for the diagnosis of Staphilococcal Scalded Skin Syndrome in Adults. Cutis 1983; 31:431-4.
5 Habif FT. Clinical Dermatology. St. Louis: Mosby, 1985:173-74.

Anthrax

Roberto Arenas

Anthrax is an infectious zoonosis which is systemic and severe in both wild and domesticated animals. Humans acquire it accidentally through transcutaneous contact, inhalation or ingestion. In the skin it can cause malignant edema or anthrax (malignant pustule). The latter is characterized by one vesicle surrounded by a necrotic zone where *Bacillus anthracis* is found. The inhaled form is rare and given its rapid progression and the limited response of the causative agent to treatment, it is nearly always lethal. For this reason Anthrax has been considered useful as a biological weapon. In its experimental form in Rhesus monkeys, it principally causes meningitis and hemorrhage of the mesenteric lymph nodes (Lab Invest 1995; 73 (5): 691-702).

GEOGRAPHIC DISTRIBUTION

It has a worldwide distribution and is a local problem in Pakistan, India, Iran, Mongolia and South Africa. It is infrequent in Australia, Mexico, Central America, South America and in the countries of the Mediterranean. It is rare in Europe and in the United States. In 1993, an epidemic in northern Canada affected 172 bison, 3 elk and 3 black bears (Can J Vet Res 1995; 59 (4): 256-64). It is believed that there are more than 100,000 cases of human anthrax per year, but the reported frequency is lower; it has dropped because of adequate sanitary measures and despite the lack of effective treatment. There are zootic outbreaks in hot seasons, but it can also be epidemic. It is found in both sexes, although it predominates in adult males, especially agricultural workers, stock farmers, veterinarians and butchers. The mortality rate in untreated cases is 5-20% and in treated cases it is almost null.

ETIOLOGY

The causal microorganism is *Bacillus anthracis*, a capsulated, strictly aerobic bacterium whose high virulence is communicated by means of the capsule and an exotoxin. It can survive for 20 or more years as highly resistant spores. These vegetative cells of *B. anthracis* have specific nutritional and physiological requirements. Spores are found in alkaline soils, high humidity and high organic content—specifically with high levels of calcium (Can Vet J 1995; 36 (5):295-301). The spores contaminate the grasses eaten by herbivores, which have a mortality rate of 80%.

Even the smallest trauma facilitates inoculation which can be cutaneous (95-98%), respiratory or digestive. Humans acquire the infection by cutting up or ingesting the flesh or handling the hides of contaminated animals, principally cattle (47.8%), sheep or goats. Transmission by insects, from person to person, or by contact with wool, bones or hair is rare.

CLINICAL FEATURES

The incubation period is 1-3 days and the clinical manifestations are anthrax (this term is also used to designate a furunculosis caused by *Staphylococcus*) and malignant edema. Anthrax, carbuncles or malignat pustule predominate in the exposed areas of the skin, principally the face and the upper extremities; rarely is it multiple. A reddish spot first appears, upon which emerges a pink, translucid vesicle with pearly aspect. This becomes a vesicle with serous and serosanguinous fluid surrounded by a well-defined necrotic area. There are vesicles on the periphery (Chaussier's sign) as well as painless and hardened edema (Fig. 30.1). Lymphangitis and adenitis are infrequent. In 3-4 days fever, headache, arthralgias and malaise occur. In patients with severe illness death may ensue, nearly always occuring in conjunction with a serious preexisting disease such as diabetes. Most of the

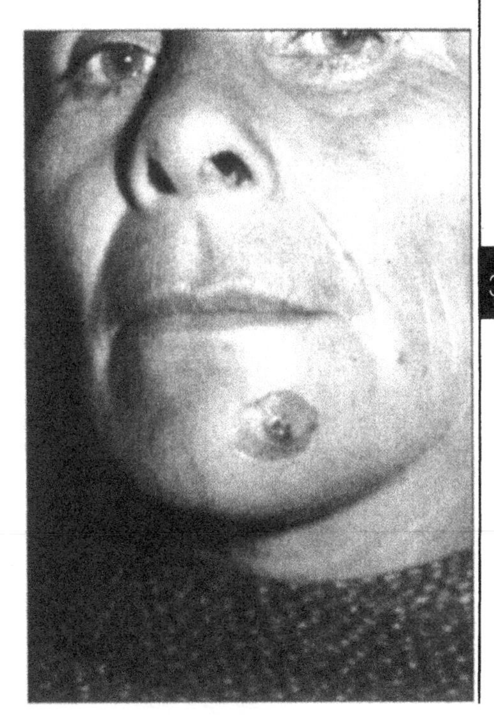

Fig. 30.1. Anthrax due to *B. anthracis*.

patients only manifest localized symptoms, and remission is observed in 2-3 weeks. A bullous variety and malignant edema have been described. Other varietes with pulmonary or intestinal localizations are rare, severe and nonspecific.

LABORATORY DATA

Biopsy is not necessary. Ulcerated epidermis covered in necrotic tissue is found. There may be spongiosis and sometimes an intraepidermic vesicle with neutrophils. In the dermis, edema, dilated vessels and polymorphonuclear infiltration which can reach the cellular tissue, are observed. It is possible to identify the bacilli by staining with hematotoxylin and eosin or Gram. Leukocytosis and accelerated erythrocyte sedimentation are present. The bacilli found in the lesions are Gram-positive, 1-1.5µm by 4-8µm in size. Culture is obtained in media for bacteria such as Mueller-Hinton, [Azide-Blood]1, and Agar-100 (Staphylococcus) in 24-72 hours. The organism can be distinguished from *B. cereus* by immunoflourescence.

TREATMENT

Procaine penicillin G, 800,000 U every 12 hours, and in severe cases, IV penicillin in high doses is indicated. Alternatives are sulfamethoxypyradazine, 500 mg to 1 g/day, or trimethoprim-sulfamethoxazole, 80/400 mg twice a day for 10 days or until the disappearance of the lesions. Tetracycline or erythromycin, 1 g/day, may be used with doxycycline or cephalosporin. Empirically, patients have used caustic solutions or burned the lesions with good results. Some authors recommend surgical excision of the necrotic area. There is a vaccination for animals and it may be used in humans for either prophylaxis or treatment. To induce a strong and stable immunity, the complete complex of antigens should be used: protector antigen, lethal and edema factors (*J Biotechnol* 1996; 44 (1-3); *Infect Immun* 1995; 63(4): 1369-72). An antianthrax globulin is also commercially available. Animals who express the disease or have died because of it should be incinerated and the area should be decontaminated with formaldehyde.

Selected Readings

1 Arenas R. Dermatología. Atlas, diagnóstico y tratamiento. México. Interamericana/ McGraw-Hill. 1996: 272-73.
2 Dragon DC, Rennie RP. The ecology of anthrax spores: Tough but not invincible. Can Vet J 1995; 36 (5): 295-301.
3 Fragoso-Uribe R, Villicaña-Fuentes H. Antrax en dos comunidades de Zacatecas, México. Bol Of Sanit Panam 1984; 97(6):526-533.
5 Grasa MP, Bizcarguenaga J, Gimenez H et al. Carbunco cut·neo: a propósito de cuatro casos. Actas Dermo-Sif 1988; 79(1):39-42.

Tularemia

Roberto Arenas

Deer fly fever, O'Hara's disease, Utah fever, lemming fever or yatobyo is a generally lethal infection in wild rodents caused by *Francisella tularensis* transmitted accidentally to humans. It causes an ulceroglandular complex that is characterized by a chancre constituted by a necrotic papule on the distal part of an extremity and regional adenopathy accompanied by malaise and fever.

GEOGRAPHIC DISTRIBUTION

It occurs throughout the world, involves both sexes equally and predominates in teenagers and adults. It is more common in hunters of wild rodents. It occurs most commonly in restricted geographic areas of North America, Russia and Japan. It has been found in some countries of Europe such as France after the introduction of hares used for sport. In 1991 two epidemics occurred in Turkey that resulted in 98 cases and in 1996 an epidemic in Sweden involved 676 cases.

ETIOLOGY

The causal microorganism is *F. tularensis* (*Bacterium, Pasteurella*), a gram-negative, pleomorphic, encapsulated, obligate aerobe coccobacillus from 0.3-0.7 µm. It is intracellular and is transmitted to the offspring of the vectors. The natural vectors are ticks (*Haemaphysalis leporis palustris, H. concinna, Dermacentor andersoni, D. variabilis, C. occidentalis, D. reticulatus, Ixodes ricinus*), flies (*Chrysops discalis, Stomoxys calcitrans, Ceratophyllus acutus*) and lice (*Haemodipus ventricosus, Polyplax serratus*). In endemic areas it is found in 2-5% of these populations, but they may transmit it to other arthropods. The most common reservoirs are wild rodents such as rabbits and hares. In humans the bacteria penetrate through small trauma in the skin and the infection can occur upon skinning or slaughtering the animals, by the bite of a vector or by contamination with their excrement, wild rodent bites, or by accidental contamination in the laboratory. Two biovariants have been identified: Type A or *tularensis*, which predominates in North America and type B or *palaearctica*, in Europe. The immunological changes are not well known, although cellular immunity plays an important role. Increased levels of soluble receptors for interleukin-2 and decreased levels of receptors for transferrin have been observed. Cytokines such as alpha-tumor necrosis factor and gamma-interferon (*Infect Immun* 1996; 64(8): 3288-93; *Microbiology* 1996; 142

Tropical Dermatology, edited by Roberto Arenas and Roberto Estrada. ©2001 Landes Bioscience.

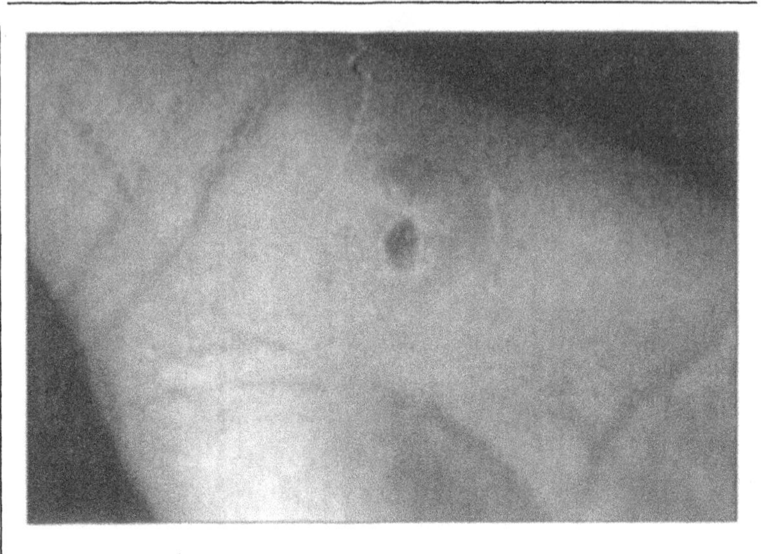

Fig. 31.1. Tularemia, chancre.

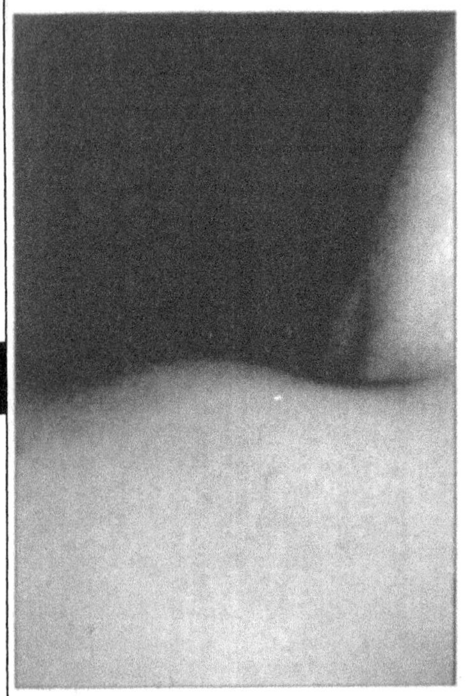

Fig. 31.2. Tularemia, axillary adenopathy.

(Pt 6): 1369-7) and an important antigen of the membrane (lipopolysaccharide) which induces protective immunity (*Vaccine* 1995; 13(13): 1220-5) seem to be essential to control experimental disease.

CLINICAL PICTURE

The incubation period is 10 days. In endemic areas an asymptomatic form occurs which is diagnosed serologically. There are ulceroglandular, oculoglandular, purely glandular, typhoid and pulmonary forms. The most common form is the ulceroglandular (60%). At the site of inoculation, usually an exposed area on the distal upper extremities, a painful papule appears that grows rapidly, becomes painless, ulcerates and is covered by a necrotic area (Fig. 31.1). Soon thereafter epitrochlear and axillary adenopathy occur which can produce a sticky, clotted and white-chocolate pus (Fig. 31.2). There may be linear nodular lymphangitis. These manifestations can be (3-25%) accompanied by a generalized papular eruption, erythema multiforme or erythema nodosum (tularemides). There is malaise, high fever, and, during the initial stages with 90% of cases present with mild and transitory bronchopulmonary involvement which is seen on X-ray. In the oculoglandular form (1%), the primary lesion occurs in the conjunctiva while in the purely glandular form a generalized adenitis occurs. The typhoid and pulmonary forms are severe and are manifested by gastroenteritis and broncho-pneumonia. They are acquired through ingestion of infected meat or food that has been contaminated by the excrement of rodents or inhaled through the respiratory tract or from the dissemination of a local infection.

LABORATORY DATA

Histopathology reveals a central zone of necrosis surrounded by histocytes, epithelioid cells, lymphocytes, polymorphonuclear cells, plasma cells and eosinophils. The microorganisms are best stained with Giemsa. Patients with serious forms of the disease and those with erythema nodosum have increased IgG, IgA and IgM, as well as reactive C proteins, complement (C3c, C4) and an elevated erythrocyte sedimentation rate. Agglutination reactions are positive (in 13% Proteus agglutinin OX19 at ratios higher than 80) as are complement fixation, immunoflourescence and ELISA tests. Polymerase chain reaction (PCR) can be used for diagnosis. (*Am J Trop Med Hyg* 1996; 54(4): 364-6). The intradermal reaction with tulargen and tularine is positive at 48 hours. Culture should be done at 37°C in blood agar or chocolate media enriched with cysteine and egg yolk. Subcutaneous or intraperitoneal inoculation in animals, principally in guinea pigs and rabbits, generally causes death.

TREATMENT

Tetracycline 2 g/day, streptomycin 1 g/day or minocycline, 200 mg/day are indicated until remission of the lesions, or for 2-4 weeks. Spiramycin, 2-3 g/day orally divided in two or three doses, or gentamicin, 80 mg, intramuscularly three times a day for 10 days also are effective. There has been little experience with trimethoprim/sulfamethoxazole and flouroquinolones (ciprofloxacin, norfloxacin) while chloramphenicol, erythromycin and other macrolides, rifampin, imipenem/cilastatin cefotaxime, moxalactam, ceftazidime and ceftriaxone are less commonly used (*Clin Infect Dis* 1995; 20 (1): 174-5). The treatment period can be shortened by puncturing the lesions. Protective measures should be taken for exposed persons. In some countries a vaccine is available which should be administered annually.

Selected Readings

1 Akdis AC, Kiliçturgay K, Helvaci S et al. Immunological evaluation of erythema nodosum in tularaemia. Br J Dermatol 1993; 129:275-79.
2 Arenas R. Dermatología. Atlas, diagnóstico y tratamiento. México:Interamericana/McGraw-Hill, 1996:304-305.
3 Cerny Zdenek. Skin manifestations of tularemia. Int J Dermatol 1994; 33(7):468-47.
4 Gurycova D, Kocianova E, Vyrostekova V et al. Prevalence of ticks infected with Francisella tularensis in natural foci of tularemia in western 5 Slovakia. Eur J Epidemiol 1995; 11(4):469-74.
6 Scott G. Tularaemia. In: Cook GC, ed. Manson's tropical diseases. 20th ed. Philadelphia:Saunders, 1996:899-905.

Rhinoscleroma

Gisela Navarrete

Rhinoslceroma or respiratory scleroma is a granulomatous infectious disease which has a chronic and progressive course and is caused by *Klebsiella rhinoscleromatis.* The most common site of involvement is the nasal mucosa, but it can affect all the areas of the respiratory tract and can involve any organ.

GEOGRAPHIC DISTRIBUTION

It occurs throughout the world and is endemic to Central Africa, South America, the Middle East, South, Central and Eastern Europe. It has been reported in Russia, Poland, Czechoslovakia, Yugoslavia, Italy, Germany, Romania, Hungary, China and Indonesia as well as in Egypt, Uganda, Morocco and Kenya. In America it is found from the United States to Chile. In developed countries it has increased due to immigration and AIDS (*Clin Infect Dis* 1993; 16(3): 441-2). It is present, above all, among those in low socioeconomic levels, and some familial cases have been reported. It affects both sexes and it is most commonly seen between 20 and 40 years of age.

ETIOLOGY

It is caused by an Enterobateriaceae, *K. rhinoscleromatis*, an encapsulated, gram-negative, aerobic coccobacillus from 2-3μ in length. It has low infectivity, seems to be transmitted during the secretory phase of rhinitis and is believed to have a long incubation period.

CLINICAL FEATURES

It begins in the nasal mucosa in 96% of cases (*East Afr Med J* 1993; 70 (3): 186-88) and gradually invades neighboring structures such as the septal cartilage, choanae, rhinopharynx, larynx, trachea and bronchia. It can also affect the wings of the nasal alae, cheeks, upper lip, gums, soft and hard palates, uvula, orbits, lachrymal ducts, Eustachian tube and the tympanic cavity.

Tropical Dermatology, edited by Roberto Arenas and Roberto Estrada. ©2001 Landes Bioscience.

There are three distinct stages in its development:

1) Rhinitis (exudative or catarrhal),
2) Infiltrative or proliferative (nodular or granulomatous), and
3) Cicatricial (fibrous or sclerous).

Initially there is a hyalin secretion which later becomes mucopurulent, abundant and fetid, forming adherent crusts. It lasts from weeks to months. The nasal pyramid is deformed by the presence of nodules that form large granulomatous masses. These are visible through the nasal fossae, and they may cause obstruction and respiratory difficulty (Figs. 32.1, 32.2). If ulceration occurs, the central part of the face is destroyed and if infiltration of the larynx occurs there is dysphonia and dyspnea. There may be epiphora and ectropion, atrophy of the alveoli and the loss of teeth. Finally there is fibrosis, and when it affects the Eustachian tube and the middle ear both hypoacusia and otalgia occur. Secondary infection and hemorrhage can occur. Death due to asphyxia may ensue.

LABORATORY DATA

In the proliferative stage an hemotoxylin and eosin biopsy is characterized by an abundance of bacteria. There is dense plasmocytic infiltration, as well as a few Russell's bodies, large vacuolated histiocytes or Mikulicz's cells, from 100-200 μm, with peripheral nuclei and abundant Frisch's bacilli (*K. rhinoscleromatis*) that are more evident with PAS, Giemsa or silver stains (Fig. 32.2). In the later phase only fibrosis is observed. Ultrastructural studies

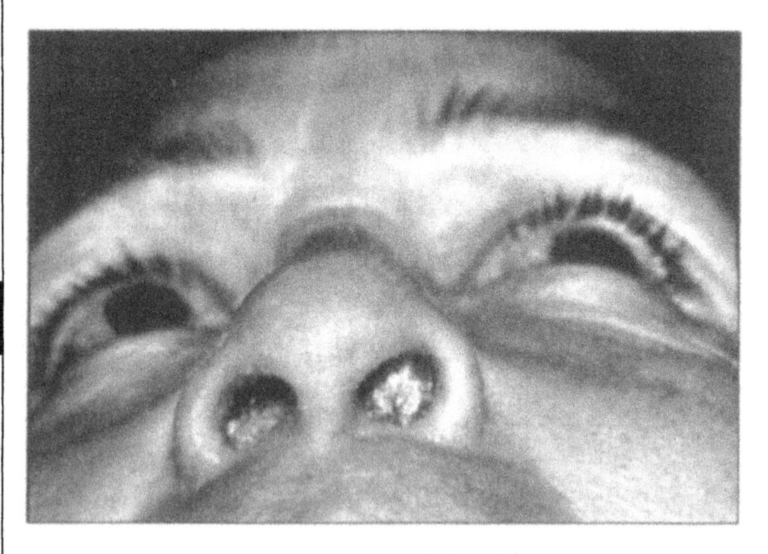

Fig. 32.1. Rhinoscleroma, proliferative phase (Courtesy of Olga De Celis).

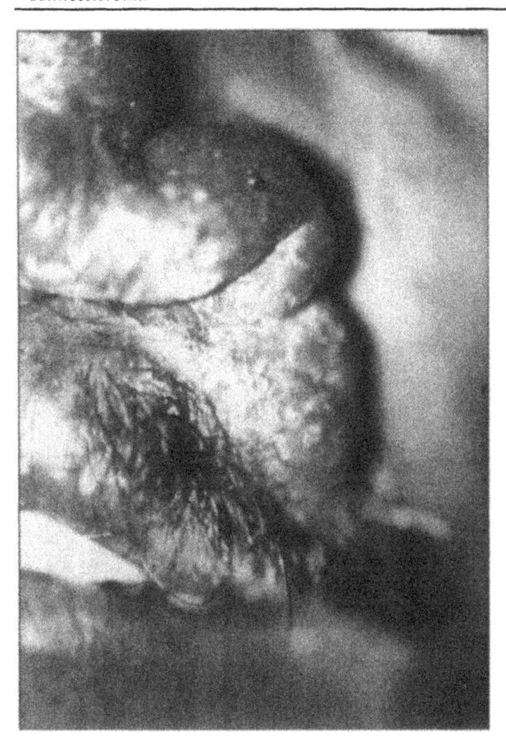

32.2. Rhinoscleroma granulomatous masses.

reveal so-called A and B granules. The organism is readily cultured from secretions or tissue in media of agar, gelatin, potato, broth, milk, glucose, maltose and other sugars. It is distinguished from other *Klebsiella* by means of fermentation and characteristic somatic O antigens and capsular K antigens. The latter antigen can be identified by means of immunoperoxidase. Complement fixation is valuable for diagnosis and to monitor the result of treatment. Endoscopy, radiographic studies and computerized axial tomography are all helpful.

TREATMENT

Streptomycin, 1 g/day and tetracycline 2g/day, have been used individually or together. Rifampin has also been used and currently fluoroquinolones such as ciprofloxacin are preferred at high dosages and for long periods (750 mg b.i.d.) (*Lancet* 1993; 342:122). Also recommended is minocycline 200 mg/day. The inhibitory properties of trimethoprim-sulfamethoxazole, clavulanate/amoxicillin, chloramphenicol, cephalexin, cefuroxime and cefodoxime have been demonstrated in vitro. Medical treatment has been combined with surgery, cryosurgery, radiotherapy and recently with CO_2 laser surgery.

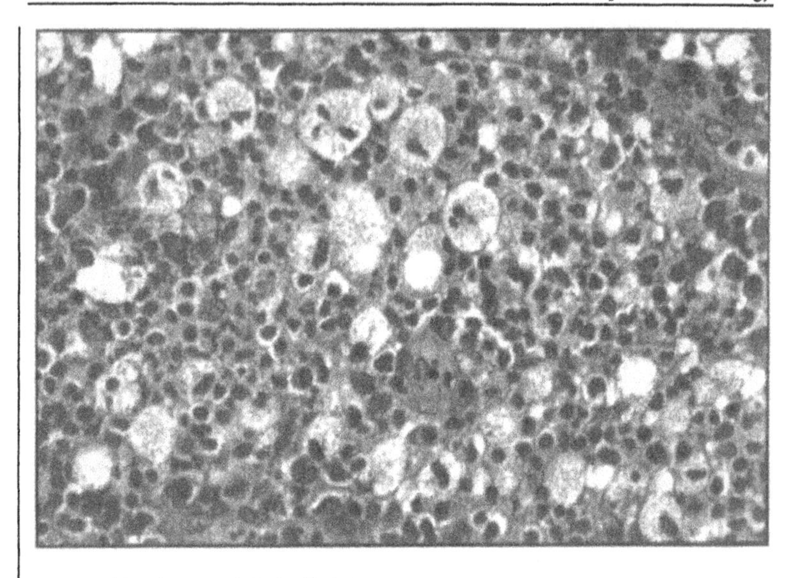

Fig. 32.3. Rhinoscleroma, Mikulicz's cells (H.E. 100X).

Selected Readings

1 Sedano HO, Carlos R, Koutlas IG. Respiratory scleroma: a clinicopathologic and ultrastructural study. Oral Surg Oral Med Oral Pathol Oral Radiol Endod 1996; 61(6): 665-71.
2 Stiernberg CH, Clark W. Rhinoscleroma. A diagnostic challenge. Laryngoscope 1983; 93:866-70.
3 Paul C, Pialoux G, Dupont B, et al. Infection due to *Klebsiella rhinoscleromatis* in two patients infected with immunodeficiency virus. Clin Infect Dis 1993; 16(3):441-2.
4 Wabinga HR, Wamukota W, Mugerwa JW. Scleroma in Uganda: A review of 85 cases. East Afr Med J 1993; 70 (3):186-88.

Lyme Borreliosis

Roberto Arenas

Lyme disease, or erythema chronicum migrans, is a multisystemic infection caused by *Borrelia burgdorferi* and transmitted to humans by a tick vector of the genus Ixodes. It affects the skin, joints, heart and the nervous system. It has an early phase which is manifested by migratory erythema, disseminated secondary lesions and lymphocytoma, and a later phase with atrophic chronic acrodermatitis and sclerodermiform lesions.

GEOGRAPHIC DISTRIBUTION

It has been known in Europe since the beginning of the century and it predominates in Austria, Switzerland, Germany, France, Denmark, Sweden and Great Britain. In the United States it predominates in the Northeast, Midwest and the Pacific Coast (*Science* 1982; 216:1317-19). By 1990, 7,997 cases had been reported; 6% were children. Isolated cases are found in Canada, Africa, Austria, Mexico and South America. The geographic distribution is related to the presence of the vectors, especially in the United States, *I. damini*, *I. pacificus* and *I. scapularium*, and in Europe, *I. ricinus* and *I. persulcatus*. Prophylaxis is difficult due to the ecological complexity of many epizootic and enzootic areas and due to the lack of effective control of the vectors.

ETIOLOGY

It is caused by *B. burgdorferi*, a gram-negative, flagellated and mobile bacteria. The organism is transmitted to humans by the bite of an hematophagic arthropod. The organism is a spirochete with diverse serotypes and variable pathogenicity which contains abundant protein particles that induce a specific immunological response. In the chronic disease IgM antibodies are present and in acute disease, IgG.

CLASSIFICATION

1) Early:
   Localized (Migratory erythema and lymphocytoma by *Borrelia*) and disseminated (Multiple migratory erythema, acute meningopolyneuritis, arthritis, carditis and affection of other organs).

Tropical Dermatology, edited by Roberto Arenas and Roberto Estrada. ©2001 Landes Bioscience.

2) Late (Chronic):
   Chronic atrophic acrodermatitis, neurological, rheumatological and other manifestations.

CLINICAL FEATURES

From 3-30 days (average of one week) after the bite, the early phase occurs, characterized by an asymptomatic lesion, migratory erythema (erythema chronicum migrans in 60-80%) appearing on the torso (38%) and on the lower (38%) and upper extremities (11%), usually in the axillae, groin, thighs and glutei. In children it predominates on the head and neck. It begins as a erythematous macule or papule that peripherally expands and produces an annular lesion. The border expands in days and the lesion persists for weeks or months; it is indurated or elevated (Fig. 33.1), the center is blue or purple. Sometimes there are vesicles (6%), ulcers and necrosis. Annular lesions are also observed due to hypersensitivity (6-48%). General symptoms are present such as headache, neck pain, myalgia, arthralgia, fever, photophobia, anorexia, dysesthesia and lymphadenopathy (50%). There may be an early disseminated infection that produces secondary or multiple lesions of migratory erythema and lymphocytoma (1%). This last condition is manifested by a nodular lesion from 1-5 cm that often is accompanied by lymphadenopathy most commonly seen in the outer ear, areola, scrotum and nose. There may also be meningitis, cranial neuritis and peripheral neuropathy (10%), auriculoventricular blockage, and myocarditis (10%), and in North America

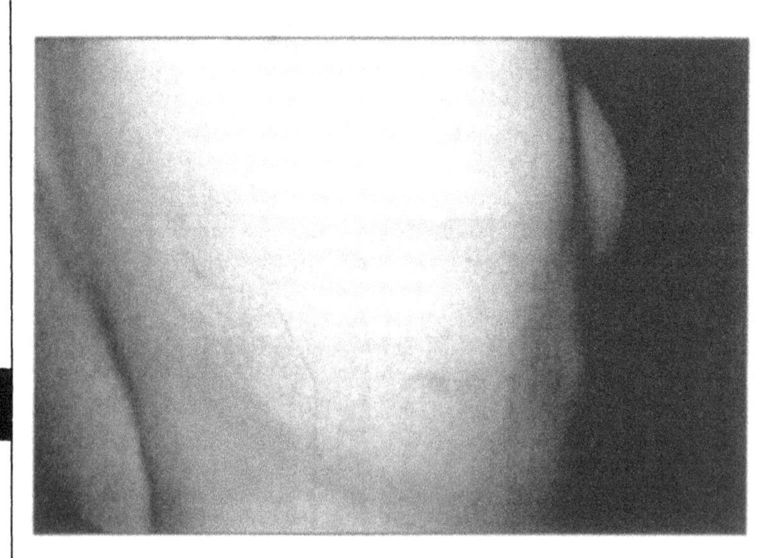

Fig. 33.1. Erythema chronicum migrans.

monoarticular and pauciarticular arthritis is observed (60%). Weeks or years later infection manifests as chronic and atrophic lesions called acrodermatitis (10%). This affects the extremities, face and torso, and predominates in women (65-80%). Other sclerodermic lesions (pseudoslcerodermia) may exist, and in 50% there is peripheral neuropathy. Arthritis, encephalomyelitis and marked fatigue may also occur.

LABORATORY DATA

Biopsy of chronic erythematous lesions reveals a superficial and deep lymphohistiocytic infiltration with plasma cells. Spirochetes are scarce and are identified with silver stains such as Warthin-Starry (40%). In lymphocytoma there is a dense B cell infiltration. In acrodermatitis there are telangiectasia and degeneration of the elastic fibers and collagen, as well as atrophy. Using indirect methods, immunoflourescence, ELISA or immunohistochemical studies involving monoclonal and polyclonal antibodies, specific antibodies can be identified. Western blot is also used, as is the T-cell proliferation test. Direct methods of antigen detection include culture, the capture of antigens and immunohistological studies, as well as DNA detection by means of polymerase chain reaction (PCR).

TREATMENT

Tetracycline, 1 to 2 g/day; doxycycline, 200 mg/day; amoxicillin, 2 g/day; erythromycin 1 g/day, all for 21 days; oral penicillin V, 1 g three times daily for 10 days. In children penicillin V, 50 mg/kg; amoxicillin 30 mg/kg, divided into three doses for 10 days; cefotaxime 100 mg/kg, divided into two doses, for 14 days; ceftriaxone, 50-80 mg/kg, for 14 days. For a neurological disease, ceftriaxone, 2 g/day IV for 10 days; chloramphenicol, 1 g/day IV, for 10-21 days are indicated (*Ann Intern Med* 1991; 114:472-481).

Selected Readings

1 Asbrink E, Hovmark A. Lyme. Borreliosis. Clinics in Dermatology 1993; 11(3):329-429.
2 Berger BW. Dermatologic manifestations of Lyme disease. Rev Infect Dis 1989; 11(suppl 6):S1475-81.
3 Burgdorfer W, Barbour AG, Hayes SF et al. Lyme Disease. A tick-borne spirochetosis. Science 1982; 216:1317-19.
4 Costello CM, Steere AC, Hayes SF et al. A prospective study of tick bites in an endemic area for Lyme disease. Rev Infect Dis 1989; 159:136-139.

H. Treponematosis and Genital Ulcers

Syphilis

Pinta

Yaws (Pian, Frambesia)

Endemic Syphilis (Bejel)

Chancroid (Soft Chancre)

Granuloma Inguinale (Donovanosis)

Lymphogranuloma Venereum

Syphilis

Sergio Eduardo Gonzalez-Gonzalez

Syphilis is a chronic, infectious disease also known as lues. For the most part, it is sexually transmitted although it can also be passed through the placenta and by blood transfusion. After penicillin was introduced, the incidence of syphilis decreased, but increased again in the late 1960s. When associated with AIDS, the clinical course of syphilis differs from the classic presentations.

GEOGRAPHIC DISTRIBUTION

Distribution is worldwide. It predominates in big cities where sexual promiscuity is high, especially among the lower economic classes.

ETIOLOGY

Treponema pallidum belongs to the Spirochaetae. It is indistinguishable morphologically, chemically or immunologically from the treponemas that cause pinta, frambesia and endemic syphilis. It cannot be cultured. It is 6-15 μm long and 1.5 μm wide. It has several regular spirals and a rotatory movement.

CLINICAL PICTURE

Primary syphilis: The first manifestation or chancre appears 9-90 days (3 weeks average) after infection. It is an erythematous papule on the site of inoculation (commonly in genitals) that grows and ulcerates rapidly. It is 1 cm in diameter, well-circumscribed and indurated at the base, hard and painless. Within a week after appearance of the chancre, bilateral, painless regional adenopathy develops unless there is a secondary bacterial infection (Fig. 34.1). The chancre disappears without any treatment in 3-6 weeks without leaving a scar.

Secondary syphilis: The manifestations of the lues present 3-6 weeks after the appearance of the chancre. The cutaneous lesions are asymptomatic. They are generally accompanied by systemic manifestations such as headache, anxiety, anorexia, weigh loss and fever. The initial skin manifestations are like roseola; they are pink, macular, lenticular lesions, generally of 0.5-1 cm in diameter, and surrounded by a collar of scales. They appear mainly on the trunk, and the proximal part of the arms. They can spread over the entire body. When they involve the

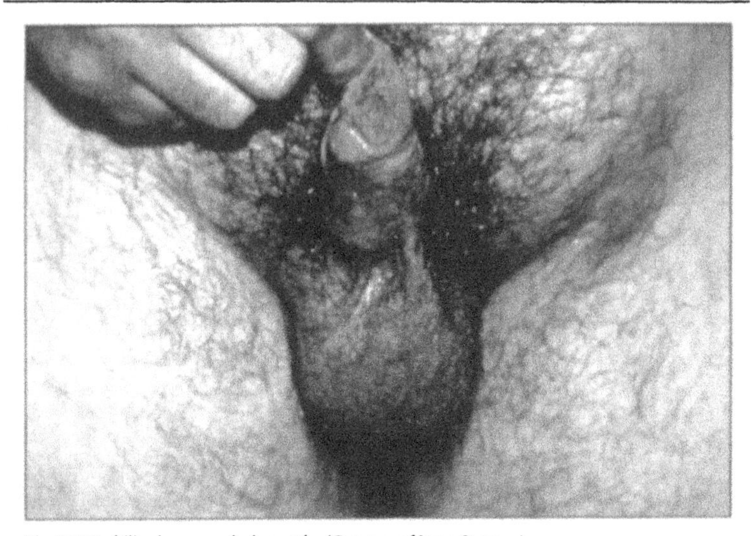

Fig. 34.1 Syphilis, chancre and adenopathy (Courtesy of Jorge Ocampo).

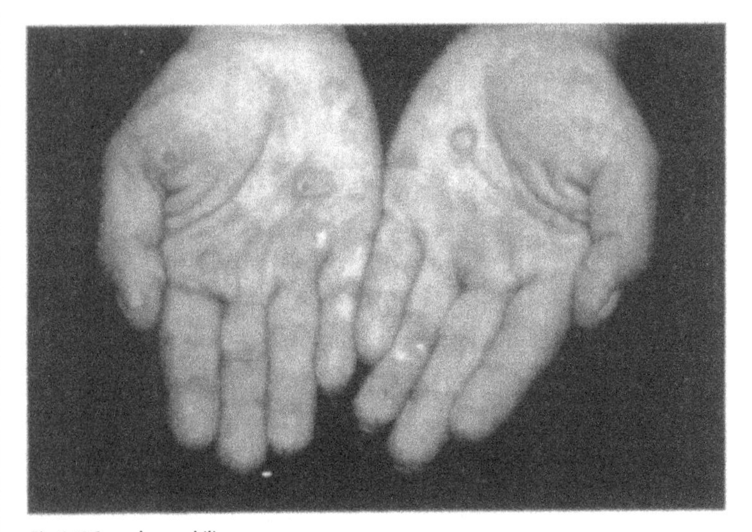

Fig. 34.2 Secondary syphilis.

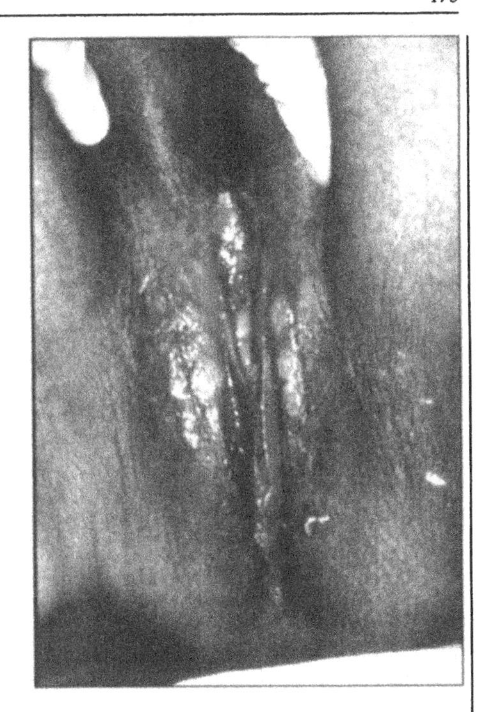

Fig. 34.3. Condylomta lata.

palms and soles, they are very pathognomonic for secondary syphilis (Fig. 32.2). The papules may coalesce and appear crusted, follicular, annular, circinate, and papuloerosive that in humid and hot climates resemble viral warts—condyloma latum or planus condyloma—on genitals, breast and intergluteal folds (Fig. 34.3). Alopecia in hairy skin and eyebrows appears as a "mouse bite" (Fig. 34.4). White or gray plaques appear on the oral mucosa and nails. Generalized lymphadenopathy is common. All lesions of secondary lues are rich in treponema.

Late syphilis: After two years without treatment syphilis can take several forms. It may heal spontaneously, it may pass into latency for the rest of life, or it may cause three types of clinical manifestations. Late syphilis may cause destructive gummatous lesions in bone, mucosae and skin (Fig. 34.5) that may give way to cardiovascular syphilis or to neurosyphilis. Cardiovascular syphilis is manifested by angina, coronary stenosis, aortic insufficiency or aortic aneurysm. Neurosyphilis may be asymptomatic with only CSF changes or it may manifest with neurovascular lesions, causing generalized paralysis, dorsal tabes or ocular syphilis.

Prenatal syphilis: Generally after the third month of pregnancy, syphilis is transmitted to the fetus through the placenta. It can cause a miscarriage before the fourth month. Early prenatal syphilis is observed from birth up to 2 years of age. The first manifestation is so-called syphilitic pemphigus with blisters, mainly on the palms and soles; perianal condyloma planus; fissures around the mouth; hepatosplenomegaly; periostitis, and osteochondritis in extremities. Late prenatal

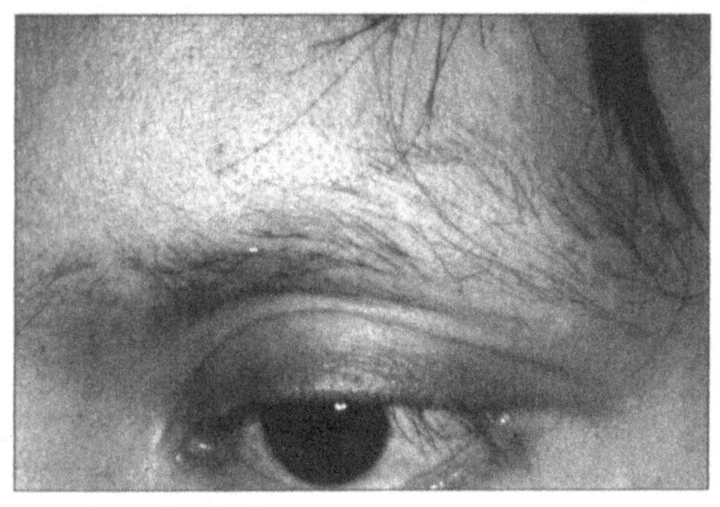

Fig. 34.4. Alopecia as "mouse bites."

syphilis is manifested by interstitial keratitis, perforation of the palate, telescope nose and less frequently neural deafness.

Syphilis and HIV: This association results in frequent anomalous findings, e.g., false negative serologic tests, failure of a serologic response to treatment, marked cutaneous reactivity attributed to a polyclonal stimulation of the B lymphocytes by HIV, failure to respond to conventional treatment, rapid progression from early to late syphilis, reactivation of syphilis by vaccinations and malignant syphilis.

LABORATORY DATA

Dark field microscopic examination with or without fluorescent antibodies and biopsy with silver stains or fluorescent antibodies are useful. Reaginic tests are not very specific and the most used are the VDRL (Venereal Disease Research Laboratories) and the RPR (Rapid Plasma Reagin). Treponema tests are specific; the most often used are FTA-abs (fluorescent treponemal antibody absortion) and the MHA-TP (microhemaglutination assay for antibodies to *Treponema pallidum*). The test FTA-abs IgM has been developed with fractionated blood (19S). It is the most sensitive and specific test for the diagnosis of prenatal syphilis.

TREATMENT

Primary and secondary syphilis: The treatment of choice for all those type of syphilis is parenteral penicillin G (*Clin Infect Dis* 1995; 20 (Suppl 1): S23-38). In adults, benzathine penicillin G, 2.4 million U in a single IM dose is indicated. In

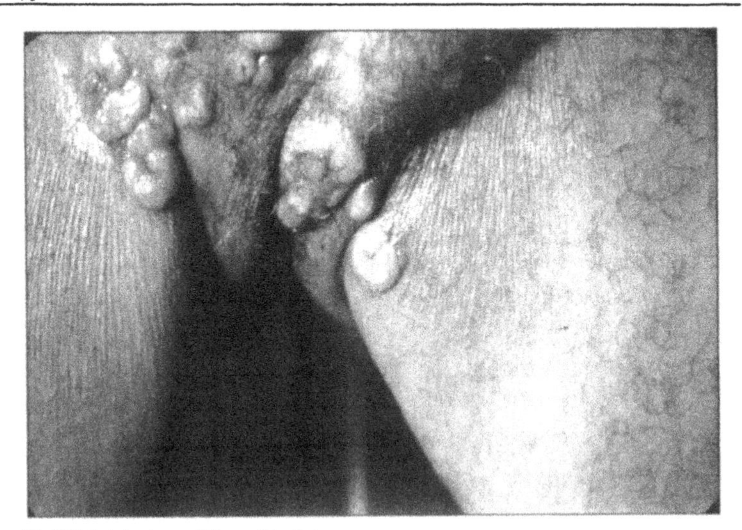

Fig. 34.5. Late benign syphilis, nodular lesions.

children with acquired primary or secondary syphilis: benzathine penicillin G 50,000 U/Kg in a single IM dose. In late syphilis: benzathine penicillin G 2.4 million U IM per week for 3 weeks. In children: benzathine penicillin G 50,000 U/kg IM in three weekly doses. With penicillin allergy of tetracycline or erythromycin, 500 mg every 6 hrs for 2 weeks is indicated (in latent syphilis up to 4 weeks), or doxycycline 100 mg every 12 hrs for 2 weeks are recommended. In late syphilis benzathine penicillin G 2.4 million U IM in three doses over 1 week. In neurosyphilis aqueous penicillin G 2-4 million U IV every 4 hrs for 10-14 days. An alternative regimen is 2.4 million U of procaine penicillin daily, plus oral Probenecidae, 500 mg every 6 hours for 10-14 days. The CSF should be examined every 6 weeks. Syphilis and HIV, primary and secondary: benzathine penicillin G 2.4 million U IM. Follow-up in 1, 2, 3, 6 ,9, and 12 months. Latent syphilis and HIV: In patients with both infections, CSF examination is recommended before treatment. If it is normal, benzathine penicillin G 7.2 million U IM in three doses. Syphilis and pregnancy: The treatment corresponds to the stage of the illness. Tetracycline and doxycycline are contraindicated. Erythromycin is not effective for treatment of the infected fetus. Infants of a mother with untreated syphilis or with evidence of relapse or re-infection after treatment must be treated.

If the mother has physical evidence of active disease, radiologic evidence, reactive VDRL on CSF, non-treponemic tests reactive for at least 4 times the title of the mother or IgM anti-treponemic specific, suggested treatment: aqueous penicillin G 100,000-150,000 U/kg/day (50 000 U/Kg every 12 hrs the first 7 days of life and every 8 hrs thereafter for 10-14 days). Procaine penicillin 50,000 U/kg IM daily in a single dose for 10-14 days. The Jarisch-Herxheimer reaction is an acute, febrile hypersensitivity reaction that presents in the first 24 hours after the onset of treatment. It is accompanied by fever, malaise, headache, joint pain, nausea, and tachycardia. It is more common in early syphilis, but more serious in late syphilis.

Selected Readings

1 Adimora AA, Hamilton H, Holmes KK et al Sexually Transmited Diseases, 2nd Ed. New York, McGraw-Hill, 1994: 1-9, 63-86335, 365-77.

2 Centers for Disease Control and Prevention. 1993 Sexually Transmited Diseases Treatment Guidelines. MMWR 1993

3 Holmes KK, Mardh PA, Sparling PF et al. Sexually Transmited Diseases, 2nd Ed. New York, McGraw-Hill, 1990: 205-11, 213-9, 221-30, 231-46, 247-50, 251-62, 771-801, 821-42, 927-34, 935-39.

3 Rolfs RT et al. Treatment of syphilis, 1993. Clin Infect Dis 1995; 20(Suppl 1): S23-38.

Pinta

Roberto Arenas

Pinta is a leukomelanodermic cutaneous disease autochthonous of Latin America that has almost disappeared. The course is chronic and benign. In its early stage it produces erythematous scaley plaques, and in the late stage it produces dyschromic lesions. It is contagious, non-venereal and it is caused by *Treponema herrejoni* (*T. carateum*).

GEOGRAPHIC DISTRIBUTION

It used to be found only in intertropical regions of Latin America: Mexico, Central America, Panama, Colombia, Venezuela, Peru, Ecuador, Bolivia, Guayanas and Antilles. In the last 20 years it has been endemic to the western Amazon region in Brazil where 265 cases have been reported: 10% have been children (*An Bras Dermatol* 1979; 54:215-237).

ETIOLOGY

It is caused by *Treponema herrejoni* (*T. carateum*) transmitted person-to-person or probably by an insect vector. Most cases have been reported in adults. It is not transmitted by sexual intercourse. There is no cross immunity with syphilis. The treponema penetrates the skin, and 1 week to 3 months later a pinta chancre appears that lasts 1-5 months; 5-12 months thereafter disseminated lesions or pintids appear which last several months. In some occasions, they are related to the initial lesion. These two first stages comprise early pinta. Late pinta is relentlessly progressive and causes permanent, dyschromic lesions. The pigmented changes may be a post-inflammatory effect or due to the inhibition of melanocytes by the treponema.

CLINICAL FEATURES

The pinta chancre is usually a single lesion and it appears on the legs, feet, arms, forearms and, less frequently, on the face. It is a 1-3 cm pink, slightly scaly papule that rapidly forms a round or oval, scaly, erythematous plaque with sharp edges surrounded by a hypochromic halo. Pintids are localized on the trunk and extremities. They are asymmetric and are not found in folds or on genitals. They

Tropical Dermatology, edited by Roberto Arenas and Roberto Estrada. ©2001 Landes Bioscience.

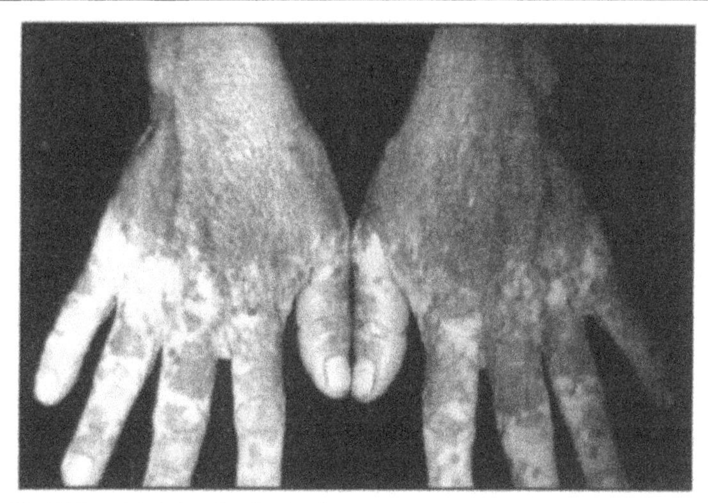

Fig. 35.1. Late pinta, leukomelanodermic lesions.

are papulosquamous plaques, smaller than the initial lesion. They tend to be dyschromic, and they disappear without leaving a trace or only with residual hyperchromic spots. There can be epitroclear, cervical and inguinal adenopathy and systemic symptoms. Late lesions are leukomelanodermic, permanent spots. They are usually disseminated and very symmetric. They predominate on elbows, knees, ankles, hands, feet and anterior surface of the wrist and trunk (white pinta). Lesions spare flexion folds, interdigital areas, the interscapulovertebral region, genitals, face and hairy skin. On the wrist, the achromic triangle is a characteristic lesion. There are large achromic spots and hyperchromic lenticular or felideform spots (Fig. 35.1). Cases of black pinta are less frequent. They appear on areas exposed to sunlight, e.g., the face, the decolletage, dorsal surface of forearms, hands, legs, feet and bony prominences. These lesions are gray or black and involute without leaving scars. In chronic cases they are dry and atrophic. There can be hyperkeratosis on the palms and soles or on elbows, knees, ankles and dorsum of hands and feet. Some patients present with depigmentation of the hair, thickness of nails and striae.

LABORATORY DATA

On biopsy atrophy of the epidermis is observed and there is a loss of sebaceous and eccrine glands. There can be an abundance of melanic pigment in the epidermis and superficial dermis or it may be scant, with vasodilatation and lymphocytes and plasma cells infiltrates. The treponema can be visualized with silver stains. On dark field microscopy *T. herrejoni* may also be observed. The VDRL and FTA-abs are strongly positive.

TREATMENT

Benzathine penicillin 1.2 million U every 8 days up to a total of 6 or 8 million U; sometimes 2.4 million U are adequate. If there is a penicillin allergy, then tetracycline or erythromycin 500 mg every 6 h for 10 days is administered.

Selected Readings

1 Castro LG et al. Nonvenereal treponematosis (Correspondence to the Editor). J Am Acad Dermatol 1994; 31(6): 1075-1076.

2 Dominguez-Soto L, Hojyo-Tomoka MT, Vega-Mejije E, Arenas R, Cortes-Franceo R et al. Pigmentary problems in the tropics. In Parish LCH, Millikan LE. Dermatologic Clinics. Philadelphia: Saunders 1994; 12(4):777-784.

3 Koff AB, Rosen T et al. Nonvenereal treponematoses: Yaws, endemic syphilis and pinta. J Am Acad Dermatol 1993; 29:519-535.

36 Yaws (Pian, Frambesia)

Roberto Arenas

This disease is also known as bouba, parangi and paru. It is an infectious and chronic non-venereal disease. It is transmitted by direct contact and caused by *Treponema pallidum* var. *pertenue.* It is more common in children, and it is characterized by a primary ulcerated lesion, secondary macular lesions, papules and plaques and late destructive lesions.

GEOGRAPHIC DISTRIBUTION

Pian occurs in Africa, Asia, Central America, South America and Pacific Islands in areas where sanitation is poor and the climate is humid and rainy. It affects 40 million people, mostly young people less than 15 years of age; 100 million children are at risk. Ten years ago in Central Africa positive serology was found in 15% of children tested. In the province of Santiago in Ecuador serology was positive in 90% of the adult population and 10% had active lesions. Thanks to some massive campaigns, this disease has been eradicated in some areas.

ETIOLOGY

T. pallidum subsp *pertenue* is the etiological agent and cannot be distinguished from *T. pallidum* by techniques such as Western blot, Southern hybridization or techniques of immunoblot. It is transmitted from person-to-person by direct contact through breaks in the skin or by insect bites. The absence of neurological involvement, unlike the treponematosis of syphilis, is a mystery.

CLINICAL FEATURES

Yaws is usually extragenital. If untreated it persists for decades. There are three stages, two early (primary and secondary), and a late one or tertiary. They are separated by periods of asymptomatic latency. After an incubation of 3-5 weeks, the primary lesion appears at the site of inoculation (main bouba, main Yaws). It generally appears in feet, legs or buttocks. It is a papulonodular lesion of 1-5 cm of diameter, indurated and painless that grows and ulcerates. The surface is reddish (fambresia) and is covered with a yellowish crust. Satellite lesions appear. It heals spontaneously in several weeks or months leaving a hypopigmented area

Tropical Dermatology, edited by Roberto Arenas and Roberto Estrada. ©2001 Landes Bioscience.

surrounded by a dark halo. Rarely the primary lesion is not present. There can be fever, joint pain and regional adenopathy.

The secondary stage appears weeks or months thereafter; it is accompanied by headache, fever, malaise and joint pain, with disseminated lesions smaller than the initial lesion. There are macules, papules or plaques (daughter yaws or "pianids") that ulcerate and are eczematous ("pianoms" or "framboesias"). They are seen on joints or around the nose and mouth. Sometimes they are annular or circinate, appearing as macerated condylomas in axilla and groin. There is a painful palmar and plantar hyperkeratosis (rough yaws) and periungual papillomas causing paronychia. There can be generalized lymphadenopathy and osseous involvement with painful osteoperiostitis and polydactylitis with asymptomatic abnormalities of the CSF. The secondary lesions are recurrent in the axilla and perianal region, and they can enter in a period of latency.

The late stage appears from 5-15 years after the primary infection, and it develops in 10-15% of patients. There are nodules that darken and ulcerate, palmar and plantar hyperkeratosis and hypertrophic lesions with bone and joint abnormalities. Chronic osteitis causes so-called "Sabre tibia." Nasal deformities (goundou) or nasal and palatal perforation (gangosa) occur. Neurological and ophthalmologic involvement is controversial. Yaws in this stage spares skin folds.

LABORATORY DATA

Dark field microscopy and serologic tests do distinguish the etiologic agents of syphilis, Bejel and pinta. This is accomplished with rapid plasma reagent (RPR) VDRL, fluorescent-treponema antibodies (FTA-abs), treponema immovilization (TPI) and *T. pallidum* hemaglutination (TPHA) (Chapters 34, 35 and 37). IgG EIA is sensitive, but it has a low specificity in endemic zones (*J Clin Microbiol* 1995; 33(7):1875-8). Histopathology demonstrates ancanthosis, papillomatosis, epidermal edema and intaepidermal microabscesses with neutrophils. In the dermis there is a dense infiltrate composed by plasma cells, lymphocytes, histiocytes, neutophils and eosinophils, and some endothelial proliferation. Treponema is observed with silver stain. On electron microscopy they can also be seen in the epidermis and in macrophages. (*Genitourin Med* 1991; 67(%):403-7).

TREATMENT

Arsenic and bismuth are only of historic interest. The preferred treatment is penicillin, and the dose recommended by WHO since 1980 is benzathine penicillin 1.2 million U in adults and 0.6 million U in children less than 10 years of age. With penicillin allergy tetracycline can be used, 1-2 g/day for 5 days, erythromycin or chloramphenicol 8-10 mg/kg 4 times/day for 15 days. The new tetracyclines and analogs of erythromycin have not been tested.

Selected Readings

1 Engelkens HJH, Jrdanarso J, Oranje AP et al. Endemic treponematoses: Part I. Yaws. Int J Dermatol 1991; 30:77-83.
2 Engelkens HJH, Niemel PLA, Van Der Sluis JJ et al. The resurgence of yaws: World-wide consequences. Int J Dermatol 1991; 30:231-8.
3 Koff AB, Rosen T et al. Nonvenereal treponematoses: Yaws, endemic syphilis, and pinta. J Am Acad Dermatol 1993; 29:519-35.
4 Meheus A, Antal GM et al. The endemic treponematoses: Not yet eradicated. World Health Stat Q 1992; 45(2-3):228-37.
5 Schmid GP et al. Epidermiology and clinical similarities of human spirochetal diseases. Rev Infect Dis 1989; 11:S1460-8.

Endemic Syphilis (Bejel)

Roberto Arenas

This disease is also known as njovera, dichuchwa, belesh, and bishel. It is a non-venereal, chronic, treponematosis. It is recurrent in infancy and caused by *T. pallidum* subsp *endemicum*. It is characterized by mucocutaneous lesions: macules, plaques, papules, ulcers and scars as well as osseous and destructive cartilaginous lesions.

GEOGRAPHIC DISTRIBUTION

This disease predominates in Third World countries where hygiene and sanitation are poor. It is endemic in the North Africa, Southeast Asia, the Arabian peninsula and the eastern Mediterranean. It predominates in rural areas with a dry and hot climate. It is observed in young people less than 15 years of age, and in both sexes. It has been eradicated from Bulgaria and Bosnia, but in some parts of Africa 22% of individuals are seropositive and 7.5% of children 5-7 years old have active lesions (*Bull Soc Pathol Exot* 1988; 81:827-31). A few cases have been reported from Europe.

ETIOLOGY

T. pallidum subsp *epidemicum* is a treponema morphologically and antigenically similar to *T. pallidum*. It is 10-13 µm long by 0.15 µm wide. It replicates in 30 hours and it contains an external membrane (it synthesizes an envelope of mucopolysaccharides), an electrodense envelope of peptidoglycans and a cytoplasmic membrane. It is almost exclusively present in children and only infrequently in susceptible adults. The form of transmission is not well known: from person-to-person by skin contact, mucous secretions and probably by fomites. AIDS may cause reactivation of the latent treponematosis and at the same time facilitates the transmission of the infection by HIV.

CLINICAL FEATURES

The initial lesion appears 2-3 months after inoculation. It occurs in nasopharyngeal mucosa at the site of inoculation. There is a papule or painless ulcer accompanied by regional adenopathy. Secondary lesions last 6-9 months. They are composed of a macular or papular eruption with multiple erosive and crusty lesions that predominates in the limbs. There can be anogenital condylomata. There is generalized adenopathy, and there can be osteoperiostitis of the long bones that causes nocturnal pain. Tertiary lesions appear 6 months to several years after the primary lesion. They appear in the nasopharynx, larynx, skin and bones. They are destructive gummas that ulcerate and leave atrophic scars, often geometrical. Lesions of the palate and nose are destructive and disfiguring. There is also synovitis, uveitis, choroiditis and chorioretinitis.

LABORATORY DATA

Serology and dark field microscopy are similar to syphilis and yaws (Chapters 34-36). On radiology, osteoperiostitis can be seen, especially involving the tibia. Skin biopsy shows parakeratosis, acanthosis, spongiosis and dermal inflammatory infiltrates with lymphocytes, epithelioid and plasma cells, melanophages, and vasodilatation.

TREATMENT

The treatment is similar to yaws (Chapter 36). In patients with destructive lesions, plastic surgery and rhinoplasty are necessary (*Plast Reconstr Surg* 1984; 74:589-602). The development of a vaccine is the hope of the future. There have been massive and effective campaigns to combat this disease, and in some places it has been eradicated. But in other places (*Med Trop* 1989; 49:237-44), especially in Central and West Africa, the efforts have not been successful. Control is complicated by the absence of natural immunity, periodic contagiousness and the presence of subclinical illness.

Selected Readings

1. De Schryver A, Meheus A et al. Revue: Les treponematoses endemiques en sont toujours pas eradiquees. Med Trop 1989; 49:237-44.
2. Erdelyi RL, Molla A et al. A burned-out endemic syphilis (bejel): Facial deformities and defects in Saudi Arabia. Plast Reconstr Surg 1984; 74:589-602.
3. Gazin PP, Meynard D et al. Enquete clinique et serologique sur le bejec au nord du Burkina Faso. Bull Soc Pathol Exot 1988; 81:827-31.
4. Koff AB, Rosen T et al. Nonvenereal treponematoses; Yaws, endemic syphilis, and pina. J Am Acad Dermatol 1993; 29:519-35.
5. Meheus A, Antal GM et al. The endemic treponematoses: Not yet eradicated. World Health Stat Q 1992; 45(2):228-37.

Chancroid (Soft Chancre)

Sergio Eduardo Gonzalez-Gonzalez

Chancroid is a sexually transmitted disease (STD) that is characterized by ulcers and adenopathy. It is caused by a Gram-negative bacillus described by Ducrey in 1889.

GEOGRAPHIC DISTRIBUTION

It is endemic in tropical and subtropical countries. There are sporadic outbreaks in the USA, Europe and other non-tropical countries.

ETIOLOGY

Haemophilus ducreyi is a gram negative, anaerobic, facultative coccobacillus. Its pathogenicity is not well known. There are avirulent and other virulent forms relatively resistant to phagocytosis.

CLINICAL FEATURES

After incubation from 12 hours to 7 days, at the site of inoculation—usually the genital area—one or several papules surrounded by peripheral erythema appear. These rapidly become pustular and erode, leaving an ulcer with sharp edges without induration and with the characteristic of being autoinoculable (Fig. 38.1). In women the symptoms are minimal and many times go unnoticed. In men, the lesion is generally painful. In approximately half of the patients there is painful, unilateral inguinal adenopathy. The skin over the lymph nodes is erythematous, and a gumma develops which spontaneously drains, leaving scars. Systemic effects are usually mild.

A secondary anaerobic infection can occur producing gangrenous ulcers and extensive destruction of genital tissue. There has been disagreement regarding the role of this infection in the transmission of AIDS. Infection in newborns has not been reported.

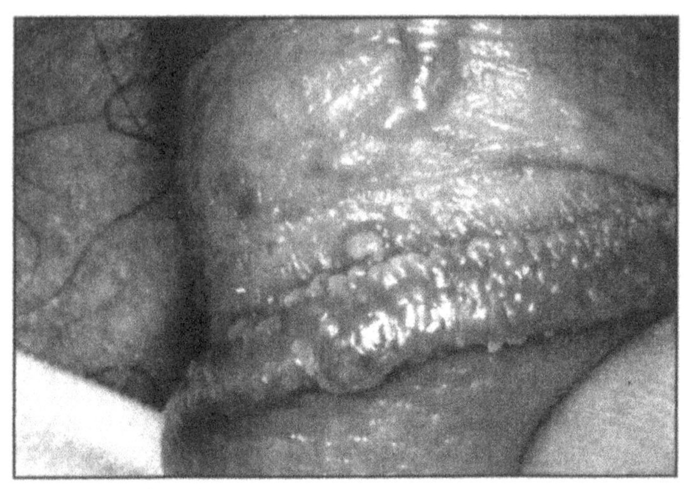

Fig. 38.1. Chancroid, initial lesion in the glans.

LABORATORY DATA

Direct exam or smears from the lesion occasionally reveal the Gram-negative extracellular organisms. Culture, usually in agar enriched with hemoglobin, confirms the existence of *H.ducreyi.* The colonies appear in 48-72 h. Indirect immunofluorescence with monoclonal antibodies can be useful in diagnosis. ELISA and PCR also are useful in diagnosis.

TREATMENT

Antibiotic resistance is common and has been documented with ampicillin, sulfonamides, tetracycline, chloramphenicol, and kanamycin. These drugs can be administered in a single dose: Azithromycin 1 g or ciprofloxacin 500 mg orally, ceftrioxon 250 mg or spectinomycin 2 g IM (*Clin Infect Dis* 1995; 21(2): 409-14). The following can be given orally in multiple doses: amoxicillin 500 mg every 12 hours for 3 days, erythromycin 500 mg every 6 hours, or ofloxacin 400 mg b.i.d both for 7 days. Another alternative is fleroxacina 400 mg in a single dose or 400 mg/day for 5 days in HIV positive patients (*Am J Med* 1993; 94(3A):85S-88S). In order to avoid scar retraction, it is important to aspirate the affected lymph nodes; they should not be drained.

Selected Readings

1. Adimora AA, Hamilton H, Holmes KK et al. Sexually Transmitted Diseases, 2nd ed. New York:McGraw-Hill, 1994:87-92.
2. Center for Disease Control and Prevevention. 1993 Sexually transmitted diseases treatment guidelines. MMWR 1993.
3. Homes KK, Mardh PA, Sparling PF et al. Sexually transmitted diseases, 2nd ed. New York, MgGraw-Hill, 1990: 263-261, 711-716.
4. Martin DH, Sargent SJ, Wendel GD Jr et al. Comparison of azithromycin and ceftriazone for the treatment of chancroid. Clin Infect Dis 1995; 21(2):409-414.

Granuloma Inguinale (Donovanosis)

Clemente Moreno-Collado

Granuloma inguinale (GI) or donovanosis is a chronic, granulomatous disease that affects the skin of the anogenital and inguinal regions. It is usually acquired by sexual contact and is characterized by ulcers that are moderately painful and progressively destructive. It is the least common of the venereal diseases. It is a sexually transmitted disease (STD) in which malignant transformation can be observed when diagnosed in young individuals. Its relation to penile cancer is unclear. Because it causes ulcerous lesions, its control constitutes a method to decrease the transmission of HIV (*Genitourin Med* 1995; 71:27-31). Recently single-dose treatment with antibiotics has been successful.

GEOGRAPHIC DISTRIBUTION

The disease was described by McLeod in 1882 in southeast India. It occurs almost worldwide, but mainly in the tropics, in small endemic foci (*Ann Acad Med Singapore* 1995; 24:569-578). Most cases are reported in Southeast Asia, New Guinea, the Carribean, and Central and South America. In North America and Europe, outbreaks are reported, especially in urban centers with immigrants from Third World countries. It mainly affects young males.

ETIOLOGY

The causative agent is *Calymmatobacterium granulomatis*, a pleomorphic gram negative bacillus, 1.5-2 µm, bipolar and surrounded by an argyrophilic capsule. It is related to *Klebsiella rhinoscleromatis* with which it shares antigenicity. It is difficult to culture by conventional methods, and the disease has not been reproduced in animals.

CLINICAL FEATURES

The incubation period is 10-50 days. The initial lesion is a firm nodule or papule that grows rapidly and ulcerates with well-defined, elevated and laced margins, painless or tender and not accompanied by adenitis. The lesion can grow, penetrate deeply and disseminate by autoinoculation. It is granulomatous at the base, bright red and bleeds easily (Fig. 39.1). As the ulcers extend, fibrosis and

Tropical Dermatology, edited by Roberto Arenas and Roberto Estrada.

vegetative epithelial hyperplasia develops. The initial lesion can be observed in the pubic area, genitals, perineum, groin or perinanal region, especially in homosexuals. The lesion in men affects the foreskin, glans, perineum or scrotum. In women, the lesion affects the labia, the pubis and contiguous areas. The forms described are ulcerovegetative, nodular, hypertrophic and scarred. Inguinal involvement does not start in the lymph node as with lymphogranuloma venereum (LGV) but from a granulomatous periganglionic lesion called a "pseudobubo" that becomes necrotic and ulcerates (Fig. 39.2). Keloidal scars with peripheral ulcers develop in months or years. Elephantiasis of the external genitals with rectovaginal or vesicovaginal sinus tracts, as well as urethral stenosis, can occur. Unlike LGV, rectal stenosis and buboes are not observed or are rare in GI. This disease does not have a tendency to heal spontaneously. Progression is slow with intermittent and irregular extension, in some occasions over several years. In 3% of the cases there can be extragenital lesions, mainly on the face (Genitourin Med 1991; 67:441-452).

LABORATORY DATA

The causative agent can be identified by smears of the granulation tissue of the lesion or from a small fragment of the edge of the ulcer stained with Wright or Giemsa which demonstrates Donovan bodies, generally, in the cytoplasm of macrophages. However, the best method for diagnosis is identification of Donovan bodies with Warthin-Starry stain. Donovan bodies appear as small, straight

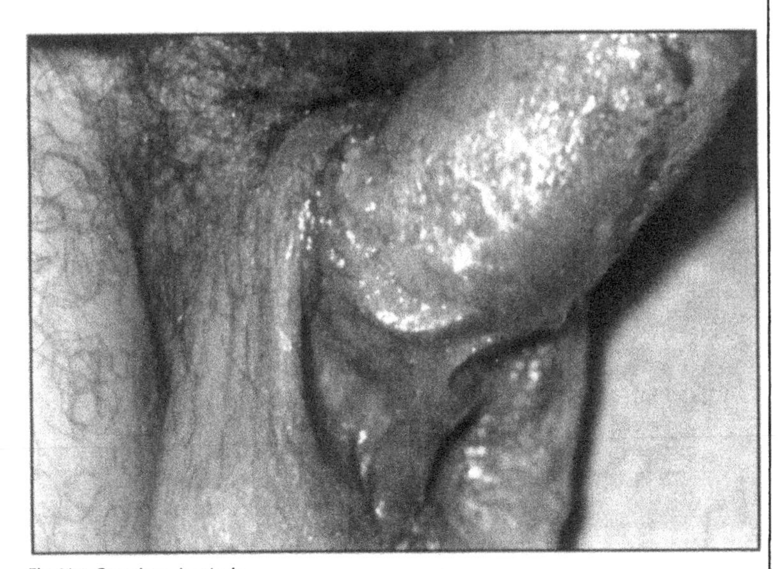

Fig. 39.1. Granuloma inguinale.

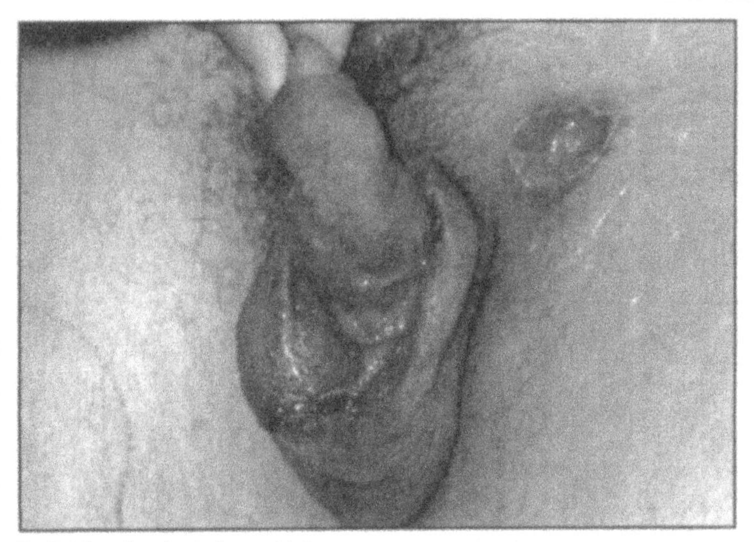

Fig. 39.2 Granuloma inguinale, pseudobubo.

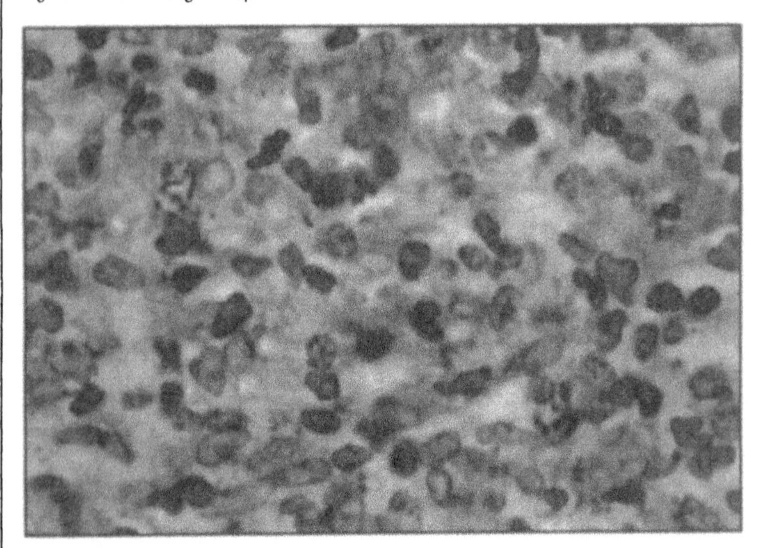

Fig. 39.3 Donovan bodies (Warthin-Starry, 100X).

or slightly curved rods that have the greatest affinity for stain at the ends where chromatin is condensed to give the appearance of "safety pins" (Fig. 39.3). There is a dense granulomatous infiltrate with polymorphonuclear cells, plasma cells and vascular neoformation. There are fewer plasma cells than in syphilis, and microabscesses are smaller and not as regular as in the LGV.

TREATMENT

The treatment of choice is an antibiotic soluble in lipids that reach high concentrations in the interior of cells and that are effective against Gram-negative bacilli such as tetracycline 500 mg qid orally for 3 weeks. Formerly, streptomycin was used, in a total dose of 20-30 g IM for 5-10 days (or 1 g IM every 6 hr for 5 days). Also, trimetroprim-sulfamethoxazole 160/800 mg bid for 2 weeks has been used with good results, adequate tolerance and few side effects. In pregnant women, the treatment of choice is erythromycin. Successful single-dose antibiotic treatment has been reported (*Genitourin Med* 1996; 72:17-19).

SELECTED READINGS

1 Hart G. Donovanosis in Holmes KK, Mardh PA, Sparling PF et al. Sexually transmitted diseases, 2nd ed. New York, McGraw-Hill, 1990:273.
2 Mulhall BP, Hart G, Harcourt C et al. Sexually transmitted diseases in Australia: a decade of change. Epidemiology and surveillance. Ann Acad Med Singapore 1995; 24:569-578.
3 O'Farrell N. Global erradication of donovanosis: An opportunity for limiting the spread of HIV-infection. Genitourin Med 1995; 71:27-31.
4 Richens J. The diagnosis and treatment of donovanosis (granuloma inguinale). Genitourin Med 1991; 67:441-452.

Lymphogranuloma Venereum

Clemente Moreno-Collado

Lymphogranuloma venereum (LGV), or lymphogranuloma inguinale, is a sexually transmitted, systemic disease caused by serotypes L1, L2, and L3 of *Chlamydia trachomatis.* It occurs in three well-recognized stages. Males suffer from urethritis. Woman are asymptomatic reservoir, and babies born of infected mothers can have the disease. LGV, often mistaken for syphilis, was first described in 1913 by Durand, Nicolas and Favre (*Bull Med Soc Med Hosp* 1913; 35:274).

GEOGRAPHIC DISTRIBUTION

LGV is very common in the tropics. It occurs sporadically in industrialized countries, especially in epidemics in urban places of Europe and the USA (*Schweiz Med Wochenschr* 1993; 123:1250-1255). It is more common in men than in women. As with other STDs most frequently affected are sexually active individuals in low socioeconomic classes. Occasionally cases have been reported in children infected via sexual or by nonsexual transmission (*Genitourin Med* 1993; 69(3): 213-221).

ETIOLOGY

The Chlamidiae family is a group of obligate intracellular bacteria. *C. trachomatis* is classified into 15 immunotypes or serotypes that have different pathogenic properties. The types A, Ba, and C cause trachoma, an endemic ocular infection that is a frequent cause of blindness. Types D-K are sexually transmitted, and they frequently cause urethritis, cervicitis, endometritis, and salpingitis, as well as ocular infections and Reiter's syndrome by contamination with vaginal secretions. Serotypes L1, L2 and L3 predominantely infect lymphatic tissue. They cause tissue destruction and are the most common etiological causative serotypes of LGV.

CLINICAL FEATURES

The lesions of LGV occur in three stages. After a period of incubation from 1 week to 3 months, the primary lesion appears as a papule, vesicle or small erosion; it is rarely observed by doctor or patient because of its evanescent nature. It can be accompanied by urethritis or cervicitis. In the homosexual man it can be present as a rectal infection with bloody diarrhea and tenesmus. The primary lesion, usually is asymptomatic on the glans, balanoprepucial furrow, scrotum or urethra in men, and in women on the inner surface of the labia, posterior vaginal wall and cervix (Fig. 40.1). It disappears spontaneously. Rarely the lesion persists for several weeks; then it is observed to persist with the second stage—the most characteristic of the illness—called the "inguinal syndrome" which is observed 2-6 weeks after the first lesion. Adenopathy, usually inguinal, begins as an acute inflammation. The lymph nodes enlarge, become tumescent and later slightly indurated. They are nontender and only mildly painful on walking. Fever and malaise can occur. As the disease progresses adjacent lymph nodes become involved forming a hard or doughy mass in the groin that can include the femoral nodes. The overlying skin is included in the process; it becomes indurated, violaceous and looks like orange peel. In time the mass softens in certain areas and appears multilocular. It drains to the surface through numerous sinus tracts; it does not have a tendency to scar. At this stage the lesion is known as a "bubo" which characteristically is situated above and below the inguinal ligament, and the zones are separated by a linear depression known as the "sign of the groove"—the most characteristic manifestation of LGV (Fig. 40.2). Intra-abdominal adenopathy occurs mainly in women as lymph nodes of the internal iliac fossa become involved. It is moderately painful, but it suppurates or drains through fistulas. In some cases other lymph node groups can be affected, e.g., the axillary and cervical. Erythema nodosum is observed in 2-10% of individuals. In the third stage proctitis, rectal stenosis, perirectal abscesses, genital edema and fistulas occur. Proctocolitis is accompanied by fever, pain, tenesmus, and lymphedema of the perirectal tissues with formation of abscesses, fistulas, ulcers, and scars. When this occurs in women it is known as esthiomene. These

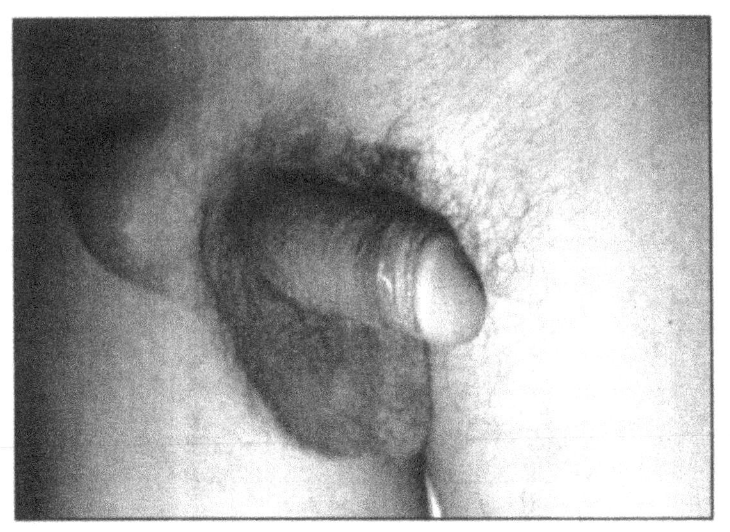

Fig. 40.1. Lymphogranuloma, primary lesion and adenopathy.

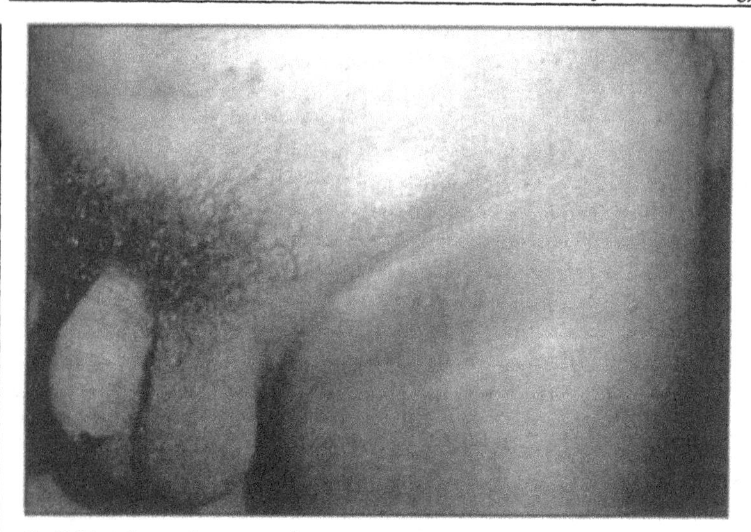

Fig. 40.2. Lymphogranuloma, groove sign.

lesions of the digestive tract and genitals were described by O. Jersild in 1926 as the "genito-anorectal syndrome". A common feature of this syndrome, especially in women, are painless or slightly tender, destructive ulcers with poorly-defined margins and a fibrous base. Systemic complications like meningitis, conjunctivitis, pneumonia and pericarditis are rare.

LABORATORY DATA

A positive test for *Chlamydia,* as well as culture using HeLa-229 or McCoy cells, confirm the diagnosis. Formerly the diagnosis was established by the Frey test, the intradermal injection of killed microorganisms in chicken embryo. This test, commercially called Lygranum, is no longer used. Also, the diagnosis of LGV was based in the detection of elevated titers of antibodies and complement. However, it is difficult to obtain serial blood to demonstrate changes in titer. Histologically the initial lesion is characterized by acute inflammation with epithelial loss. The base of the ulcer contains fibrin and polymorphonuclear cells, and the deep tissues have an inflammatory infiltrate composed of lymphocytes, histiocytes and plasma cells. In the nodes starry microabscesses surrouded by macrophages, epithelioid cells, lymphocytes and neutrophils can be observed. These abscesses, though rare, suggest the diagnosis. Also *Chlamydia* produces characteristic inclusion bodies demonstrated by Papanicolau stain (Fig. 40.3). The most precise diagnostic method involves fluorescent monoclonal antibodies in clinical specimens of frotis of the primary lesion, of the aspirated bubo or from frozen tissue biopsies. The frotis is fixed in acetone and much the section of frozen section tissue is placed in chamber; the commercial reactivate is added (Siva Microtrak) for 15 min, washed with distilled water, and left to air dry. In the positive cases, an immunofluorescent pattern that resembles a starry sky (Fig. 40.4) is seen. It is essential, as always when dealing with suspected cases of STD, to do luetic and HIV serology tests.

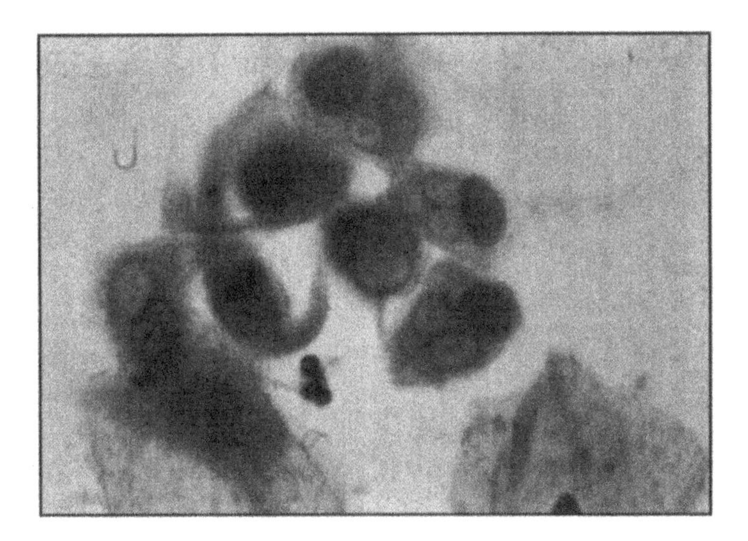

Fig. 40.3. Cytology of *Chlamydia trachomatis* (Papanicolau, 100X).

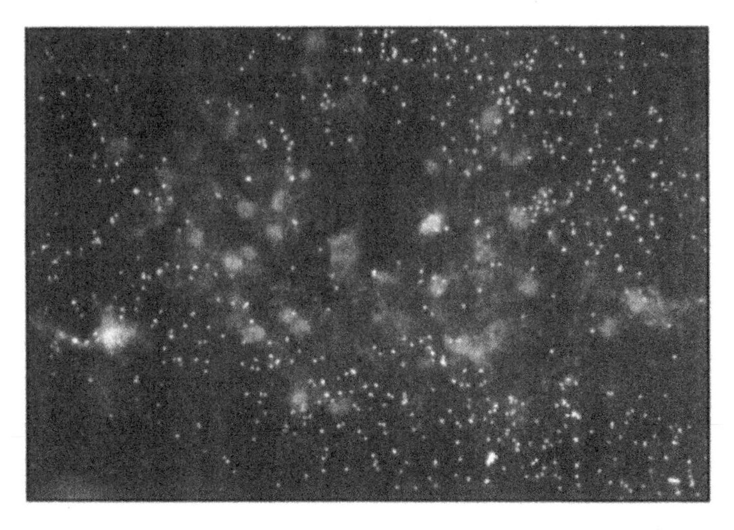

Fig. 40.4. Immunofluorescence, "starry sky".

TREATMENT

In some cases the illness is self-limited. If diagnosed and properly treated at an early stage, all damage is reversible. The first drug used to effect was sulfonamide. Subsequently tetracycline, 500 mg qid for 3 weeks and doxycyline and erythromicin have been reported to be effective (*Semin Dermatol* 1994; 13(4):269-274). Recent reports of single dose antibiotics are of interest (*New Engl J Med* 1992; 327:921-925).

Surgery is an auxiliary therapeutic measure. Fluctuant and soft bubos can be drained; this can improve the course. Drainage should be performed from the inferior part of the bubo upwards with a large, thick needle inserted through uninvolved skin to avoid the formation of fistulae.

SELECTED READINGS

1. Buntin DM. The 1993 sexually transmitted disease treatment guidelines. Semin Dermatol. 1994; 13(4):269-274.
2. Goh BT, Forster GE. Sexually transmitted diseases in children: chlamydial oculo-genital infection. Genitourin Med 1993; 69(3) 213-221.
3. Stamm W, Holmes KK. Chlamydia trachomatis infections of the adult. In Holmes KK, Mardh PA, Sparling PF and Wiesner PJ. Sexually transmitted diseases, 2nd ed. New York. McGraw-Hill. 1990.
4. Thomas GJ, Osborn MF, Munday PE, Evans RT, Taylor RD. A 2-year quantitative assessment of chlamydia trachomatis in a sexually transmitted diseases clinic population by the Micro Trak direct smear immunofluorescence test. Int J STD AIDS 1990; 1(4):264-267.

I. Parasitic Dermatosis

Pediculosis

Scabies (Sarna)

Larva Migrans (Larva Migrans Syndrome)

Tungiasis

Myiasis

Leishmaniasis

Cutaneous Amebiasis

Trypanosomiasis

Onchocerciasis

Trichinosis

Dracunculosis

Cysticercosis

Pediculosis

Roberto Estrada

Pediculosis is an ecoparasitosis caused by hematophagous wingless insects (lice) of the gender Anoplura. In humans infestation is caused by *Pediculus humanus capitis* or head louse, *Pediculus humanus humanus* or body louse, and *Phtirus pubis* or pubic louse. *Pediculus h. humanus* and *capitis*, may in addition transmit infectious diseases such as endemic typhus, relapsing fever and other rickettsial diseases (*Lancet* 1993; 342:1213-15), especially among victims during war time (*Isr J Med Sci* 1993; 29:371-73). Both varieties seem to be closely related and may interbreed; nevertheless, different species must be considered since their feeding habits, survival requirements and habits are different (*Primary Care. Clinics in Office Practice.* Saunders 1989:551-68).

ETIOLOGY

The head louse is a 2-4 mm flat insect. The female has a larger abdomen than the male. It has three pairs of legs that end in claws which stick to the hair and clothing. The body louse has a thin skin which requires humidity and the heat of its host in order to survive. Out of this environment it dies in approximately 24 hours (Fig 41.1). It is mobile and passes easily from one individual to another when people sleep in close contact or when children play most commonly at school. The louse moves on its back using the hair as ladder crossbars. The head louse assumes a similar color to the host's hair. The color tones range from a very black to gray whitish. White ones observed at a distance are empty and are not viable. The life cycle lasts a month during which the female lays up 7-10 eggs called nits which firmely attach to hair by a special glue that she expels with the eggs (Fig. 41.2). Incubation is about 8 days. If we consider that the average hair growth is 1 cm a month, the distance from the scalp at which we will find the nits indicates the time of the parasitation. The parasite undergoes three skin changes before becoming an adult in 8-12 days. It feeds every 4-6 h, and when it does, it injects thrombin inhibitor factor Xa at the site of the bite (*Med Entomology and Acarology J Insect Physiol* 1996). The most important infectious risk factor in this disease seems to be the number of people living in one house, and mainly those that sleep in the same bed. The distinction between *Pediculus h. humanus* and *P. capitis is* difficult to establish. The former is slightly larger and its color is lighter. In reality, they share many characteristics although their differences are important. For example, body lice live on fabric, especially in people who do not change their clothing, a vital factor in their survival. They lay their eggs in the clothing fabric in

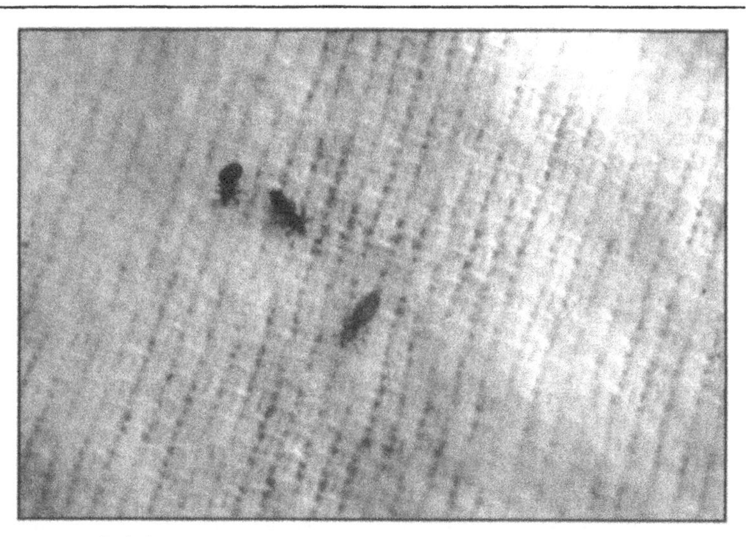

Fig. 41.1. Pediculosis capitis.

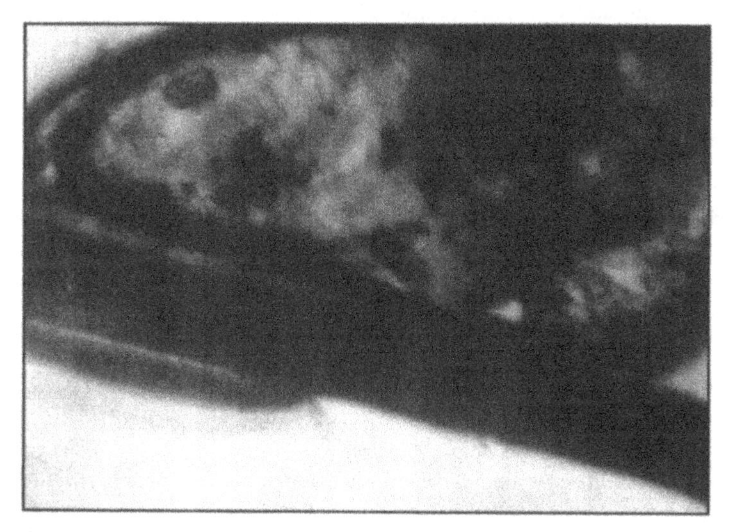

Fig. 41.2. Nits.

the areas where they can get warmth and humidity from skin from which the lice feed immediately after hatching. It may also be the vector for endemic typhus caused by *Rickettsia prowasewkii*, trench fever caused by *Rickettsia* (*Rochaliamea*) *quintana*, murine typhus caused by *Rickettsia typhi* (mooseri) and the relapsing fever caused by *Borrelia recurrentis*. In this last, transmission is by scratching and introducing the feces of the parasite—not by the bite as was formerly believed. These diseases are particularly important in crowded conditions and during war—time (*Rats, Lice and History*. Boston, Little Brown 1935). Pediculosis pubis is caused by *Pthirus pubis*. It is 2-3 mm (Fig. 41.3) and is transmitted sexually.

GEOGRAPHIC DISTRIBUTION

Epidemics of head lice occur in schools in developed countries (*Maternal and Child Health* 1983; 8:51), while it is endemic in the schools of developing countries. Generally, infestation is related to poverty and the lack of hygiene. We have found out that up to 33% of the school children in suburban and rural areas may be infested. The families of infested children usually also are infested. Children are more prone to be infested at home than at school (Proceedings of the Royal Institution of Great Britain 1983:55). In a study done in 944 school age children, we found that knowledge about its infestation reduces the risk of contagium (Paredes S, Estrada R, Chavez G et al. *Int J Dermatol* 1997; 36:826-830). In our series men were less often infested than women. Long and/or dirty hair was a risk factor in our patients. This contradicts the study which reports infestation to be common in children with clean and short hair (*Head Lice Advice*. BLM Publications Croydon, 1986).

CLINICAL FEATURES

Pediculosis capitis. One of the first indications of infestation is intense itching. Pruritus indicates infestation of about two months' duration. The lice are difficult to see in clean individuals who have only a minor infestation. Yet they can be abundant and easily, seen in malnourished individuals with poor hygiene. Nits can be confused with dandruff and may be distinguished from it with a magnifying glass. Also dandruff falls from the hair easily whereas nits firmly attach to it. It should also be distinguished from seborrhea, psoriasis, the shafts which cover the hair in the pityriasis sicca or from the residual particles of hair spray. On examination of hairy skin, it is possible to see lichenification and severe scratching marks and erythema, especially in the occipital or retroauricular regions (Fig. 41.4). If a white cloth is place under the head of an infested child and a fine-toothed comb is drawn through the hair, lice, easily nits or ova, and a black powder-like lice feces will fall. Impetigo, folliculitis or furunculosis are frequently

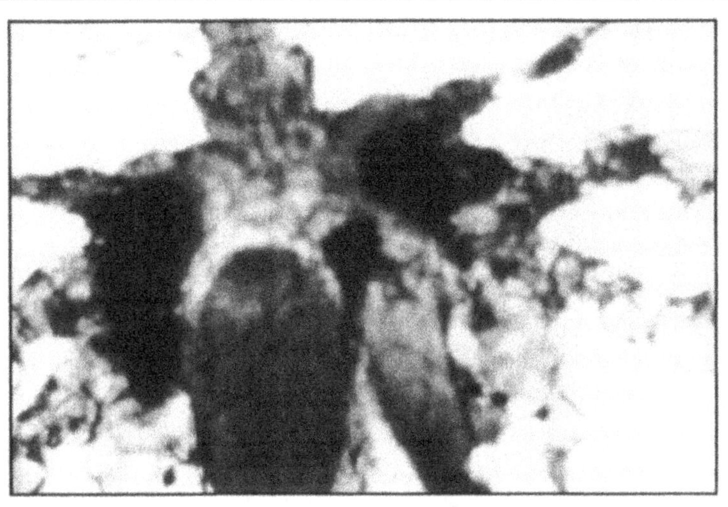

Fig. 41.3. *Phtirus pubis.*

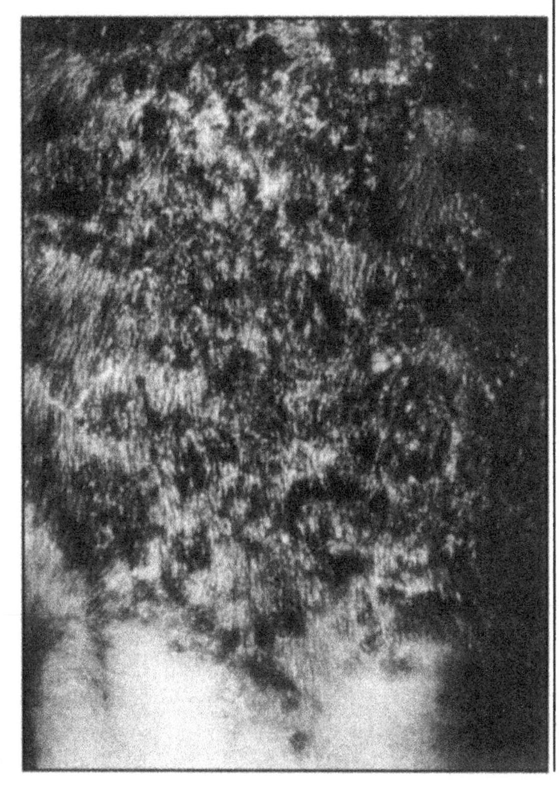

Fig. 41.4. Pediculosis capitis.

associated with this infestation. Furunculosis can be accompanied by fever, malaise and regional adenopathy.

Pediculosis corporis. Papular lesions, scabs, hives, and scratch marks are seen. As with pediculosis capitis, symptoms begin only after several weeks of infestation or sooner in cases of reinfestation (Fig. 41.5). Pruritis is the main symptom, although with secondary infection there may be pain, fever, malaise and lymphadenopathy. To confirm the diagnosis, meticulous examination of the clothing will reveal lice and nits.

Pediculosis pubis. Pubic lice cause severe pruritus that is difficult to ignore. Examination reveals the black powder-like louse feces as well as mites firmly adherent to hair in the pubic and abdominal areas. Small blue-black dots caused by the irritating secretion of the lice's bite (macula cerulea) are characteristic. In hirsute individuals infestation may spread to the perianal and gluteal area, axilla, chest, back and umbilicus. It may also involve the beard and mustache. Children can become infested by sleeping with their parents or other affected people, not necessarily by sexual transmission. In children the common areas of involvement are the eyebrows and eyelashes (pediculosis ciliaris). It can cause blepharoconjuntivitis with epitheliopathy (*Br J Ophtalmol* 1993; 77:815-16), as well as other ocular inflammations. The patient has intense pruritus, burning or pain. (*Arch Dermatol* 1973; 107:916-17).

LABORATORY DATA

Wood's lamp examination will reveal bluish, pearl-colored nits.

TREATMENT

Here are the most important drugs of choice: Permethrin and its derivatives (lotion rinse to 1%) occupy the most prominent place in the treatment of pediculosis because they are effective and only mildly toxic. This ointment is applied in the affected areas for 5 minutes and then rinsed off well. One single application is sufficient although it can be repeated in 10 days to treat the viable ova that may have hatched since the initial application. It has greater ovicidal effect than other antiparasitics agents. Cases resistant to other antiparasitics can be treated successully with permethrin (*Pediatr Dermatol* 1986; 3:344-8). It is safe in pregnant and nursing women as well as in young children. Malathion or lindane 1% is an acceptable option for treatment although it cannot be used in pregnant and nursing women or in young children. Excessive application induces resistance and/or toxicity, especially if it is applied after a warm bath which favors its absorption. Toxicity may include local irritation, CNS irritation and aplastic anemia (*CID* 1995; 20(Suppl 1):104-109). For pediculosis ciliaris permethrin and lindane are applied with swabs and parasites and eggs are extracted with

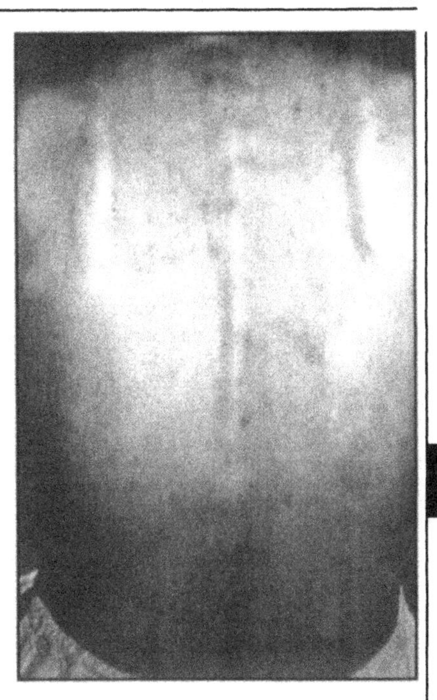

Fig. 41.5. Pediculosis corporis.

tweezers after applying Vaseline to the eyebrows and eyelashes for several days. Children do not cooperate with this treatment; usually only a single application is tolerated. Also, physostigmine, 0.25-1%, has been used, but it can cause dermatitis and visual disturbances, especially in young children. Mercuric oxide 1% can be used but causes a similar irritation. Insecticides can provoke various skin reactions: rash, urticaria and irreversible hair loss. Toxicity of insecticides to the bone marrow and to the kidney are well known. Yet, unfortunately, ignorance about the toxic effects of insecticides, their relatively low cost and easy availability make nonsupervised treatments quite common. The use of trimethoprim/sulfamethoxazol, 80/400 mg/12hr/3 days is moot because it frequently causes adverse cutaneous reactions (*Dermatologia Rev Mex* 1989; 33 (5):298-299). The effectiveness of treatment is enhanced in the following ways:

1) Just as in scabies, the entire family must be treated simultaneously.
2) For a few days after treatment, brush the hair with a fine-toothed comb to remove nits. Vinegar or acetic acid, formerly recommended, do not seem to have any effect on the nits.
3) Clean the bed sheets, combs, hair brushes, hats, and objects that could have the parasite and its ova.
4) Do not overtreat although a second application of lindane is recommended in one week to eliminate the ova that hatched in the interval.

5) In body lice the application of the medication must be accompanied by the elimination of infected clothing.
6) Pubic lice are sometimes resistant so three or more applications of any of medication can be necessary. The entire family may be treated.
7) Shaving the pubis and axilla eliminates the parasites, but the removal of hair can injure skin and generally is not recommended.

Selected Readings

1 Brown S, Becher J, Brady W. Treatment of ectoparasitic infections: Review of the english language literature, 1982-1992. Clin Infec Dis 1995;20(Suppl 1):104-9.
2 Mumcuoglu KY, Galun R, Miller J et al. Human lice. Medical Entomology and Acarology. Internet. Http.www.md.huji.ac.il/depts/parasitology/p3-7.html).
3 Maunder JW. The appreciation of lice. Proceedings of the Royal institution of Great Britain 1983; 55.
4 Ruiz-Maldonado R, Parish LC, Beare JM. Tratado de dermatologia pediatrica (Pediatric Dermatology) Interamericana/McGraw-Hill, 1992:593-6.
5 Sundnes KO, Haimanot AT. Epidemic of louse-borne relapsing fever in Ethiopia. Lancet 1993; 32:1213-15.

Scabies (Sarna)

Roberto Estrada

Scabies is a parasitosis caused by *Sarcoptes scabiei* var. *hominis* that occurs in 30-year cycles with up to 15 years between epidemics. It is characterized by its ease of transmission as well as by intense pruritus that increases at night. Although its origin goes back to ancient times, it does not seem to have existed in America before the Spanish Conquest.

GEOGRAPHIC DISTRIBUTION

Scabies occurs throughout the world. It is more common under conditions of poverty and where hygiene is poor; however, during epidemics it affects all who are exposed regardless of their social position or hygiene, age, sex or health. It is epecially common in nursing homes, even when the general conditions are adequate (*Int J Dermatol* 1991; 30:703-6). In the last epidemic outbreak in Mexico that reached its most intense activity about 10 years ago, infestation was most frequent in rural areas.

ETIOLOGY

The causative agent is an arthropod that specifically parasitizes humans. There are also species that infest dogs and other animals. There is little tendency for cross-species parasitations except in cases of immunosuppression as with AIDS. The life cycle of the mite is approximately 20 days. The adult female is 0.3-0.4 mm and is responsible for the clinical manifestations. The fertile female tunnels under the corneal layer and deposits 2-3 eggs daily which hatch in 3-4 days. The larvae come to the surface and mature in a hair follicle. They return to the skin surface and begin a new cycle (Fig. 42.1). The male is smaller than the female. The determinant factor in the contagion is the duration and extension of the contact area with an affected person. Scabies can be transmitted sexually; however, an adults in general have less predisposition except during wars or while residing in jails, asylums or nursing homes. Migration of the parasite requires at least 5 minutes. For this reason a hand shake or a hug is not considered a risk factor.

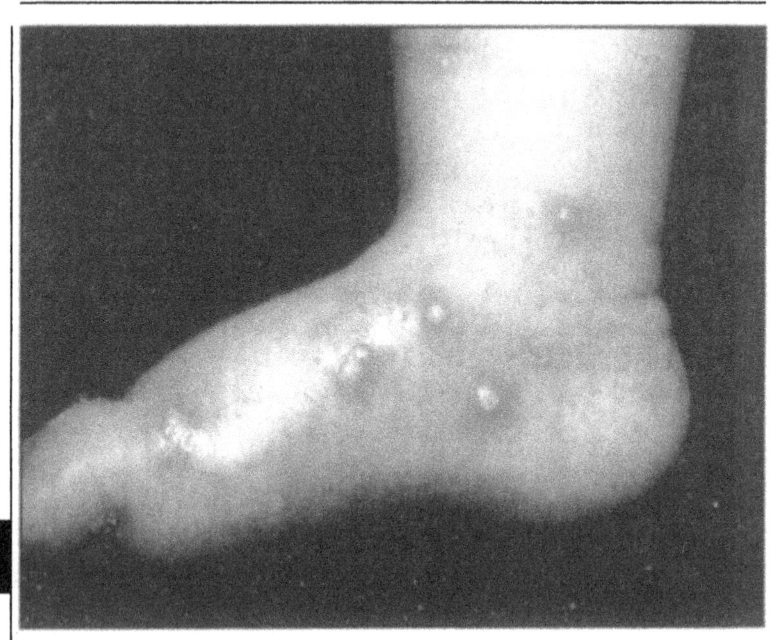

Fig. 42.1. Scabies in children.

CLINICAL FEATURES

It affects interdigital folds of the hands, inner aspects of the wrist, gluteus, regio axillaris folds, elbows and knees. The vesicular lesions under which the parasite is found are easiest to see on the hands and genitals. The penis, nipples in women, and the soles in babies, are the common areas of involvement (Fig. 42.1) Although the face is rarely affected and is generally not treated, some therapeutic failures have been attributed to scabies of the retrauricular folds. We have also seen babies with paronychia caused by mites, and there are reports that mites hide beneath fingernails (*JAMA* 1984; 252:1318). Vesicles and papules are the response to the parasite and to their eggs and feces. There are scabs and excoriations due to the intense scratching (Figs. 42.2 and 42.3). The incubation period is 3-4 weeks during which the intense pruritus is present, especially at night. With re-infections this period is shorter. Subsequently lichenification and scaling appear on hands, elbows and knees. Despite the abundance of lesions, the total number of mites does not exceed 10. In individuals with Norwegian or crusted scabies, there are many mites, but symptoms may be diminished making the diagnosis difficult and favoring epidemic outbreaks among hospital personnel and the patient's family. This form of scabies is manifested by abundant lesions, especially involving elbows, knees, feet, hands and fingernails (Fig. 42.4). It has been reported in Down's syndrome, diabetes, leprosy, dementia, syringomyelia, stroke,

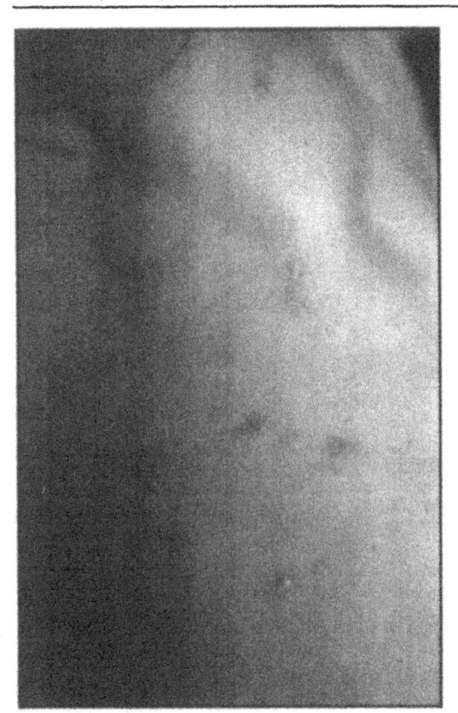

Fig. 42.2. Scabies, vesicles and furrow.

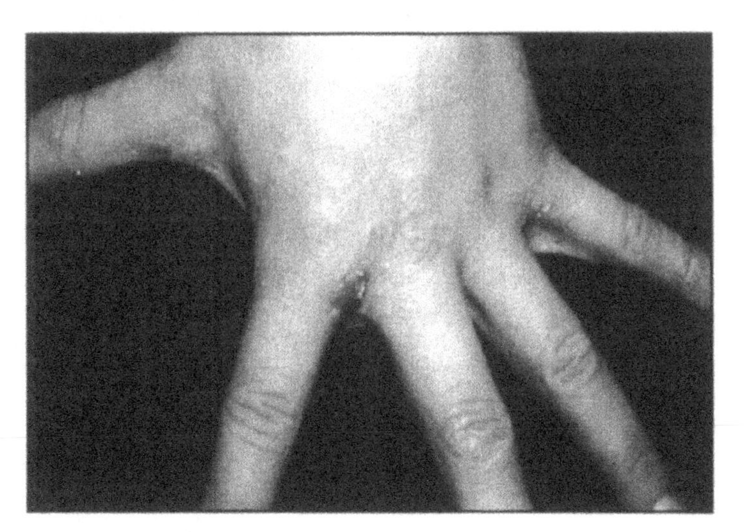

Fig. 42.3. Scabies, interdigital lesions.

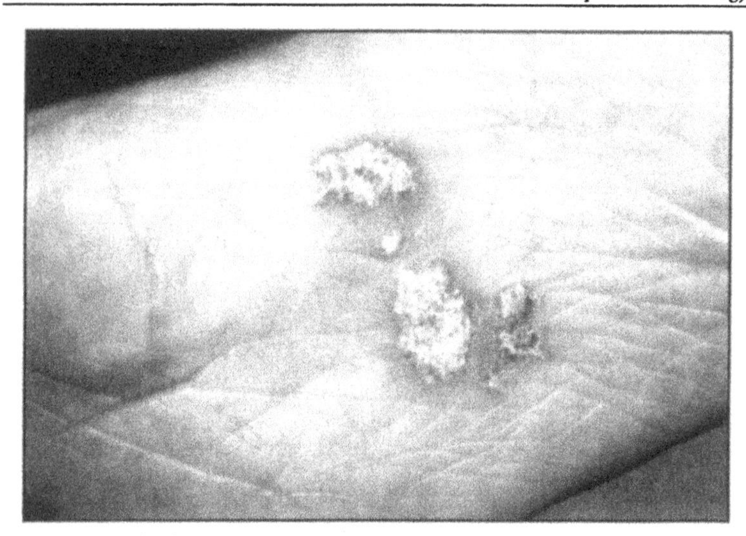

Fig. 42.4. Crusted scabies.

leukemia, neoplasms, and AIDS (*Int J Dermtol* 1990; 29:258-65). The intense scratching is the cause of frequent complications such as impetigo and contact dermatitis. In chronic cases nodular lesions are found in the penis. These lesions have been attributed to allergic hypersensitivity to the mite and its feces. In clean people there are cases in which the clinical lesions are very discrete. In those cases a more detailed examination is required to detect vesicles or the characteristic furrows seen most frequently in folds and on the penis.

LABORATORY DATA

A drop of mineral oil is applied in the suspicious area and is scraped with a scalpel blade. The material is examined microscopically to detect mites, eggs or feces. There are alternative methods. A surface biopsy can be performed after applying a drop of cyanoacrylate. Scotch tape can also be applied to the involved skin and the adherent material examined. Tunnels can be seen after applying India ink to the lesion and then cleaning the skin with alcohol. The ink is retained inside the tunnel which becomes visible. If a solution of tetracycline is applied in the same way, the tunnels can be seen under a Wood's lamp. Sometimes biopsy stained with hematoxylin and eosin is useful (Fig. 42.5). In scabies cases with abundant crusting, the diagnostic value of direct exam is very useful.

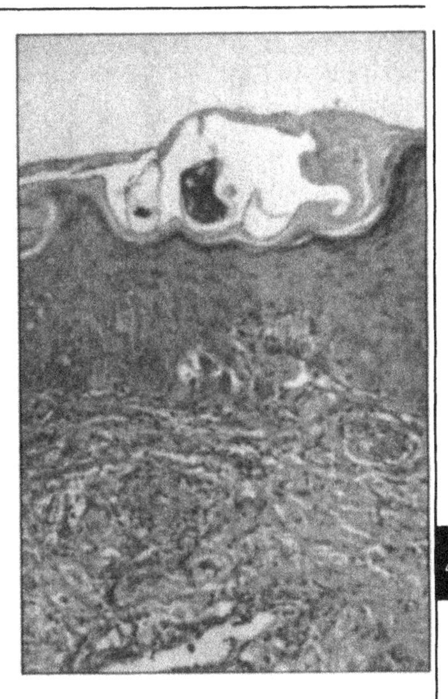

Fig. 42.5. Mites in epidermal tunnel (H.E. 20X).

TREATMENT

In many countries benzyl benzoate is used. However, because of its potential for contact dermatitis, it has given way to new antiparasitics that are effective on a single application, are less irritating, cosmetically acceptable and safe even in children. Benzyl benzoate, 25% emulsion, is applied over the entire body for three consecutive nights. It is washed off the following morning. It is inexpensive, but it is very irritating, especially in children for whom it must be diluted before topical use. Lindane lotion 1% is usually effective after a single application. It is inexpensive, readily available, and cosmetically acceptable. It is applied at night and washed off in a morning bath 12 h later. If overused it is toxic to the central nervous system. It is not recommended for small children in whom resistance has been reported (*J Am Acad Dermatol* 1991; 24(3): 502-3). Permethrin cream, 5%, is administered in a single application, after washing the skin. It is left on over night for 8 h and then washed off in a bath. It has low toxicity, is cosmetically acceptable and it is indicated when other scabicides have failed. It is more expensive than lindane and more difficult to obtain in rural areas. Crotamiton cream 10% is applied for three nights followed by a daily morning bath. It is cosmetically acceptable but it is less effective than lindane and permethrin and more expensive. Oral ivermectin, 200 mcg/kg day, in a single dose has been used in our country with good results (74% cures) and very low toxicity (*Gac Med Mex* 1993;

129(39):201-5). It is not yet commercially available. It is an adequate alternative for the control of epidemic outbreaks. In rural areas where medications are not available or are too expensive to purchase, vaseline or petrolatum with sulfur 5% can be applied for four consecutive nights followed by a daily morning bath. Its main advantages are the low cost, availability and tolerance in children. Disadvantages include a fetid odor, it is cosmetically unpleasant, it is messy and if it overused it dries and irritates the skin. The medication should be repeated for one week if there are persistent lesions on the hands, feet and genitals. It is important to treat the complications before scabies such as pyoderma (*Lancet* 1991; 337:1016-18). Patients must be instructed about measures to prevent re-infestations. The main causes of relapse are failure of concomitant treatment of family, close friends, sexual contacts or those sharing quarters with patients; improper application of medication; and re-exposure of small children with infested persons. Misinformation given to the patient by the doctor about the medication or about the illness is also a cause of re-infestation. The patient should be instructed that symptoms can persist for several days after treatment. Pruritus must be treated to avoid dermatitis. The administration of medication should be described in written instructions so that accurate information can be available to the patient's family.

Selected Readings

1 Burkhart CG. Scabies: An epidemiologie reassessment. Ann Intern Med. 1983; 98:498-503.
2 Denenholz DA, Crissey JT. Infestaciones cutaneas. (Cutaneous infestations) In: Ruiz-Maldonado R, Parish LC, Beare JNL, eds. Tratado de Dermatologia Pediatrica (Pediatric Dermatology) Mexico:Interamericana-McGraw-Hill, 1992:585-96.
3 Hermanss JF, Trinh Le. Dermatoses dues aux arthopodes. In: Pierard GE, Caumes E, Franchimont C, Arrese-Estrada J, eds. Dermatologie Tropicale. Bruxelles:AUPELF, 1993:395-399.
4 Parish LC, Witkowski JA, Millikan L. Scabies in the extended care facility. Int J Dermatol 1991; 30(10):703-706.

Larva Migrans (Larva Migrans Syndrome)

Roberto Estrada

Larva migrans or creeping eruption is an infestation of the skin by a nematode larvae. These worms normally parasitize the intestine of other animal species. The nematode cannot complete its life cycle in man. As the worm migrates in the epidermis, it produces a characteristic lesion. They lack the capacity to cross the basal layer in the epidermis due to lack of essential collagenases or due to the host immune response (*Actas Dermosif* 1987; 78:751-52).

GEOGRAPHY DISTRIBUTION

It is a tropical disease as the worm requires high humidity and temperature for its development. It affects both sexes and all ages.

ETIOLOGY

The following types of worms cause cutaneous larva migrans: *Uncinaria*, *Gnathostoma* and *Necator*. Some fly larvae of the varieties *Gasterophylus* and *Hypoderma bovis* cause similar clinical pictures. Most cases are caused by *Uncinaria*. The principal agents are *Ancylostoma braziliensis* and *A. caninum*. Fewer cases due to *A. ceylanicum*, *A. stenocephala*, *Bunostomum sp*. and *Necator suillus* have been reported. All of these worms normally parasitize other mammals and reach human skin by accident. Their life cycle begins with the expulsion of eggs of the parasite in the feces of the host. On humid soil or sand, they hatch giving place to rhabditi-form larva which change twice before forming filariform larva which penetrate unprotected skin. By the circulatory system, the larvae lodge in the small intestine. Eggs as well as the larvae can be transported by flies, increasing their dissemination (*East Afr Med J* 1989; 66(5(:349-52). The human skin is penetrated in mud, dirt or, especially, sand. In some cases we suspect it has been acquired by contact with concrete or tile floors washed with contaminated water. Filariform larvae have been found on these surfaces. They are capable of invading the skin through follicular pores or sweat glands where they could remain for a long time and cause folliculitis (*Arch Dermatol* 1991;

127(4):547-9). Usually, they migrate through the epidermis at speeds that vary from milimeters to centimeters per day. Rarely they penetrate the dermis. Nevertheless, eosinophilic enteritis caused by *A. caninum* has been described. It is caused by hypersensitivity to antigens produced by the parasite (*Lancet* 1990; 335(8701):1299-302).

In *Gnathostoma* infestation, the principal organisms are *G. spinigerum*, *G. hispidus*, *G. niponnicum* and *G. doloresi*, and to a less degree *G. procyonis*, *G. turgidum* and *G. miyasaki*. They are distinguished by the form and disposition of the spicules that cover their body as well as by their cephalic bulb. In their third developmental stage they are generally parasitize the digestive tracts of dogs, domestic and wild cats, raccoons, pigs, and wild boars. In their second developmental stage they parasitize fish, salamanders, frogs and snakes. The reside in the stomach wall where they form cystic tumors with orifices through which they expel their eggs. These appear in the host excreta and release the larva of the first phase. These larva are ingested by copepods crustaceans of the gender *Cyclops* in which they enter the second larval stage. The third stage occurs in the musculature of fresh water fish or amphibians like frogs, toads and even birds and small mammals that feed on these crustaceans. The nematode becomes an adult after passing through these temporary hosts. In the permanent host they mature in the muscle or subcutaneous tissue to which they arrive after perforating the stomach wall. For some time they reside in the kidney until they finally return to the stomach wall where the cycle begins again. They infest humans through ingestion of raw or semi-raw fish (Oriental sashimi or Latin American "ceviche") (*Dermatologia Rev Mex* 1995; 39(2):77-80) or incompletely cooked chicken that is infested by the parasite. Although it is not usual, it is thought that man can act as a permanent host. Periods of up to 3 years are required (*Rev Lat Amer Microbiol* 1970; 12:83-91). Nevertheless, most cases are detected and eliminated in the early stages because of the intense symptoms that they cause. In the larva mygrans syndrome *Strongyloides* infestation is included. The principal agent of anguillulidos is *Strongyloides stercolaris*. It is a common parasite of humans, although cases of *S. myopotami* and *S. procyornis* have also been reported. These last can not complete their life cycle in humans. Normally they parasitize otters and raccoons. In man *Strongyloides* infests the duodenum. It is approximately 2.5 mm long. It can reproduce asexually (parthenogenesis) in the host or sexually. Its cycle is complicated, and it begins with eggs eliminated in the feces of the host. In the soil the rhabditi-form larvae transform into the filariform larvae that are capable of penetrating the human skin producing lesions very similar to *Ancylostoma*, and arrive in the digestive tract through the circulation. The internal (endogenous) asexual cycle does not have to be outside the host to develop. This is a form of autoinfection. In this cycle the rhabditi-form larvae transformed into filariform in the digestive tract. This explains the prolonged presence of the anguillulidos for 30 years or more in endemic areas.

CLINICAL FEATURES

Larva migrans caused by *Uncinaria*: Skin penetration is usually asymptomatic. The infestatation presents with the formation of a pruritic papule associated with an erythematous track that follows an erratic course (Fig. 43.1). Vesicles or blisters can accompany the initial lesion or its track. This lesion is usually intensely pruritic, and it can cause burning or pain. The most commonly involved areas are the soles, hands, gluteus, and back. A rare case had abundant lesions in the face of a patient that slept in direct contact with the sand (Fig. 43.2). Frequently the infestation occurs in tourists who have visited tropical areas (*Clin Infect Dis* 1995; 20(3):542-8). Some cases resolve spontaneously in 4-8 weeks.

Larva migrans caused by *Gnathostoma* (Gnathostomiasis): The clinical manifestations vary depending on the involved organ; the digestive and genitourinary tract, kidney, lungs, brain, eyes and ears may be effected. Skin is the most frequently involved organ and the easiest to detect:

1) The inflammatory form or migratory panniculitis is of variable intensity. It is characterized by erythematous, edematous, circular or irregular, discreetly elevated plaques. The surface is warm, painful or burning, with and orange-like aspect to the skin, and they can displace from 1-5 cm daily (Fig. 43.3). The lesions disappear spontaneously (weeks, months or years) or with treatment, and they periodically reappear in areas near or distant to the previous site. The most frequently involves the trunk and abdomen, then upper and lower limbs, neck and face.
2) The superficial or serpiginous form presents as a irregular, sinuous track with a mild inflammatory reaction (Fig. 43.4).

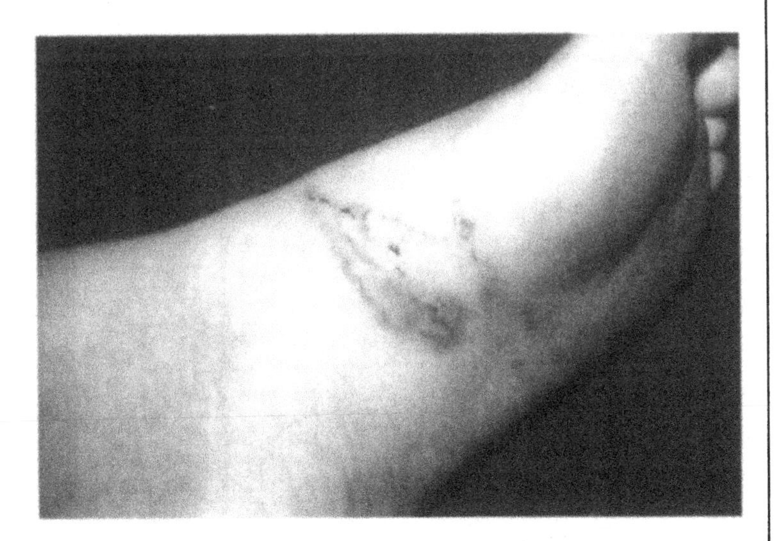

Fig. 43.1. Larva migrans on the feet.

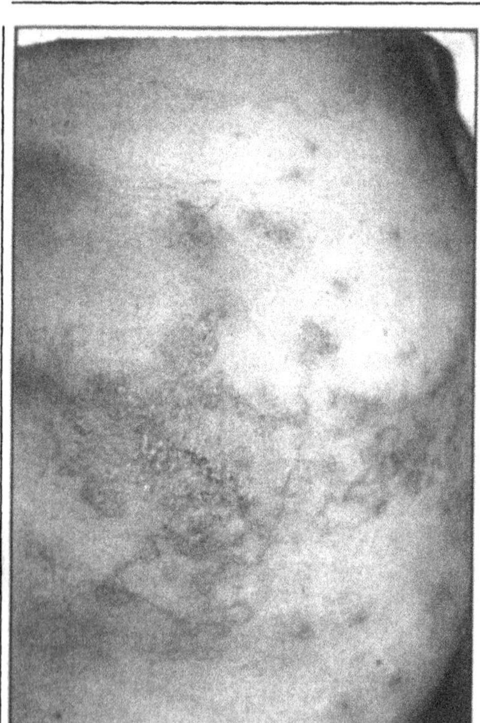

Fig. 43.2. Disseminated larva migrans.

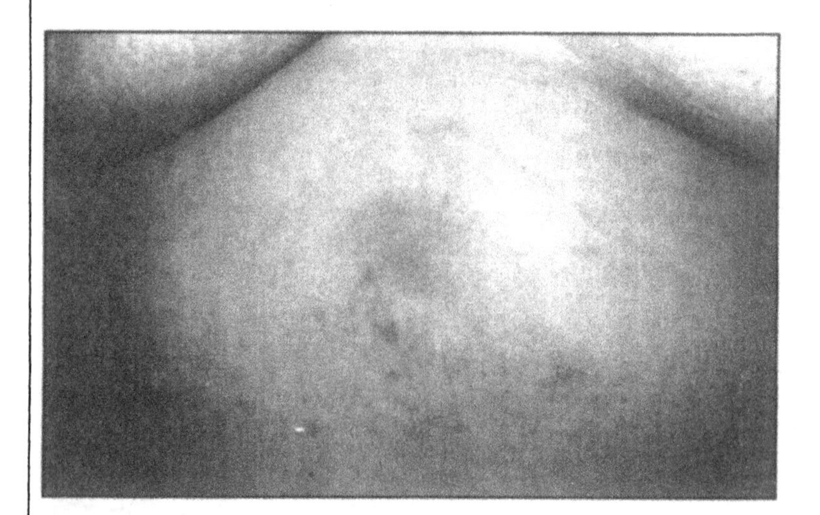

Fig. 43.3. Migratory panniculitis due to *Gnathostoma*.

Fig. 43.4. Superficial Gnathostomiasis.

The pseudofurunculous form presents as small, superficial inflammatory plaques with central necrosis. As the parasite is expelled the infestation resolves. Larva migrans caused by *Strongyloides* (anguillulidos): The systemic phase is manifested by Larva currens, rapidly developing (5-15 cm/hr) serpentine skin lesions that may spontaneously disappear in hours (*Arch Dermatol* 1988; 124:1826-30). They frequently present around the anus and in the gluteal area and can affect the lumbar, pelvic and thoracic regions. The cutaneous lesions are accompanied by intense pruritus and sometimes by a papular, pseudourticarial eruption. In immunodepressed patients or in those with prolonged steroid therapy there can be accelerated proliferation of larva and adults with massive visceral invasion. Larva migrans caused by fly larvae: This is also known as rampant or migratory myasis. The gender *Gasterophylus* is the main etiological agent, and the species *G. intestinalis*, *G. haemorrhoidalis* and *G. precorum* are, among others, the most frequently involved. These are habitual parasites of the stomach and rectum of horses. In humans the larva make tunnels into the epidermis and form linear tracks that progress from 1-2 cm daily. Vesicles and blisters form along the track. Larval activity and pruritis are more intense at night.

LABORATORY DATA

Biopsy is little help since the inflammatory reaction remains behind the actual location of the parasite. Nevertheless, it can be attempted after initial treatment since the medication seems to immobilize the organism. The common histologic findings are an intense, perivascular, inflammatory infiltrate composed of lymphocytes, plasma cells, polymorphonuclear cells and abundant eosinophils throught the entire dermis. In Gnathostomiasis there is moderate leukocytosis with eosinophils above 20%, especially with visceral involvement. A reliable skin test is not available. Biopsy can be performed after treatment with albendazole since it seems to stimulate the migration of the *Gnathostoma* to the skin surface (*Southeast Asian J Trop Med Public Health* 1992; 23(4):716-22).

TREATMENT

The following medications are used: Albendazole 20 mg/kg/day for 3-6 days (400 mgs/3 days); Thiabendazole 25-50 mg/kg/day in a single dose repeated in one week or daily doses administered on three consecutive days. These medications usually cause intense gastric discomfort and are not used in many countries. Ivermectin 200 mcg/kg in a single dose. This is a very promising medication, but it is not yet available. Some authors prefer the topical use of thiabendazole in cream or emulsion 10% or 15% three times a day for a week. Cryotherapy with liquid nitrogen, solid carbon dioxide or ethyl chloride. Surgical excision may be performed when the location of the larva is very obvious. Prophylaxis includes the use of shoes in sandy or contaminated areas; avoid sitting or lying down in direct contact with the beach sand; the periodic antiparasitic treatment of household dogs and cats; and the control of street dogs, especially in the area of beaches. Up to 96% of these dogs are parasitized by *Uncinaria* (*Parasite* 1996; 3(2):131-4). The results of the treatment of *Gnathostoma* are variable. Sometimes it is frustrating for both the patient and the doctor. The following can be used: Mebendazole 100 mg/12hr, or albendazole using the mentioned dose for 7-10 days or for up to 15 days. If the larva migrates to the surface, surgical excision is also indicated. Thiabenazole at the above mentioned dose, orally and topically. Praziquantel 30 mgs/kg/day for 7 days. Diethylcarbamazin 5 mg/kg/day for 10 days. The indications for ivermectin are not yet clear. These medications can be administered in association with the topical treatments already mentioned. The most important prophylactic measure is to limit the consumption of freshwater fish in raw or semi-raw form. Cryotherapy or the direct extraction of fly larvae with a sterilized needle may constitute an adequate treatment (*Piel* 1989; 4(7):309-18).

Selected Readings

1. Caumes E, Carriere J, Guermonprez G et al. Dermatoses associated with travel to tropical countries. Clin Infect Dis 1995; 20(3):542-8.
2. Caumes E. Nematodoses Intestinales. In: Pierard GE, Caumes E, Franchimont C, Arrese-Estrada J, eds. Dermatologie Tropicale. Bruxelles: AUPELF, 1993:342-53.
3. Farah FS, Klaus SN, Frankenburg S. Protozoan and Helmintic infections. In:Fitzpatrick TB, Eisen AL, Wolff K et al. Dermatology in General Medicine. 4th ed. New York:McGraw-Hill, 1993:2769-87.
4. Miller AC, Walker J, Jawoeski R et al. Hookworm follicuitis. Arch Dermatol 1991; 127(4):547-9.

5 Prociv O, Croese J. Human eosinophilic enteritis caused by dog hookworm *Ancylostoma caninum.* Lancet 1990; 335(8701):1299-302.

6 Suntharasamai P, Riganti M, Chittamas S, Desakorn V. Albendazol stimulates outward migration of *Gnathostoma spinigerum* to the dermis in man. Southeast Asian J Trop Med Public Health. 1992; 23(4):716-22.

Tungiasis

Giancarlo Albanese

Over the last few years, due to the increasing number of people spending holidays abroad in particularly attractive geographical areas, we are starting to see patients who have contracted tropical diseases in endemic countries. One example that confirms this new tendency is Tungiasis (in different countries: chigoe infestation, nigua, bicho dos pes, moukardam), which was formerly restricted to the equatorial zones of the world.

GEOGRAPHICAL DISTRIBUTION

Tungiasis is a typical tropical disease caused by the sand flea *Tunga penetrans* (jigger, chigger or chigae). This infestation is widespread in the tropics and is usually contracted in Central and South America, the Caribbean, India, Pakistan and tropical Africa. By way of example, we report a survey carried out on 5,595 primary school children in Lagos State, Nigeria (*Acta Trop* 1981, 38:79-84). It showed that most of the children were over-loaded with parasitic infestations which included malaria (37.7%), schistosomiasis (13.4%), ascariasis (74.2%), trichuriasis (75.8%), hookworm (29.5%) and tungiasis (49.5%). Tungiasis is rarely diagnosed in Italy or in other European countries or North America, so it can really be considered a "tourism-transmitted" disease, contracted by tourists in endemic areas. French authors have reported examing 269 patients in a tropical disease unit in Paris over a 2-year period (*Clin Infect Dis* 1995; 20:542-548). The average age of these patients was 30 years, 137 patients were male, 76% of the patients were tourists and 38% had visited sub-Saharian Africa. In 61% of cases, cutaneous lesions had appeared while the patient was still abroad, whereas in 39% they had appeared after the patient's return to France. Diagnosis was definite in 260 cases; 137 of these (53%) involved imported tropical diseases. The most common diagnoses were cutaneous larva migrans (25%), pyodermas (18%), pruritic arthropod-reactive dermatitis (10%), myiasis (9%), tungiasis (6%), urticaria (5%), fever and rash (4%) and cutaneous leishmaniasis (3%).

ETIOLOGY

Tungiasis results from the cutaneous infestation of humans by the gravid female flea, *Tunga penetrans*. The parasite penetrates the skin and completes

Tropical Dermatology, edited by Roberto Arenas and Roberto Estrada.

pregnancy in the infestation site. Before dying, it discharges many eggs into the ground. It is a blood-sucking ectoparasite of mammals and is usually found on dogs, pigs and cows. Infestation may be contracted from sandy soil in areas inhabited by these animals.

CLINICAL FEATURES

The lesion appears as an asymptomatic, non-inflammatory, translucent nodule, single or multiple, with a central dark spot (Fig. 44.1); it is located in sites that favor the penetration and growth of the parasite (usually between the toes, under the nails and on the soles of the feet, rarely elsewhere). The usual black dots may then become inflammatory nodules. In fact, the fully developed lesion resembles an abscess with a black center and a discharge of pus; sometimes discomfort and disability are reported. Tungiasis occasionally leads to secondary infections and may be the cause of long-term inflammation or erysipelas. Complications include severe itching, pain, inflammatory reaction leading to autoamputation of digits, pyoderma (cellulitis) and tetanus.

LABORATORY DATA

If the dermatologist is familiar with tungiasis, it is easy to diagnose by considering the typical clinical features. If necessary, however, diagnosis can easily be confirmed by removing the top of the lesion and examining it directly

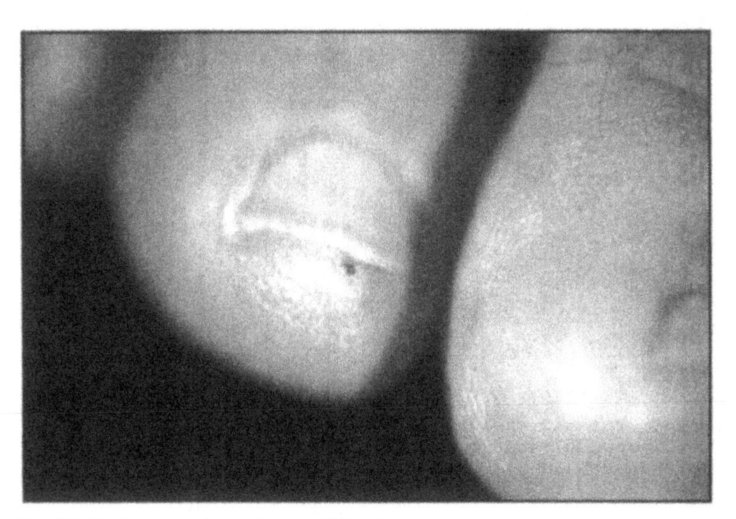

Fig. 44.1. *Tunga penetrans* in a typical site of infestation.

with a microscope. This will immediately reveal a cluster of eggs (Fig. 44.2) or parts of Tunga.

TREATMENT

Spontaneous recovery does sometimes occur, but the disease usually needs to be treated. In an early infestation, treatment consists of excising the lesion and removing the gravid female flea with a sterile needle. It is then advisable to carefully clean the cystic cavity in the infestation site and to protect the patient against pyoderma. Healing is usually complete without sequelae, but Tetanus cannot be ruled out. Tetanus prophylaxis may be indicated, depending on the wound and its location.

As regards this problem, it is interesting to refer to a study reported by authors from the Congo (*Dakar Med* 1989: 34:44-48). Just like other tropical areas, this country is not free of tetanus. Over a period of 11 months, the authors treated 44 cases where tetanus entered by various means. The removal of a chick-flea, for example, leaves a hole through which the tetanus spore can penetrate: this accounted for 25% of the cases (11 cases). In the group under observation, no death attributable to tetanus occurred. To prevent tungiasis, it is essential to recommend the use of protective clothing and shoes when tourists plan to spend their holidays in endemic areas.

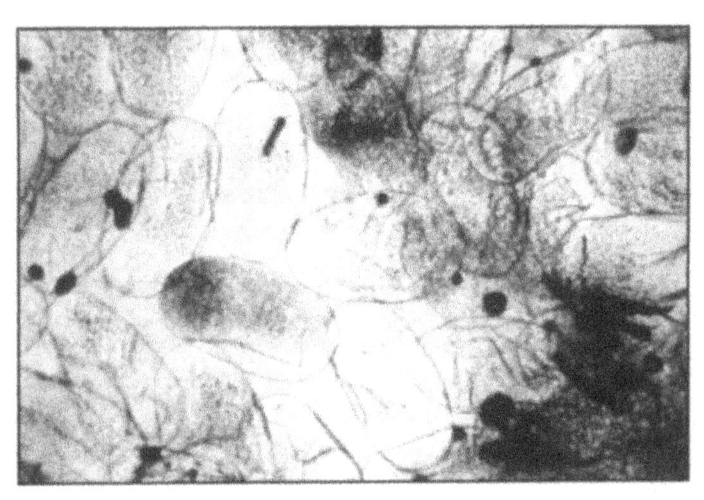

Fig. 44.2. Direct examination: A cluster of eggs from the top of a lesion (100x).

Selected Readings

1 Caumes E, Carriaere J, Guermonprez G et al. Dermatoses associated with travel to tropical countries: A prospective study of the diagnosis and management of 269 patients presenting to a tropical disease unit. Clin Infect Dis 1995; 20:542-548.
2 Ejezie GC. The parasitic diseases of school children in Lagos State, Nigeria. Acta Trop 1981; 38:79-84.
3 Obengu I. Tungiasis and tetanus at the University Hospital Center in Brazzaville. Dakar Med 1989; 34:44-48.

Myiasis

Roberto Estrada

This parasitosis is caused by larvae of dipterous insects that reside in the tissue of warm-blooded animals. It affects humans only occasionally. Cutaneous myiasis (from the Greek mya = fly) is a primary dermatosis (*Int Dermatol* 1995; 34(9):624-6) or secondary complication of wounds and ulcers (*Piel* 1989; 4:319-24).

ETIOLOGY

Semiobligate myiasis is caused by flies of the genders Chrysomya, Calliphora, Callitroga, Musca, Phormia, Sarcophaga and Wolfahrtia. The host skin is penetrated by different ways. The larvae feed and develop in tissues and secretions of wounds, ulcers or suppurative areas. The larval phase develops in excrement, cadavers and wounds of living animals. Obligate myiasis is caused by the genders Hypoderma, Gasterophilus, Oestrus, Cordylobia and Dermatobia.

CLINICAL FEATURES

There are four types of myiasis: myiasis in wounds and ulcers, furunculoid myiasis, subcutaneous migratory myiasis and cavitary myiasis. In myiasis of wounds and ulcers, flies of obligate or facultative life cycles feed on organic decomposing matter and occasionally on healthy tissue. The inoculation is always through exposed lesions. For this reason diagnosis is evident when observing the larvae, unless their penetration had been complete. In America the flies are *Cochliomya hominivorax*, in Asia and Africa Chrysomya. Wohlfahrtia, Sarcophaga, Lucilia and Calliphora are distributed throughout the world. Geographic distribution is determined by the habitat of the flies. Nevertheless, any open, poorly treated wound can support larvae. For this reason fly larvae, especially the species *Lucilia sericata* (green fly), have been used to clean wounds under controlled conditions to avoid bacterial infection. In furunculoid myasis, the most important causal species are *Hypoderma bovis* and *H. lineatum* of European origin, *Cordylobia anthropophaga* (Tumbu fly) in Africa and *Dermatobia hominis* in America. These are the clinical forms most often seen by physicians (*Br J Dermatol* 1995; 132:811-814). Larvae cross the skin rapidly and without causing discomfort. Initially they form erythematous papules that become inflamed in a few weeks. They form a furuncule inhabited by the parasite larvae. Serosanguineous or purulent exudates issue from the orifice. Frequently there is pain, itching and adenopathy

Tropical Dermatology, edited by Roberto Arenas and Roberto Estrada. ©2001 Landes Bioscience.

with secondary infections. Without treatment this stage lasts 3-4 months and heals when the larva is expelled. The fly species is identified by the posterior spiracles that are characteristic in each species and that are visible in the central part of the lesion. *D. hominis* eggs adhere to the abdomen of hematophagic mosquitoes (Psorophora). The larvae are deposited on contact of the fly with the skin of homeothermic animals. The skin then is invaded. In contrast, *C. anthropophaga* deposits its eggs in dirt moistened with urine, in wet diapers, clothes with dry urine, or in stool. *H. bovis* and *lineatum* rarely affects man. It is seen in the winter and in individuals with extended contact with livestock.

Cutaneous migratory myiasis. Distribution is worldwide. It is caused by *Gasterophilus hemorroidalis, G. veterinus, Hipoderma bovis, H. lineatum* and *H. diana.* Development takes place in the digestive tract of horses. The eggs are deposited in animal's hair and pass to the mouth when the animal licks its hair. In the stomach, they form pupae discharged in the stool to finally mature in the dirt. When they accidently enter a human, the cycle cannot be completed. They reside in the spinous cell layer where they form serpentine tracks confused with larva migrans. There is inflammation with vesicles or blisters. Clinically the picture is similar to that caused by larvae of Uncinaria. In cavitary myiasis the most frequent causative species are Lucilia (worldwide), Wohlfahrtia and Sarcophaga (European), and rarely Cuterebra (American). While the host sleeps, eggs are deposited in the ears, mouth, nose, and even eyes. Flies are attracted by the abundant secretions or necrotic tissue. Their larvae are incapable of penetrating the intact skin, but they can penetrate mucosae. The tissue destruction caused by numerous larvae may be severe. Clinical manifestations vary depending upon the affected organ. In the nose there is itching, and pain, constant congestion, frequent epitaxis, and tissue destruction (*J Dermatol* 1995; 22(5):348-50). In the ears there may be pain, dizziness and hearing loss. In the eyes the development of ophthalmomyiasis and conjunctivitis is described. This is caused by the fly of the sheep, *Oestrus obvis* (*Ger J Ophtalmol* 1995; 4(3):188-95). Even when the clinical picture is advanced, control generally is not difficult since man is not the natural host.

LABORATORY DATA

The presence of the larvae in the wound is diagnostic (Fig. 45.1). Biopsy identifies cases in which the parasite is observed within the skin. Larvae of various species can be distinguished by posterior spiracles. If this is not adequate, mature flies can be obtained by application of meat to the lesion. Larvae invade the meat which is placed in a ventilated, closed jar with sawdust on which the pupae will fall and will permit development of mature flies for identification.

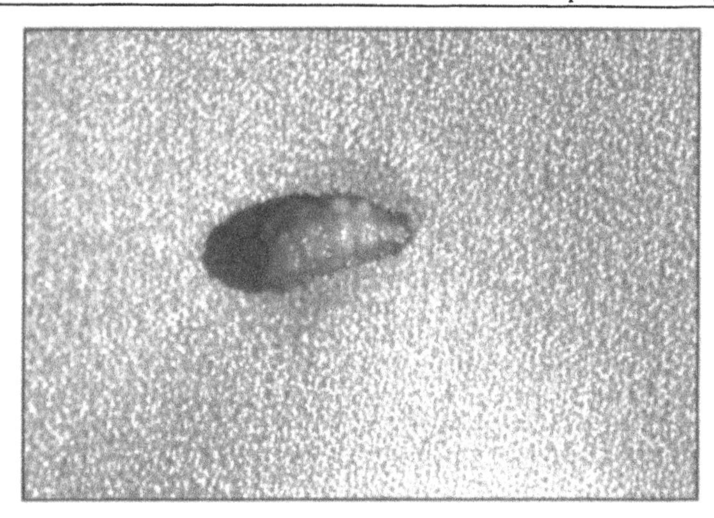

Fig. 45.1. Larva of Dermatobia.

TREATMENT

In isolated areas where there is no medication, wounds should be cleaned with soap and water. Topical antiseptics and oral antibiotics control associated bacterial infections. It is important to determine if there is immunosuppression caused, for example, by malnutrition or HIV. Vaseline can be applied to the wound to promote larval death or withdrawal from the wound. The application of paraffin compresses with chloroform or ether for at least 24 h has a similar effect (*Ger J Ophthalmol* 1995; 4(3):188-95). Many furunculoid larvae can simply be expressed, although not in the case of *D. hominis* which has hooks with which they adhere firmly to the skin. They must be removed surgically and irrigated with chloroform, 5%, in olive oil; potassium permanganate and ethyl chloride are also effective. Cavitary lesions require surgical debridement and antibiotic therapy for the associated infections. In some forms of migratory myiasis the larvae can be visualized and removed with a needle. Myiasis will be controlled when sanitation improves.

Selected Readings

1 Gordon PM, Hepburn NC, Williams AE, Bunney MH. Cutaneous myiasis due to Dermatobia hominis: A report of six cases. Br J Dermatol 1995; 132:811-814.
2 Hermanns JF, Le T. Dermatoses dues aux arthropodes. In: Pierard GE, Caumes 3 E, eds. Franchimont C, Arrese-Estrada J. Dermatologie Tropicale. AUPELF:Bruxelles, 1993; 395-413.
4 Jelinek T, Nothdurft HD, Rieder N, Loscher T. Cutaneous myiasis of 13 cases in travelers returning from tropical countries. Int J Dermatol 1995; 34(9):624-6.
5 Jeremias FJ, Soriano JC, Gimenez JM. Miasis cutanea por *Lucilia caesar* en el hombre (Cutaneous myasis by *Lucilia caesar* in men) Piel 1989; 4:319-24.

6 Mumcuoglu KY, Lipo M, Ioffe-Upspensky I et al. Maggot therapy for the treatment of a severe skin infection in a patient with gangrene and osteomyelitis. First Intern Congr Biosurgery, Prthcawl, South Wales. May 3-4, 1996:4.
7 Rook A, Wilkinson DS, Ebling FJG. Textbook of Dermatology. Oxford:Blackwell. 1982:918-20.

Leishmaniasis

Roberto Arenas

Leishmaniasis is a chronic infestation of the skin, mucosa or viscera caused by several species of intracellular protozoan of the genus *Leishmania*. They are transmitted to human beings mainly by vectors of the gender *Lutzomyia* spp. and *Phlebotomus* spp. The clinical manifestations depend on the species of the parasite, on the genetic background and the state of immunity of the host.

GEOGRAPHIC DISTRIBUTION

Leishmaniasis occurs throughout the world with the exception of Australia where it has not yet been reported. It is endemic in several parts of India, the former Soviet Union, Asia, Africa, the Mediterranean basin, and in Central and South America. It has been seen from the southern United States to Argentina. It is one of the six tropical illness important for the WHO (tuberculosis, leprosy, onchocercosis...). It is calculated that about 15 million people in 83 countries have been infested and that there are 400,000 new cases every year. It is a reportable disease in only 30 countries. In Latin America there are about 59,300 cases per year. It is more common in southern Mexico, Guatemala, Nicaragua, Brazil and Venezuela. In Costa Rica the rate is 1 per 1000 inhabitants. It occurs in forested, tropical and semiarid zones. Prevalence is much higher in dogs (20-40%) than in humans (1-2%). It is most common in men, especially in farmers, hunters, archeologists and soldiers. It is an opportunistic infection in people infected with human immunodeficiency virus.

ETIOLOGY

The parasite is a unicellular dimorphous protozoan of the gender Leishmania from the family Trypanosomtidae. *L. mexicana* and *L. braziliensis* predominate in America. In the Old World it is caused by *L. tropica*, *L. major*, *L. aethiopica* and *L. donovani*. *Leishmania* is a zoonotic disease with the exception of *L. tropica* and *L. donovani*. The invertebrate vectors of the genus *Phlebotomus* spp. (Old World), *Lutzomya* spp. (America) are involved in transmission and sometimes *Sergentomyia* spp., *Brumptomyia* spp., *Warileya* spp. and Psychodopygus. The sand flies that are the vectors live in low and humid prairies and proliferate in rainy seasons. Promastigote and Paramastigote are found in the intestine of these invertebrate vectors. The first has a flagellum and the second is a transitional stage. The

Tropical Dermatology, edited by Roberto Arenas and Roberto Estrada. ©2001 Landes Bioscience.

parasite or amastigote (Leishman-Donovan bodies) is found within macrophages of the vertebrate hosts. It presents a mitochondrial structure containing DNA and called a kinetoplast. It becomes flagellated (promastigote); this form is transmitted when the fly feeds again. Only the female fly is hematophagus. The cycle takes 53-100 days. The illness is limited to vertebrate reservoirs, like wild or domestic mammals and reptiles. The different species of *Leishmania* are morphologically identical. Their classification is controversial and is based on molecular biology. One of the most useful classification methods is based on the location of isoenzymes: *L. donovani, L. infantum, L. chagasi, L major, L. tropica, L aethiopica, L. enrieti, L. hertigi, L. mexicana* complex (*L. m mexicana, L. m amazonensis, L. m pifanoi, L. m venezuelensis, L. m aristedesi, L. m gamhami*) *L. (viannia) braziliensis* complex (*L. b braziliensis, L. b guyanensis, L. b panamensis*), and *L peruviana.*

Man is an accidental host, and the amastigotes multiply in macrophages. They may cause either a sub-clinical or self-limited infection, the localized cutaneous form which has a lymphocytic response or as a diffuse form if the infection spreads. Specific antibodies IgG can be detected in the mucocutaneous form, while in the diffuse form there may be high levels of IgA that enhance the expression of ICAM-1, HLA-DR on keratinocytes, a loss of Langerhans cells and Th1 and Th2 granuloma. In this form a lack of ICAM-1 and HLA-DR expression is observed as well as a defect in the production of cytokines; the granuloma is of the Th2 type with many parasites in the macrophages. A fundamental step for invasion of the host cell depend of surface receptors of Leishmania as lipophosphoglycan and glycoprotein 63. In experimental disease in mice, expansion of Th1 with production of interleukin-2 and gamma-intereferon leads to resolution of infection and on the other hand, Th2 proliferation and interleukin-4 production leads to progression of the disease.

Immunopathologic classification of Leishmaniasis:

I. Immunoallergic form:
   1. Abortive (subclinical),
   2. Cutaneous (single lesion, lymphangitic, disseminated nonallergic),
   3. Tegumentary (mucocutaneous).

II. Immunoanergic forms:
   1. Cutaneous diffuse (disseminated nodular),
   2. Visceral (kala-azar),
   3. Secondary cutaneous (from visceral) or leishmanoid.

CLINICAL FEATURES

The cutaneous form has different geographic variants. It is caused by *L. tropica, L. major* and *L. aethiopica* in the Old World (Oriental sore) and in the New World it is caused by numerous species and subspecies, especially *L. mexicana mexicana, L. m. amazonensis* and *L.(viannia) braziliensis.* New and Old World forms are clinically similar, but New World infections are more severe and chronic. Both affect areas exposed to insect bites such as the face, trunk, and extremities. The incubation period varies from 1-4 weeks, but it may be several years. The lesion is a painless erythematous nodule, 1-10 cm in diameter. It can be psoriasiform or hyperkeratotic (dry or urban form), or it can ulcerate in 1-3 months (moist or rural form). It heals spontaneously in 6 months to 4 years. It leaves a depressed and dyschromic scar with telangiectasias (Fig. 46.1). Regional lymphangitis is rare. In endemic areas 33% of the patients present with re-infections and autoinoculation is also possible. The recurrent, lupoid or tuberculoid form is atypical. It may last many years and does not respond well to treatment. It is confused with lupus vulgaris (Fig. 46.2). Leishmaniasis recidiva cutis may represent a reactivation of an initial infection, probably due to the persistence of parasites and is considered by some authors as synonymous with lupoid leishmaniasis. In the Andes one or several lesions that resolve spontaneously (Uta) are seen in children; it is caused by *L. peruviana*. In Panama there is an ulcerative form (Bejuco's ulcer) caused by *L. panamensis*. When it affects the ears it causes the cutaneochondral form or chiclero's ulcer (which occurs most frequently in the workers who harvest chicle for chewing gum). In Mexico it is seen in 60% of the cases and is caused by *L. m. mexicana* and is usually associated with mild lesions. It begins with an insect bite-like lesion that becomes an infiltrated plaque or chronic ulcer. It is frequently painful. It can heal spontaneously after a long time and leave a disfiguring

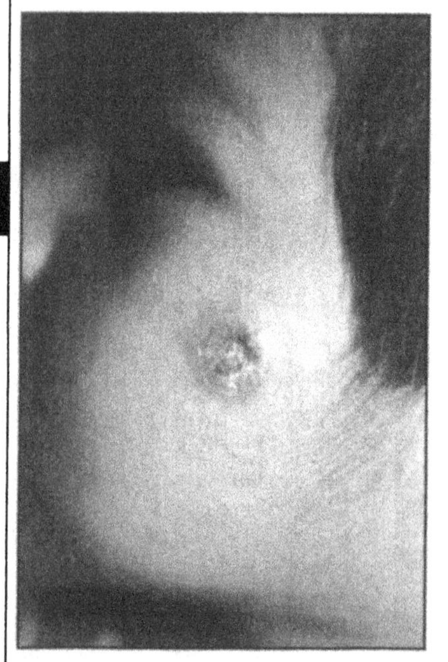

Fig. 46.1. Cutaneous leishmaniasis.

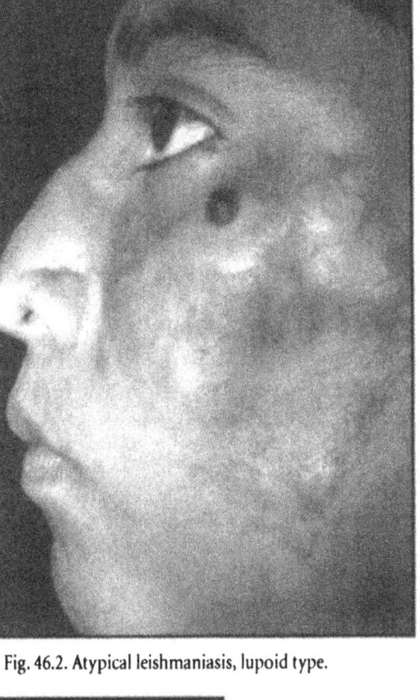

Fig. 46.2. Atypical leishmaniasis, lupoid type.

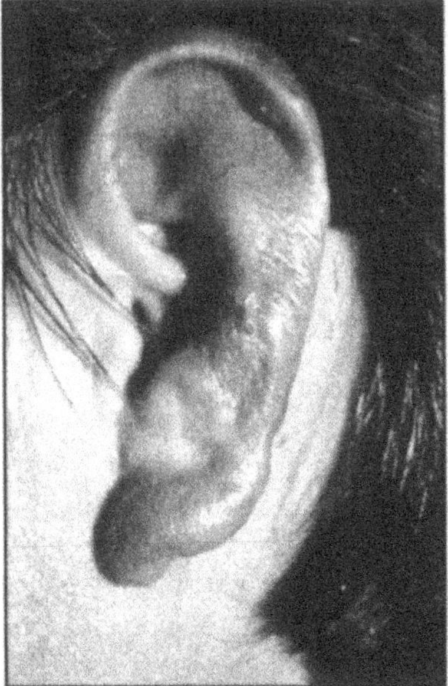

Fig. 46.3 Cutaneochondral leishmaniasis or chiclero's ulcer.

notch-shaped scar (Fig. 46.3). Disseminated cutaneous leishmaniasis or tegumentary (nodular disseminated) leishmaniasis, tends to cover the entire body, but mucosae involvement is infrequent. It is caused by several species and subspecies: *L. m. amazonensis*, *L. aethiopica* and in Venezuela by *L. mexicana* (*L. m. pifanoi*). It usually involves the exposed areas, the pinna of the ears, cheeks, ciliary regions and extremities. The folds and hairy skin are usually free of involvement (Fig. 46.4). It is characterized by firm, gray-red, smooth or verrucous nodules and plaques that can ulcerate. Lymphedema, lymphadenopathy, malaise and fever may be present. Mucocutaneous leishmaniasis (American leishmaniasis or espundia) is commonly limited to South America. It affects youngsters and is caused by *L. braziliensis braziliensis*, *L. aethiopica* and *L. mexicana*. There is a primary cutaneous lesion at an exposed site, generally on the lower extremities. It is nodular and can ulcerate or assumes a vegetative form(multiberry or frambesoid aspect). There can be lymphangitis and adenitis. After several years the nasal septum is affected (tapir nose) with obstruction, epistaxis or perforation. The lips, gums, pharynx and larynx (Fig. 46.5) may also be involved. Eighty percent of untreated primary cutaneous lesions progress to the mucocutaneous form even after 30 years.

An atypical variant of disseminated cutaneous leishmaniasis is caused mainly by *L. chagasi* and less frquently by *L. mexicana*. It has been observed in endemic areas of visceral leishmaniasis in Central America, especially in Honduras. The patients are usualy children with chronic, asymptomatic and non-ulcerated papules and nodules in the exposed areas of the skin (Fig. 46.6).

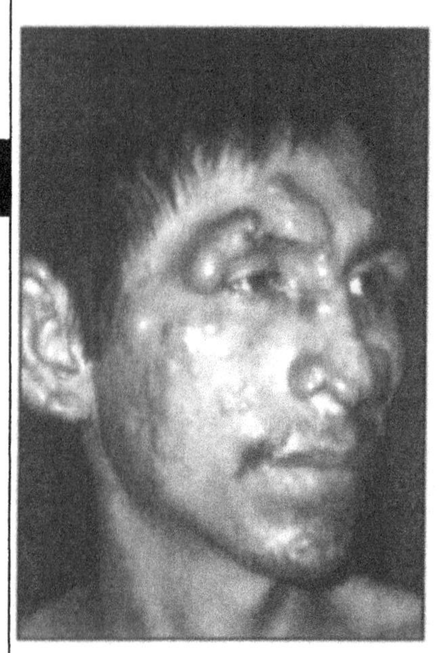

Fig. 46.4. Anergic leishmaniasis (Courtesy of Nixma Eljure).

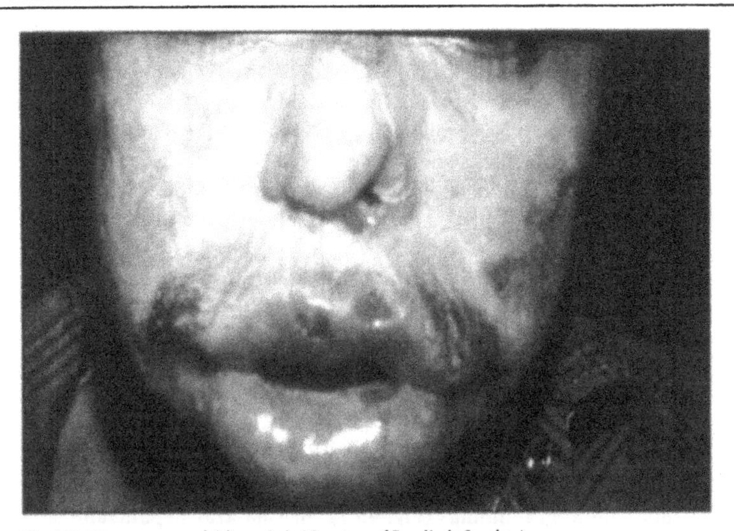

Fig. 46.5. Mucocutaneous leishmaniasis (Courtesy of Rosalinda Sanchez).

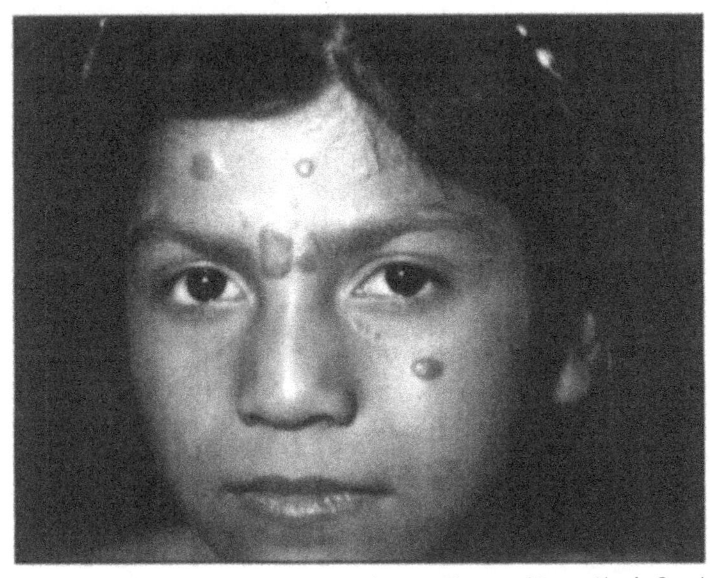

Fig. 46.6. Disseminated lesions in patients of Central America (Courtesy of Gustavo Lizardo-Castro).

Visceral leishmaniasis or kala-azar can be endemic, sporadic or epidemic. It is due to *L. donovani*, *L. tropica*, or *L. chagasi*. In India it is more common in adults while in Latin America it predominates in children and is often fatal. It causes lesions in the reticuloendothelial system manifested by adenomegaly, hepatosplenomegaly, fever, weight loss, and asthenia. In the skin there is hyperpigmentation ("black fever or fiebre negra") as well as hypopigmented areas, especially on the forehead, around the mouth, hands and midline of the abdomen. It is confused with other myeloproliferative syndromes. There is a cutaneous form, post-kala-azar, that manifests one to several years later after the visceral form is healed. Atypical cases of visceral leishmaniasis have been reported in patients with AIDS and in other cases with cutaneous lesions that resemble dermatomiositis.

LABORATORY DATA

The parasites are visualized when stained with Giemsa or Wright. Biopsy shows epidermal atrophy or hyperplasia. During the acute phase there are intense dermal infiltrates of neutrophils with rare vacuolated histocytes that contain the parasite (Leishman bodies). These are abundant in patients with HIV. In advanced cases lymphohistiocytic infiltrates predominate with a tendency to form tuberculoid granulomas (Fig. 46.7). In kala-azar hepatic and splenic needle biopsy, and bone marrow culture are recommended. Pancytopenia, hypergammaglobulinemia and moderate elevation of transaminases and alkaline phosphatase can be found. The intradermal reaction with leishmanin (Montenegro reaction) is sensitive and specific. It is positive in localized forms and negative in the presence of anergy. The intradermal injection of Leishmania antigen is not approved by the Food and Drug Administration. The culture is done on three N media (Nicolle-Novy-MacNeal) or a modification of the biphasic Evans method at 24°C for 4 weeks. The strains are distinguished by electrophoresis of isoenzymes in

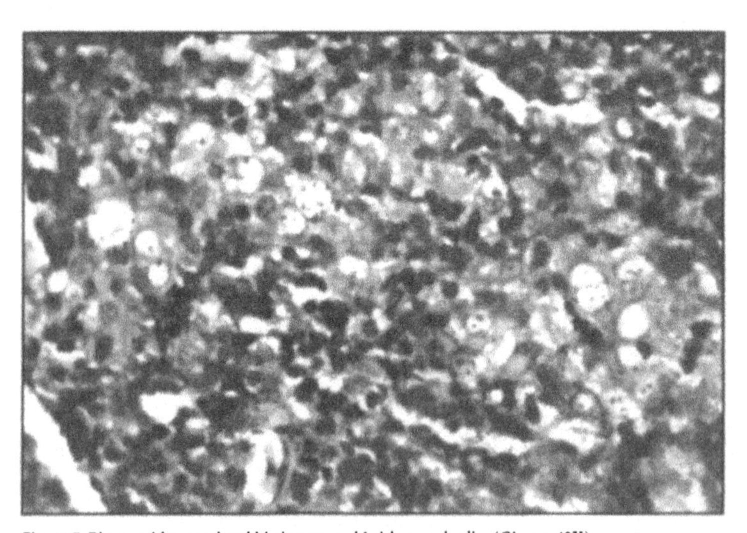

Fig. 46.7. Biopsy with vacuolated histiocytes and Leishmann bodies (Giemsa 40X).

cellulose acetate that allows identification of subspecies. Inoculation in animals such as hamsters helps in the classification of *Leishmania.* Antibodies can be detected by direct agglutination, direct immunofluorescence, complement fixation, monoclonal antibodies, and DNA probes. The following tests have been developed: Indirect immunofluorecent antibody test (IFAT), enzyme-linked immunosorbent assay (ELISA), Western blot, direct agglutination test (DAT), indirect immunoperoxidase assay (IPA) and recombinant *Leishmania infantum* proteins (rLIP2a and rLIP2b).

TREATMENT

Cutaneous leishmaniasis may heal spontaneously, yet the following treatments yield good results: trivalent antimonials administered parenterally such as repodral and anthiomalin 2-3 ml (0.02-0.03 g) on alternate days in 12-20 doses; and the pentavalents like glucantime (meglumine antimoniate) 10-20 mg/kg for 12 days to 3 weeks or until the lesions have healed. Another medication, pentostam (sodium stibogluconate) 20 mg/kg/day may be given for 20 days. For cutaneous disease antimonials may be injected intralesionally 0.2-15 ml weekly. Local antiseptics should be used. Sulfate of paromomycyn 15% and methylbenzetonio chloride 12% twice a day for 10 days and up to three weeks, or bleomycin 1% intralesionally are recommended. Also thermotherapy (local heat), cryosurgery, curettage, laser and radiotherapy have been used. In cutaneous leishmaniasis diaminodiphenylsulfone (Dapsone), 3 mg/kg/day for 3 weeks, has yielded good results. In cases caused by *L. m.mexicana* there is a good response to ketoconazole, 200-600 mg/day and to itraconazole 200-400 mg/day for one or two months (*Arch Dermatol* 1996; 132(7): 784-6). Terbinafine 250 mg in children and 500 mg in adults has been effective in cases caused by *L. tropica* (*Int J Dermatol* 1997; 36:59-60). In diffuse leishmaniasis, pentamidine 4 mg/kg is useful. It can be used given as two IM injections of 120 mg/day each in three applications, or as a daily injection in two or three series of 10 with intervals of 10 days. The side effects of the antimonials include local reaction, anorexia, nausea, vomiting, myalgia, arthralgia, an elevation of hepatic enzymes and electrocardiographic alterations (prolonged ST interval and T wave inversion). In anergic and visceral forms as well as in patients with AIDS, amphotericin B (3m/kg/day for 10 days), liposomal amphotericin B or amphotericin B lipid complex (3 mg/kg for 5 consecutive days or on alternative days with a total dose of 15 mg/kg) (*J Infect Dis* 1996; 173 (3): 762-5; *J Infect* 1996; 32(2): 133-7) or antimonial therapy for 40 days may be employed. Also rifampicin 600-1200 mg/day for more than 2 months has been used alone or with isoniazid; interferon gamma, allopurinol 20 mg/kg/day, metronidazole 250 mg three times a day for 10-15 days, trimethoprim-sulfamethoxazole 160-800 mg twice a day for 4 weeks also have been used. In visceral forms, recombinant human gamma interferon parenterally or intralesionally, or interleukin-2 have been used. The following have not been effective and generally should not be used as monotherapy: fluconazole, itraconazole (7 mg/kg daily) and terbinafine. In patients with HIV and visceral leishmaniasis, pentavalent antimonials in a single dosage of 850 mg monthly is recommended for prophylaxis. In mice infected with *L. major*, the progression of the illness has been prevented or there has been partial protection with recombinant IL-12 and CD40L (*Immunity* 1996; 4(3):283-9). In the anergic form and in kala-azar a usefull alternative is immunotherapy with BCG plus killed Leishmania promastigotes.

Prophylaxis

Eradication of the vector decreases the frequency of illness. This is accomplished by the use of insecticides, drainage of stagnant water and the early treatment of patients. The use of insect repellents based with diethyltoluamide or permethrin should be promoted, as well as thick clothing with long sleeves, pants, pavillon to sleep in and avoidance of night walks in forested areas during the time when insects are biting, generally between 18:00 and 6:00 h. The elimination of sick dogs or serologically positive dogs is also recommended. Trials are under study for the development of a vaccine (killed leishmania and BCG) that protect against all types of leishmaniasis (*Immunity* 1996; 4(3):283-9).

Leishmaniasis Treatment

Cutaneous	Mucocutaneous	Anergic	Kala-Azar
Spontaneous/Healing	Antimonials	Pentostam	Antimonials
Antimonials	Amphotericin B	Antimonials	Amphotericin B
Allopurinaol	Cyclophosphamide	Metronidazole	Azols
Dapsone	Pentamidine	Rifampicin	Terbinafine
Azols	Immunotherapy	Allopurinol	Interleukin-2
Rifampicin	Aminosidine-	Trimethoprim/Sulfa	Gamma Interferon
Paromomycyn	Sulfate	Methoxazole	Immunotherapy
Bleomycin		Amphotericin B	
Methylbenzetonio		Azols	
Terbinafine		Terbinafine	
Thermotherapy		Interleukin-2	
Cryosurgery		Gamma Interferon	
Surgery/Curettage		Immunotherapy	
Radiotherapy/Laser			

46

Selected Readings

1 Bahamdan KA, Tallab TM, Johargi H, et al Terbinafine in the treatment of cutaneous leishmaniasis: a pilot study. Int J Dermatol 1997; 36:59-60.

2 Convit J, Castellanos Putrich M, et al. Immunotherapy of localized, intermediate and diffuse forms of American cutaneous leishmaniasis. J Infect Dis 1989; 160 (1): 104-15.

3 Koff AB, Rosen T. Treatment of cutaneous leishmaniasis. J Am Acad Dermatol 1994; 31(5): 693-708.

4 Magill AJ, Grogl M, Gasser RA, et al. Visceral infection caused by *Leishmania tropica* in veterans of operation desert storm. N Engl J Med 1993; 328: 1383-1387.

5 Marsella R, Ruiz de Gopegui R. Leishmaniasis: A re-emerging zoonosis. Int J Dermatol 1998; 37: 801-14.
6 Momeni AZ, Yotsumoto S, Mehregan DR, et al. Chronic lupoid leishmaniasis. Evaluation by polymerase chain reaction. Arch Dermatol 1996; 132:198-202.
7 Noyes H, Chance M, Ponce E, et al. *Leishmania chagasi*: Genotypicaly similar parasites from Honduras cause both visceral and cutaneous leishmaniasis in humans. Exp Parasitol 1997; 85(3): 264-73.
8 Oliveira-Neto MP. Mattos M, da Silva Freitas de Souza C. Leishmaniasis recidiva cutis in New World cutaneous leishmaniasis. Int J Dermatol 1998; 37: 846-49.
9 Ponce C, Ponce E, Morrison A, Cruz A, et al. *Leishmania donovani chagasi*: New clinical variant of cutaneous leishmaniasis in Honduras. Lancet 1991; 337 (8733): 67-70.

Cutaneous Amebiasis

Roberto Estrada

This is a parasitosis caused by the protozoan, *Entamoeba hystolitica*, which may be found in the digestive tract. Through different ways, it may invade the skin where it causes painful, rapidly growing necrotic ulcers.

EPIDEMIOLOGY

In the Tropics *Entamoeba* affects about 30% of the population causing diarrhea, sometimes with systemic symptoms. The cutaneous form of amebiasis is rare. In Mexico City among 3000 dermatologic patients just 1 case was found (*Rev Med Hosp Gral* 1980; 43:33-6). Most cases affect adults of both sexes; it is rare in children although in nursing infants with diarrhea amebiasis must be considered.

ETIOLOGY

Entamoeba histolytica is the single causative agent; however, in immunodeficient individuals *Acantamoeba castellani* can cause illness (*J Am Acad Dermatol* 1992; 26:352-5). The parasite is unicellular and reproduces by binary division. The trophozoite is the invasive and reproductive form that colonizes the digestive tract. Before it is discharged with the feces, it assumes the metacystic form. It can survive in this form for several days, contaminating food and water and invading new hosts. An extreme manifestation is the invasion of the skin. The powerful amoeba's enzyme system composed of amylase, phosphomonoesterase, glutaminase, maltase, gelatinase and other proteolytic enzymes allow it to reside outside the digestive tract, in the liver, lung, brain, and skin. The organism reaches the skin barrier most often in the perianal region, especially in individuals suffering from diarrhea. Repeated rubbing when cleaning the skin causes perianal erosion and inflammation that facilitates parasitic penetration. Also the use of disposable diapers increases skin contact with contaminated feces (*Pediatrics* 1983; 71(4): 595-98). Sexual transmission has been reported. Also the skin can become involved through open surgical or traumatic wounds. When the parasite gains access to the blood or lymphatic circulations, it may reach the viscera and skin (*Int J Dermatol* 1982; 21:472-5).

CLINICAL FEATURES

Cutaneous amebiasis is characterized by rapidly growing, necrotic ulcers with severe pain, regional adenopathy, fever and malaise. The ulcers have a central granulation zone covered by purulent exudates or fibrin with necrotic tissues and an intensely erythematous halo and well-defined margins. Although classically presenting as a single lesion, one of our cases had multiple ulcers affecting the same area (Fig. 47.1). Some lesions are vegetative, verrucous or hyperkeratotic (*Clinical Tropical Dermatology* 1975:176). These are frequently misdiagnosed as malignant tumors, cutaneous tuberculosis or late syphilis. Usually lesions do not heal spontaneously. They grow rapidly and destroy extensive genital areas. When complicated by severe malnutrition, infection or immune compromise, the outcome may be lethal. Genital lesions (vulvitis, vaginitis, cervicitis, salpingitis and endometritis in women and balanitis, urethritis and prostatitis in men) can be papillomatous and may be difficult to diagnose.

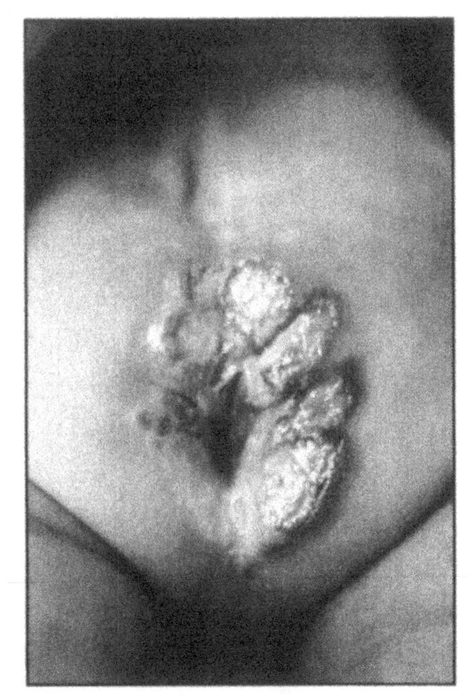

Fig. 47.1. Perianal cutaneous amebiasis.

LABORATORY DATA

Amoeba are usually found microscopically from fresh material taken from the base of the ulcer and placed on a glass slide to which warm saline solution is added to stimulate movement. On biopsy of the edge of the lesion, trophozoites may be seen. Indirect hemagglutination and counterimmunoelectrophoresis have been employed. Serologic studies are positive in invasive amebiasis (*Rev Med Hosp Gral* 1980; 43:37-9).

TREATMENT

Emetine and dehydroemetine are the drugs of choice. Emetine 1mg/kg/day for 10 days and dehydroemetine is given IM not to exceed 60 mg daily because of its cardiac effects. Oral metronidazole 30-40 mg/kg/day for 21 days or iodochlorohydroxyquinolein 0.25 g 3 times a day for 2-3 weeks can be given along with emetine and dehydroemetine. Also, aminoquinolein, oxytetracycline, erythromycin, acromycin, aminosidin and paramomycin have been used to treat cutaneous amebiasis. Because intestinal amebiasis is so common, it is convenient to do laboratory studies once or twice a year. In some places treatment of symptomatic individuals every 6 months has been adopted. Ingestion of contaminated water or food is a well known risk.

SELECTED READINGS

1 Lewis PM, Gurdip SS, Buchness MR. Diagnosis of Acanthamoeba infection by cutaneous manifestations in a man seropositive to HIV. J Am Acad Dermatol 1992; 26:352-5.

2 Ruiz-Maldonado R, Alvarez-Chacon R. Enfermedades por protozoarios y helmintos. In: Ruiz-Maldonado R, Parish L, Beare JM, eds. Dermatologia pediatrica. Mexico:Interamericana/McGraw-Hill, 1992:596-608.

Trypanosomiasis

Roberto Estrada

Trypanosomiasis is caused by flagellated protozoans of the gender Trypanosoma of which two forms are known: the American or Chagas' disease caused by *T. cruzi* and the African of which there are two varieties: *T. brucei gambiense* that causes the sleeping sickness in East Africa and *T. brucei* rodesiense in West Africa.

American Trypanosomiasis (Chagas' disease)

GEOGRAPHIC DISTRIBUTION

It affects 10-20 million individuals in rural areas of Latin America: Argentina, Brasil, Venezuela, Uruguay, Chile, Central America and Mexico.

ETIOLOGY

It is caused by *Trypanosoma cruzi* and is transmitted by hematophagic insects of the family Triatomidae. The vector feeds on contaminated blood from humans or animals. Within 2-3 weeks the trypomastigotes are deposited with feces on the skin during another blood meal. They penetrate through the conjunctivae (50% of the cases). *T. cruzei* can live intra- and extracellularly. Reservoirs besides man are dogs, cats, monkeys, pigs and rodents, among others. In the blood the microorganism is 15-20 mm long, it carries a wavy membrane and an interior flagellum which is lost when it penetrates cells of the reticuloendothelial system. Within the cells it assumes the leishmanoid form and reproduces. In 4-5 days it spreads through the lymphatics, and there is lymphadenopathy and hepatosplenomegaly. The organisms leave the cells in the flagellar form again. They re-enter in the circulation and repeat the cycle.

CLINICAL FEATURES

In the skin, at the site of the bite an erythematous, subcutaneous nodule—similar to the leishmaniasis—called a chagoma is seen. usually affecting exposed areas, e.g., face, arms, or legs. Unnoticed lesions occur in about 25% of cases. On the face, unilateral palpebral edema with painless, regional adenopathy is called the Romana-Mazza sign. This lesion may last 1-2 months. The acute phase is accompanied by fever and malaise. If cardiac or neurological lesions occur, the mortality is 10%. The cutaneous manifestations in this phase are known as squisotripanids; they consist of a morbiliform erythema marginatum-like, polymorphous or urticariform eruptions of 2-3 weeks duration. They remit spontaneously and may leave residual dyschromia. The chronic phase is characterized by organ involvement, e.g, cardiomegaly and megacolon.

LABORATORY DATA

In the acute cases, the parasite may be seen peripherally in blood or can be cultured from lymph node biopsies. In chronic cases, complement fixation, hemagglutination or immunofluorescence tests are useful. The complement (Machado Guerreiro) is very useful in epidemiological studies.

TREATMENT

In the acute phase, Nifurtimox (Lampit) 8-20 mg/kg/day orally for 3-4 months is useful as is benznidazole 5-8 mg/kg/day orally for 3-4 weeks. In the chronic phase treatment focuses on stabilization of the affected organs and symptomatic treatment.

African Trypanosomiasis

GEOGRAPHIC DISTRIBUTION

The most common known form is the "sleeping sickness" in West and Central Africa.

ETIOLOGY

It is caused by *T. brucei gambiense.* The fly vector of the gender *Glossina palpalis* known as tse-tse fly. This is a hematophagic insect of diurnal habits. Males as well as females bite and may transmit the parasite. They are attracted to dark skin, clothing, moving vehicles and the dirt of stampeding animals (*Rev Infect Dis* 1991; 13:1130-38). Man is the natural host, although the fly also feeds on animals. It affects both sexes and all ages. The flies ingest blood in which the parasites exist as trypomastigotes. They form non-infecting epimastigotes that migrate to the salivary glands of the insect, becoming infecting forms injected into the new host when the fly feeds. In the host skin, they multiply, mature and migrate to the lymph nodes. From there they involve the nervous system.

CLINICAL FEATURES

At the site of the bite a 2-3 mm nodule (chancre) forms surrounded by a whitish halo that is more common in white than in black-skin people. It is accompanied by lymphangitis and regional adenitis which spontaneously disappear in a few days. Also the initial lesion may present as a papular eruption accompanied by peripheral edema or, if on the face, as painful palpebral erythema and urticaria. This is early detected data in white patients (*J Trop Med Hyg* 1966; 69:124-131). Sometimes there are no skin lesions. There is a formation of immune complexes that are induced by the presence of the parasite; eventually it may be observed as intravascular dissemination (*Am J Med Sci* 1992; 303:258-90). Cardiac failure, arrhythmias, and infection usually are the cause of death. When the parasite enters the circulation, the patient develops fever, headache, malaise, arthralgias, dizziness, pruritus and lymphnode enlargement. Central nervous system involvement occurs months later. It presents as chronic meningoencephalitis. The individual sleeps, thus the name "sleeping sickness". There may also be ataxia, paresthesias and sudden death.

Trypanosomiasis caused by *T. brucei rodesiense* affects mainly the East African coast. Wild animals are the natural reservoir. It is transmitted by *Glossina morsitans*. In man it has an acute and fulminant course. Posterior cervical adenopathy constitutes the characteristic Winterbottom sign. The diagnosis of both forms may be difficult, especially in travelers from non-endemic areas where physicians are not familiar with this illness.

LABORATORY DATA

The erythrocyte sedimentation rate is increased 100-150 mm in the first hour. There is leukocytosis with increased monocytes, especially plasma cells (Mott cells). Gamma globulin, mainly IgM, is elevated in 85% of patients. Indirect immunofluorescence is a sensitive and specific test. Trypanosomes in the blood, in lymph node aspirate or biopsy is diagnostic.

TREATMENT

Pentamidine, 4/mg/kg for 10 days, isothionate, is effective. It is reported to have numerous side effects and is contraindicated in children weighing less than 8 kg and in pregnant women. Melarsoprol, 3.6 mg/kg/day, 2-3 iv injections every other day is also effective. It can be repeated after a 15-day interval with a maximum of 12 injections. It is effective in all stages, but similar to pentamidine, it causes several side effects. Difluoro-methyl-ornithine or eflornitine, 400 mg/kg/day for 15 days, is also effective although it is contraindicated in pregnancy, in renal insuficiency and in children under two years of age. Prophylaxis to prevent transmission, insecticides, protection against the vector and control of the animal reservoirs, must be undertaken.

Selected Readings

1 Caumes E, Danis M. Trypanosomiasis. In: Pierard GE, Caumes E, Franchimont C, Arrese-Estrada J, eds. Dermatologia Tropical. Bruxelles: AUPELF. 1993; 334:42.
2 Duggan AJ, Hutchinson MP. Sleeping sickness in Europeans: A review of 109 cases. J Trop Med Hyg 1966; 69:124-31.
3 Farah FS, Klaus SN, Frankenburg S. Protozoan and Helmintic Infections. In: Fitzpatrick T, Eisen A, Wolff K et al, eds. Dermatology in General Medicine. 3rd ed. New York:McGraw-Hill, 1987:2769-87.
5 Hadjuk S, Adler B, Bertrand K et al. Molecular biology of African trypanosomes: Development of new strategies to combat an old disease. Am J Med Sci 1992; 303:258-90.
6 Panosian CB, Cohen L, Bruckner D et al. Fever, leukopenia and a cutaneous lesion in a man who had recently traveled in Africa. Rev Infect Dis 1991; 13:1130-38.
7 Ruiz-Maldonado R, Alvarez-Chacon, R. Enfermedades parasitarias por protozoarios y helmintos. In: Ruiz-Maldonado R, Parish L, Beare J, eds. Dermatologia Pediatrica. 1a ed. Interamericana-McGraw-Hill, 1992:596.

Onchocerciasis

Roberto Arenas

Onchocerciasis is also known as oak tree sickness, erysipelas of the coast, river blindness or craw craw. It is an illness caused by the filaria *Onchocerca volvulus.*

GEOGRAPHIC DISTRIBUTION

It affects 20 million people. It is a public health problem in 31 countries of Africa from Senegal to Tanzania. Recently the Civil War in Sudan has contributed to its dissemination and seriousness. In Latin America it is mainly found in Guatemala, Venezuela, Colombia, Ecuador, Brazil and Mexico (20,000). In Oaxaca, Mexico the occurrence is 81 per 1,000 individuals. It causes blindness in 1% of cases. In endemic zones of Guatemala it occurs in 22% of the inhabitants. The endemic zones are equatorial although the illness does occur in Savannas. It affects all races; it is more frequent in adult men and farmers. It is seen in children but is exceptional under 5 years of age.

ETIOLOGY

It is produced by a nematode, the filaria *O. volvulus.* Humans are the definitive host. Microfilariae, 150-280 mm long and 5-7 mm wide occupy the anterior chamber of the eye and the skin. The parasite is thermotropic. Feeding insects take up the parasite larvae from the skin and deposit them on other individuals. In 9 months they mature into fertile, adult worms. The female is 230-700 mm and the male 20-45 mm by 0.2-0.4 mm. These forms concentrate in cystic lesions from which the microfilariae are born and pass into the skin and circulation. The insect vectors belong to the gender Simulium. In Mexico the most common species is *S. ochraceum.* The true vector is the anthropophilic female that has diurnal habits. Its activity depends on brightness, temperature (25-28°C) and humidity. It is found between 100-1500 m altitude and it can travel 5 km in 2-3 days. It breeds in moving water. The acute reaction is spontaneous or often follows treatment. It occurs in more than 80% of cases and is explained by the cytotoxic effect of the products of massive destruction of the parasite which induces elevations of serotonin and reactive protein C. There is immediate hypersensitivity with eosinophilia and an increase in IgE, induction of Il-4 and Il-5, and interferon gamma. It has been suggested that tolerance occurs in immune-compromised individuals in whom microfilariae persist (*Int J Derm* 1985;

Tropical Dermatology, edited by Roberto Arenas and Roberto Estrada. ©2001 Landes Bioscience.

24(6):349-358; *Immunol Today* 1991; 12(3):A54-8). The filarial protease contributes to the pathogenesis of chronic onchocercal dermatitis directly by enzymatic destruction of connective tissue and indirectly by eliciting an autoimmune response (*Exp Parasitol* 1994; 79(2):177-88). Destructive elastic fiber changes frequently are seen.

CLINICAL PICTURE

This disease manifests on the skin and in the eyes. Skin subcutaneous lesions occur on or near the joints, on the head and upper trunk, but may also occur on the buttocks, in the sacrococcygeal region and the lower extremities (onchocercomas). The lesions are cystic and contain adult filariae; they are spherical or oval, are asymptomatic and may perforate the skull. On the trunk and extremities so-called lichenified onchodermatitis is characterized by dry skin, ictiosiform, lichenification, pigmentation, scaling or hyperkeratosis (saurio skin) and pruritus. "Saurio skin" is the most constant characteristic, and in some occasions it is limited to only one extremity (sowdah). Later on there is atrophy, especially of lesions of the buttocks, with loss of elasticity, marked skin folds, alopecia and anhidrosis (Fig. 49.1). There is a form known as onchocercal dyschromia with alternating areas of depigmented and normal or hyperpigmented skin around the follicles (Leopard skin). It predominates in males and on shins. There may be lymphadenopathy, lymphedema and redundant inguinal and femoral skin (adenolymphocele). Elephantiasis is exceptional. The ocular involvement, occurring in 50% of cases, is caused by microfilariae in the anterior and posterior chambers and may cause choroidoretinal atrophy. The symptoms and signs are photophobia, lacrimation, burning, decreased visual acuity, conjunctival erythema and edema, keratitis (punctate and sclerosing), and iritis (Fig. 49.2). There is edema of the eyelids and blepharochalasia. The conjunctivae are vascularized and later an ochreous pigmentation appears. Leukoma, limbitis and pseudopterigium, and blindness ensue soon thereafter in a high percentage of cases. The acute stage, or so-called Mazzotti's reaction, appears spontaneously or is induced by therapy with diethylcarbamazine. It is characterized by generalized and digestive symptoms, fever, headache, anorexia, nausea, vomiting, myalgia, adenitis, abdominal pain, diarrhea and epistaxis. Cysts become more prominent, and deep erythema, edema and papules with occasional vesicles and blisters comprise what is known as acute, papular onchodermatitis (filariasic scabies). There is intense pruritus in the face, neck, upper trunk and proximal extremities; anxiety can be severe. It is not clear if the unbearable itching is just pruritus since, in addition to scratching, the patients hit their skin. With time itching decreases and the lesions become hyperpigmented. There is photophobia, conjunctival eythema, burning, or the sensation of a foreign body. The erythema can occur suddenly as a darkly cyanotic and uniform plaque (cyanotic disease or erysipelas of the coast) usually accompanied by hyperesthesia. The face is erythematous with firm, elastic

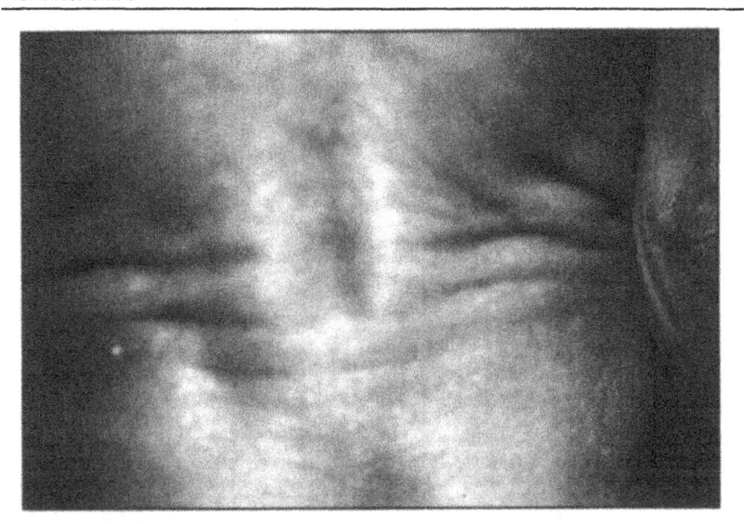

Fig. 49.1. Chronic Onchodermatitis, cutaneous folds.

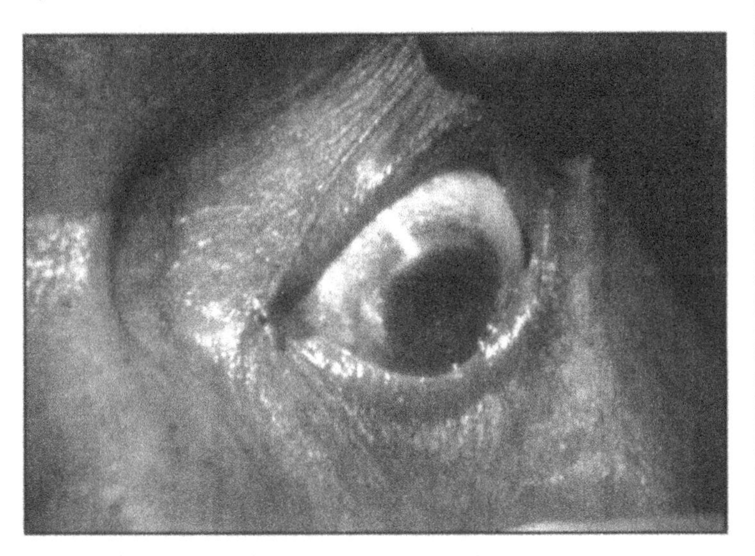

Fig. 49.2. Onchocerciasis, ocular lesions (Courtesy of R. Ortiz and V. Bravo).

edema or with erythematous pigmented and infiltrated plaques. They are numerous lesions on the eyelids, cheeks, ears and neck. The facial appearance is flaccid and leonine.

LABORATORY DATA

Slit lamp examination can reveal microfilariae in the anterior chamber, cornea and lens. Cutaneous biopsy should be obtained with a curved scissor, scalpel or razor blade; snips of 0.5 x 1 cm should include epidermis and dermis. It is not neccesary to close the biopsy sites with suture. Best sample sites are ear lobe, shoulders or supraclavicular region, and the hip. After rending the skin samples, they are observed under the microscope with saline solution to look for microfilariae. They may be fixed with alcohol and stained with Giemsa. They stain blue-violet. Exposing the skin to heat prior to biopsy induces the microfilariae to move rapidly to the surface. A negative biopsy does not rule out the diagnosis. Hematoxylin and eosin staining can demonstrate microfilariae in the dermis and hypodermis with the usual markers of inflammation (Fig. 49.3). In the acute form the changes are more marked. In the subacute form there are abundant fibroblasts, compact collagen and elastosis. The microfilariae often concentrate at a single site and become intertwined in an ovid mass that contains tubular, eosinophilic structures that stain well with Giemsa. With time they calcify and are surrounded by a granulomatous reaction with fibrosis. There is eosinophilia in peripheral blood, and the erythrocyte sedimentation rate is increased. IgE and IgG are elevated. Indirect inmunofluorescence is 98% positive. ELISA and Western blot may also be employed for diagnosis.

TREATMENT

Ivermectin (Mectizan) 150-200 μg/kg in a single dose is the treatment of choice (two tablets of 6 mg). It is a microfilaricide. It is administered twice a year, but the optimum regimen has not been established. It is able to control the disease which remains transmisibile; it also diminishes the ocular damage as it destroys larvae very slowly. The density of microfilariae in the skin and the anterior chamber is diminished; corneal opacity is also reduced (*Trans R Soc Trop Med Hyg* 1994; 88(5):581-4; *Am J Trop Med Hyg* 1995; 52(3):270-8). The pruritus, erythema, and edema that accompany the acute reaction may be mitigated with antihistaminics, antiserotonins or with thalidomide 100-300 mg/day for 3-10 days. Also, phenylbutazone or corticoisteroids can be used for brief periods. Diethylcarbamazine (Hetrazan), another microfilaricide, 250 mg/day in adults and 100 mg/day (two tablets) is given for 10 days. Two to four treatments a year are recommended. The effect is potentiated with levamizole. Metrifonate is also used, 10 mg/kg/day for six days every two weeks in four series. Both trigger Mazzotti's reaction in 88% of cases in 24 hr and may be used as a diagnostic test.

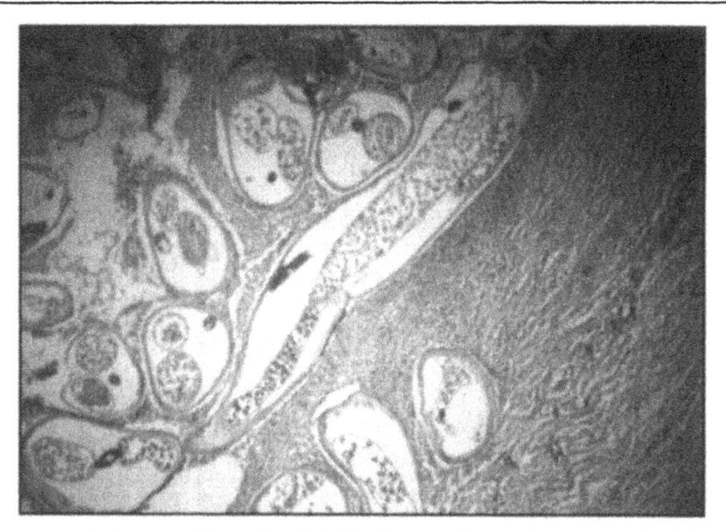

Fig. 49.3. Onchocerciasis, histopatology (H.E. 20X) (Courtesy of R. Ortiz, V. Bravo, and J. Novales).

Because of this, many patients do not follow the prescription. Prophylaxis is important. Insecticides kill larva in rivers.

OTHER FILARIASIS

A dozen filarias can cause illness in man. All have a cycle requiring a vector (Culex, Aedes) and symptoms characteristic of the nature of the adult nematodes and the immune response to the microfilariae. Prophylaxis, requiring massive chemoprophylaxis and treatment of the vectors, is almost impossible.

Lymphangitic filariasis is estimated that there are about 90 million people affected worldwide. It predominates in China, India, Indonesia and tropical Africa. It also occurs in Brazil, Guayana and the islands of the Pacific and Caribbean. In 90% of cases obstruction of the lymphatic vessels is caused by the adult worms of *Wuchereria bancrofti* and in the other cases by *Brugia malayi* and *B. timori.*

Most cases are asymptomatic, without the presence of microfilariae. Acutely there can be adenitis, centrifugal lymphangitis and recurrent fever. Lymphangitis of the genitals is not uncommon. The disease is generally recurrent. Chronic disease is characterized by adenolymphoceles, varicose veins, lymphedema and elephantiasis.

Loasis or filariasis Loa loa affects 13 million people in tropical Africa. It is acquired from *Chrysops* sp. It is manifested by cutaneous lesions and the immune reaction to destruction of the microfilariae. The incubation period is 4 months to 20 years. It is associated with pruritus of the thorax and genitalia, angioedema, and subcutaneous migration sites (Calabar swellings). Larvae in the skin form palpable, erythematous cords of less than 10 cm and that move 1 cm per minute.

The conjuntivae may be affected. It may be related to a glomerulonephritis.

Dirofilariosis occurs throughout the world and affects man in an accidental manner. It is produced by filariae of the dog *Dirofilaria repens*, *D. immitis* and the filaria of the mouse *D. tenuis*. In the skin adult filariae reside in serous-filled cysts. They are also found in lungs and heart.

Mansonelosis occurs in Asia, Africa and South America and is caused by *Mansonella perstans*, *M. ozzardi*, *M. streptocerca* and *M. rodhaini*. It is transmitted by insects of the gender Culicoides and it causes pruritus.

LABORATORY DATA

Generally there is eosinophilia. Filariae may be demonstrated in peripheral blood smears. In lymphangitic forms gel precipitation, indirect immunofluorescence and ELISA are useful. In loasis the serology is neither sensitive nor specific.

TREATMENT

In lymphangitic filariae treatment is symptomatic, e.g., bedrest, nonsteroidal anti-inflammatory agents, antihistamines and exceptionally, antibiotics and prednisolone. Symptoms of acute disease are exacerbated by antiparasitic treatment. Diethylcarbamazine is useful in the progressive form of disease, up to 6 mg/kg day in a single dose without exceeding 40 mg/day. Three courses of 10 days' duration with intervals of 10 days are recommended. Ivermectin 200 µg/kg in a single dosage is also used. In the chronic forms, cumarinics and surgical treatment are indicated. Loasis may be treated with diethylcarbamazine for 21 days as well as surgical extraction of the filariae. For dirofilariosis, mebendazole has been suggested.

Selected Readings

1 Alley Es, Plaisier AP, Boatin BA et al. The impact of five years of annual ivermectin treatment on skin microfilarial loads in the onchocerciasis focus of Asubende, Ghana. Trans R Soc Trop Med Hyg 1994; 88(5):581-4.
2 Burnham G. Ivermectin treatment of onchocercal skin lesions; observations from a placebo-controlled, double-blind trial in Malawi. Am J Trop Med Hyg 1995; 52(3): 270-8.
3 Caumes E, Danis M. Nematodoses de type filariose. In: Pierard GE, Caumes E, Franchimont C, Arrese-Estrada J, eds. Dermatologie Tropicale. Bruxelles:AUPELF. 1993:355-70.
4 King CL, Nutman TB. Regulation of the immune response in lymphatic filariasis and onchocerciasis. Immunol Today 1991; 12(3):A54-8.
5 Malatt AE, Taylor HR. Onchorcerciasis. Infect Dis Clin North Am Philadelphia: Saunders 1992; 6(4):963-77.
6 Murdoch ME, Hay RJ, Mackenzie CD et al. A clinical manifestation and grading system of the cutaneous changes in onchocerciasis. BR J Dermatol 1993; 129:260-69.

7 Rodriguez-Perez MA, Rodriguez MH, Margeli-Perez HM, Rivas-Alcala AR. Effects of semiannual treatments of ivermectin on the prevalence and intensity of Onchocerca volvulus skin infection, ocular lesions, and infectivity of Simulium ochraceum population in southern Mexico. Am J Trop Med Hyg 1995; 52(5): 429-34.
8 Stingl P. Onchocerciasis: clinical presentation and host parasite interactions in patients of Southern Sudan. Int J Dermatol 1997; 36:23-28.
9 Tawill SA, Kipp W, Luclus R et al. Immunodiagnostic studies on *Onchocerca volvulus* and *Mansonella perstans* infections using a recombinant 33 KDa O. volvulus protein (Ov33). Trans R Soc Trop Med Hyg 1995; 89(1):51-4.

Trichinosis

Rafael Herrera-Esparza and Esperanza Avalos-Diaz

This is a parasitic disease caused by *Trichinella spiralis* (filum nematode, aphasmidia class, superfamily Trichinelloidea). The family has eight groups of genes (T1-T8) and at least four of them can be identified in a species. The genome of the parasite is composed of six haploid chromosomes with a total of 2.4×10^{-8} pb (*Parasitol Today* 1992;8:299-306). The larval morphology varies according to the five stages of postembryonic development. At birth it is 100 x 5.6 microns and has a banana shape with an anterior lance-shaped end. Its growth starts after invading the muscle reaching a size up to 1300 X 400 µm. The parasite is composed of two concentric tubal structures: an external one invested by the cuticle, the vascular membranes and hypodermis, and an internal one, anteriorly containing the esophagus, with its esticosomes (exocrine glands composed of discoid cells called esticocytes). These contain antigenic material in form of alpha-0, alpha-2, beta and gamma granules. The posterior end of the internal tube contains the rectum. The adult parasite has a mouth anteriorly connected to a long digestive tract with a ventral curvature. The male is 1.5 x 0.04mm. The mouth, esophagus and neural ring are anterior structures. In the middle are the testicle and the vesicular lumen. Posteriorly are two caudal appendices for copulation. The female is 3.5 x 0.06 mm. The valve and part of the uterus are in the anterior part. The uterus in the mid-portion has larvae and embryos. Posteriorly are the seminal receptacle, the oviduct, intestine, ovary and rectum (*J Parasitol* 1990; 76:290-95).

GEOGRAPHIC DISTRIBUTION

Trichonosis occurs throughout the world and affects populations that consume raw or partially cooked meat infected with the parasite. In 1940 in the developed countries the reported incidence was 16%. This progressively decreased to 4.2% in 1970 because of enhanced sanitation and improved foodhandling technology. Its current prevalence is insignificant (*N Eng J Med* 1978; 298: 1178-80). In developing countries there are still some endemic areas where pork is clandestinely slaughtered. The meat is generally consumed in the form of half-cooked "chorizo" (sausage). Epidemics occur more frequently in spring (*Salud Pub Mex* 1985; 27:40-50).

Tropical Dermatology, edited by Roberto Arenas and Roberto Estrada. ©2001 Landes Bioscience.

CLINICAL PICTURE

It is characterized by gastrointestinal symptoms, fever, palpebral edema, muscle pain and eventually an extanthem. The natural history of the illness is related to the life cycle of the parasite: intestinal, systemic, muscular complications and convalescence. Meat infected by *T. spiralis* is ingested; larval cysts are lysed and the larvae migrate to the small intestine where they reach sexual maturity. After fertilization the females deposit newborn larvae one week after inoculation. In this intestinal phase the symptoms include nausea, vomiting, abdominal pain and diarrhea. This stage can last up to four weeks; it ends when the adult parasites are eliminated from the intestine. The newborn larvae penetrate the intestinal wall and enter the lymph and blood circulations from where they invade muscle. In this systemic phase of symptoms are fatigue and headache. Symptoms characteristic of the phase the skeletal muscle invasion include conjunctivitis, periorbital edema, myalgia and fever. In the majority of cases fever appears between day 7-10 days postinoculation. It may reach 40°C and can last up to 2 weeks. Occasionally in the third week there is an exanthem. Myalgias of the legs, trapezius, deltoids, masseters and hypopharyngeus are common. Muscles are tender and weak. Functional incapacity is reduced and walking and eating are painful. Severe cases are accompanied by myocarditis, pericarditis, cardiac failure, respiratory insufficiency, hepatic failure and death. In most patients recovery begins in the sixth week. Laboratory parameters return to normal, and the parasite is calcified (*Prensa Med Mex* 1979; XLIV:278-87). The disease must be distinguished from other myopathies.

LABORATORY DATA

In the intestinal phase it is difficult to demonstrate the cause of the illness. In the first week the white cell blood count is 10,000-20,000. There is a progressive increase in the percentage, and in the absolute number of eosinophils, up to 60%. There is also elevation of the erythrocyte sedimentation rate. Muscle enzymes SGTP, LDH, creatinine phosphatase and aldolase are elevated. Skin testing is of no value. The demonstraton of larvae in muscle is diagnostic. Larvae are encysted and surrounded by an inflammatory infiltrate of mononuclear cells. In late stages the larvae are calcified. The exanthem is a leukocytic vasculitis. Western blot (*Exp Parasitol* 1987; 63:233-36), immunofluorescence (*J Rheumatol* 1985; 12:782-84), latex particle agglutination, ELISA and PCR have been employed in the diagnosis (*Parasitol Res* 1994; 80:358-60) (Table 50.1).

TREATMENT

Thiabendazole, 50 mg/kg/day 3-4 days is effective. Nonsteroidal antiinflammatory agents may be administered as symptomatic treatment.

Table 50.1. Diagnosis of trichinosis (J Rheumatol 1985; 12:782-784.)

I. **Presumptive diagnosis**
   a. Typical clinical lesions
   b. Eosinophilia
   c. Enzymes elevation (especially A and B)
   d. Electromyographic alterations (especially A and B)

II. **Indisputable diagnosis**

Direct exam of Trichinella larvae in:
   a. Muscle: Larvae can be observed microscopically in the muscle, placed between two glass-slips, by digestion with HCl-pepsine and in tissue sections
   b. Intestine: By duodenal aspiration, during the enteric phase
   c. Blood: In parenteral phase, DNA amplification by PCR

III. **Poitive indirect test**

Circulating anti-Trichinella antibody detection:
   a. Indirect immunofluorescence
   b. Western-blot
   c. ELISA
   d. PCR
   e. Others

Selected Readings

1 Hammond MP, Bianco AE. Genes and genomes of parasitic nematodes. Parasitol Today 1992; 8:299-306.

2 Herrera R, Varela E, Morales G et al. Dermatomyositis-like syndrome caused by trichinae. J Rheumatol 1985; 12:782-84.

3 Dupouy-Carnet J, Robert F, Guillou JP et al. Identification of Trichinella isolates with random amplified polymorphic DNA markers. Parasitol Res 1994; 80:358-60.

4 Takahashi Y, Mizuno N, Uno T et al. Direct evidence that the cuticle surface of Trichinella spiralis shares antigenicity with stichocyte a-granules and the esophaus-occupying substance. J Parasitol 1990; 76:290-95.

Dracunculosis

Roberto Estrada

Dranunculosis is also known as dracontiasis, or "Medina's worm" or "Guinea worm" disease. It is a parasitosis caused by the nematode *Dracunculus medinensis.* It is a very old ailment since it is mentioned in the "Turin Manuscrupt" in the 15th century B.C. The Indian Rig-Veda also refers to it, and a recent examination of an Egyptian mummy revealed a calcified worm that corresponded to *Dracunculus.* In 1870 Alexei Fedchenko described the life cycle of the parasite and identified the *Copepods cyclops* as the vector.

GEOGRAPHIC DISTRIBUTION

The ailment is endemic of 18 nations from India, the Middle East and numerous African countries, especially on the west coast. This is perhaps the only parasitosis that may be eradicated. According to the World Health Organization, this will be accomplished by the end of this century. There has been important progress in the last decade. Before the international eradication campaign, the annual incidence was 10 million cases; it has been reduced to about 2 million. The elimination of this illness has almost been accomplished in Asia, Cameroon and Senegal. Cases in Nigeria and Ghana have fallen from 820,000 in 1989 to fewer than 240,000 in 1992 (*Am J Trop Med Hyg* 1993; 49(3):281-9).

ETIOLOGY

D. medinensis is the longest worm that affects man. The adult female may be more than 1 meter long and 2 mm in diameter. The male is smaller. The vector is the genus *Cyclops* or "water flea." These are widely distributed and are vectors for other parasitoses as well as for gnasthostomiasis. The *Cyclops* inbibes the microfilariae in the water. Within the flea it develops into the infective form. Ingested larvae reach the human intestinal tract and migrate to the lymphatics and subsequently reside in the subcutaneous tissue where they mature in 8-12 months. Mature worms migrate to the lower extremities in 90% of cases. Man seems to be the only reservoir for the adult parasite. Frequently an individual is host to more than one worm.

CLINICAL FEATURES

Until the parasite reaches the skin, the host can be asymptomatic. But when the skin is reached there is urticaria, rash, erythema, malaise and fever. Inflammation is intense where the skin opens above the worm. Vesicles or bullae appear. The pregnant female discharges microfilariae through the open area, and when the parasitized individual is in contact with water the cycle is renewed. Several weeks after the microfilariae are discharged, the parasite could die and calcify beneath the skin, if it has not been previously extracted. Occasionally a joint space is involved causing aseptic arthritis, synovitis, ankylosis and contractures. The individual often is incapacitated for long periods of time, especially if the worm has emerged on the sole of the foot. Secondary infection such as erysipelas and necrotizing cellulitis occur. The wound can be the entry point of tetanus as well. The diagnosis is clinical and obvious when visualizing the worm in individuals in endemic areas. When the parasite is calcified, it may be visualized on x-ray.

TREATMENT

Traditionally the worm has been extracted manually from the host, a little each day by traction and winding it on a stick over 1-4 weeks, although this may result in breaking the worm. No antiparasitics are really effective, although the diethylcarbamazine, which seems to act in the early stages has been used, and thiabendazole has been reported to kill the worm and decrease inflammation. The Worldwide Health Organization has employed several strategies to eradicate the illness:

1) Careful epidemiologic vigilance by local committees that report new cases, promote the sanitary education, distribute water filters and treat affected individuals;
2) The supervision of the eradication activities conducted by governmental health organizations;
3) The filtration of water sources suspected of contamination. In isolated areas that are only inhabited by nomadic shepherds or in isolated farms, water filtration through nylon fabric or pieces of tightly woven material is promoted;
4) The protection and delivery of water in itself constitutes an effective method to eliminate the disease.

Selected References

1 Bournerias I, Caumes E, Danis M. Autres nematodoses: Trichinose et dracuculose. In: Pierard GE, Caumes E, Franchimont C, Arrese-Estrada J. Dermatologie Tropicale. Bruxelles: AUPELF, 1993; 373-80.
2 Farah FS, Klaus SN, Frankenburg S. Protozoan and helmintic infections. In: Fitzpatrick T, Eisen A, Wolff K et al, eds. Dermatology in General Medicine. 3rd ed. New York: McGraw-Hill, 1987:2769-87
3 Hopkins DR, Ruiz-Tiben E, Kaiser RL et al. Dracunculosis eradication: Beginning of the end. Am J Trop Med Hyg 1993; 49(3):281-9.

Cysticercosis

Gisela Navarrete

Cysticercosis is a parasitosis produced by *Cysticercus cellulosae* (cysticerci), the larval form of the adult parasite *Taenia solium*. It affects the nervous system, eyes, skin, muscle and other organs. It was known in antiquity; Aristotle described the larva in a pig's tongue. Human infestation was described by Rumler in 1558, by Renna and Panarolus in 1558 and by Kuchenmeister in 1855. In1856 Leuckart studied the life cycle and demonstrated the relationship between the larva and the adult worm.

GEOGRAPHIC DISTRIBUTION

It occurs worldwide, but it is common in Africa, India, China and Latin America (Mexico, Guatemala, Peru, Brazil and Ecuador). It has practically been eradicated from the developed countries, and it is rare in North Africa and the Middle East.

ETIOLOGY

Cysticerci constitute the larval phase of *Taenia solium*. It is a parasite with a spherical form, white, 5 mm long by 8-10 mm wide, with a small, invaginated head. Man ingests the cysterci in raw or incompletely cooked pork or fruits and vegetables washed with contaminated water. Auto-infection occurs when an egg from a parasitized individual's perianal area is introduced into the mouth. Internal auto-infection can occur when by retroperistalsis eggs move in a retrograde direction toward duodenum or stomach where they hatch and migrate to other organs. After being digested the cysticerci localizes on the wall of the small intestine; within 5-12 weeks an adult female is 2-7 mm. The proglotides are discharged in the feces or migrate outside the intestine. The eggs that discharged from the pregnant proglotide are deposited on the soil where they can remain viable for several weeks and can be ingested by pigs, man or other mammals, liberating embryonic parasites that penetrate the intestinal wall and entering the mesenteric circulation to infiltrate muscles where after 60-70 days they become cysticerci. The cysticerci have a preference for subcutaneous tissue, muscle and the central nervous system. They calcify in the brain, and the immune response to neurocyisticercosis varies from complete tolerance to a severe inflammatory reaction.

Tropical Dermatology, edited by Roberto Arenas and Roberto Estrada. ©2001 Landes Bioscience.

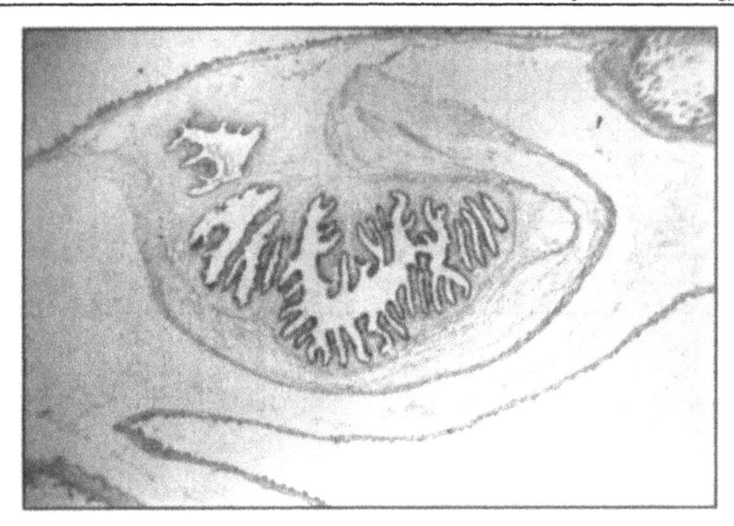

Fig. 52.1. Cysticerci in a skin biopsy (H.E. 40X).

CLINICAL FEATURES

The clinical manifestations depend on the location and number of parasites. In increasing order of frequency they affect the brain, eyes, skin, muscle, heart, lung and abdominal viscera. In cysticercosis of the brain, involvement can be localized to the meninges, ventricles, parenchyma or may involve many sites. The neurological manifestations are headache, vomiting, disequilibrium, epilepsy, meningoencephalitis, decreased mental acuity, intermittent obstructive hydrocephaly, transitory hemiparesias and even death. In ocular cysticercosis, uveitis, iritis, retinitis, choroidal atrophy, conjunctivitis and orbital muscles cysts may be develop. In the skin, the manifestations are more frequent on the trunk and upper extremities. Single or multiple, oval or round, 0.5-2.0 cm x 0.5-1.0 cm tumors can be palpated. They are skin-colored, smooth and well circumscribed. They may remain unchanged for many years or may calcify (*Arch Dermatol* 1978; 114:107-108). Prognosis depends on the site of involvement. Skin and muscle lesions are benign. They tend to calcify and are only of cosmetic importance.

LABORATORY DATA

Biopsy is the most simple diagnostic procedure. A cyst in which parasite resides makes the diagnosis. The parasite has an invaginated head and hooks. The connective tissue around the cyst is moderately inflamed (Fig. 52.1). In the cerebral and ocular cysticercosi, imaging studies are necessary to make the diagnosis. ELISA has a sensitivity of 80%.

TREATMENT

When the skin lesion is single, surgical excision is indicated. With multiple cutaneous lesions various medications have been used: albendazole 15 mg/kg/day for 8-30 days (*Int J Dermatol* 1995; 8:574-579), metrifonate (*Arch Dermatol* 1981; 117:507-509), and praziquantel 50 mg/kg/day for 8-15 days (*Sal Pub Mex* 1982; 24:679-682). This last medication has been administered in cerebral and ocular forms of the illness. Its use in children is controversial. Side effects relate to the inflammatory response elicited by the dying or dead parasite. Glucocorticoids are indicated, 1 mg/kg/day. Prophylaxis includes elimination of contaminated pork and the sanitary disposition of human feces.

SELECTED READINGS

1 Baily GG. Cysticercosis. In: Cook GC, ed. Manson's tropical disease. 20th ed. Philadelphia:Saunders, 1996:1509-16.
2 Ostrosky-Zeichner L, Garcia-Mendoza E, Rios C, Sotelo J. Humoral and cellular immune response within the subarachnoidal space of patients with neurocysticercosis. Arch Med Res 1996; 27(4):513-17.
3 Schantz PM, Sarti E, Plancarte A et al. Community-based epidermiological investigations of cysticercosis due to *Taenia solium*: Comparison of serological screening tests and clinical findings in two populations in Mexico. Clin Infect Dis 1994; 18:879-885.
4 Schmidt DK, Jordan HF, Schneider JW, Cilliers J. Cerebral and subcutaneous cysticercosis treated with albendazole. Int J Dermatol 1995; 8:574-579.

J. Virosis

Herpes Simplex/Aphthous Ulcer

Varicella/Zoster

Viral Wart/Focal Epithelial Hyperplasia

Molluscum Contagiosum

Dengue

Aquired Immune Deficiency Syndrome (AIDS)

Herpes Simplex/Aphthous Ulcer

Roberto Arenas

This is an infection caused by the herpes simplex virus (HSV) 1 and 2 that affects the skin and oral and genital mucosae. It is characterized by groups of vesicles on a red base that resolve spontaneously but are recurent. There is no specific treatment.

GEOGRAPHIC DISTRIBUTION

It affects all races, both sexes and all ages. Although it rarely appears before the age of four months. The mean age for the first episode of genital herpes is 20-25 years. Oral-labial involvement is the most frequent manifestation. The incidence of genital herpes has increased and approximately 20% of women are carriers. In the United States from 1978-1990 the incidence increased to 32%, and it is estimated there are 20 million affected individuals with 300,000-500,000 new cases per year. Fifity-five million are estimated to be seropositive for HSV-2.

ETIOLOGY

The herpes virus belongs to the Herpesviridae family of DNA viruses. Two types have been distinguished, herpes simplex virus-1 and -2. The former (HSV-1) is not sexually transmitted, but it has been demonstrated in 20% of genital infections. The latter (HSV-2) is sexually transmitted, but it has been isolated in oral lesions, probably transmited by oral sex. In general, individuals with antibodies against HSV-1 do not have genital lesions. This is not the case of those with antibodies to HSV-2. Different strains of HSV-1 and HSV-2 have been found through analysis of nucleotide sequences. The life cycle of the herpes virus is manifested by infection, latency and cellular transformation. It is introduced to a susceptible person by direct contact. This is followed by an asymptomatic invasive phase, a replicative phase within the host cell and then cell lysis. This cycle is complete in 5-6 h. From the epithelial inoculation the herpes extends along sensory nerves to the neuronal ganglia where it remains latent. The mechanism of reactivation is not known, but it usually occurs after a febrile illness, physical or emotional stress, sexual contact, menstruation, trauma and heat. The host response to the infection is complex: humoral, cellular and also nonimmunological responses are involved. Antibodies do not prevent viral reactivation and the highest titers are reported in individuals with frequent episodes of recurrences. Intact cellular immunity is important in the prevention of severe illness. Increased succeptibility to infection of HSV-2 seems to be related to

Tropical Dermatology, edited by Roberto Arenas and Roberto Estrada. ©2001 Landes Bioscience.

decreased exposure to HSV-1 during childhood. Its viral oncogenicity is moot, but it has been observed that women with genital herpes have a higher incidence of cervical and uterine cancer. The varicelliform eruption is related to atopic dermatitis. Ulcerations caused by herpes increase the risk of infection from human immunodeficiency virus (HIV). Concomitant herpes infection in some way enhances the speed with which HIV causes immunodeficiency.

CLASSIFICATION

1) Stomatitis: labial herpes or herpetic gingivostomatitis
2) Genital: balanitis or herpetic vulvovaginitis
3) Other: Herpetic proctitis and perianal herpes, keratoconjunctivitis, perinatal herpes and disseminated, panadizo, and herpectic eczema.

CLINICAL FEATURES

The incubation period in the primary infection is 2, 3, and up to 20 days. There are premonitory symptoms 24 h before in 50-70% of the patients such as paresthesias or a burning sensation. The episodes are characterized by one or several vesicles grouped in bunches over an erythematous base accompanied with pruritus or mild burning. Sometimes the lesions become pustular, ulcerate and form meliceric crusts (Fig. 53.1). Regional adenopathy and generalized symptoms may occur. The onset is acute and resolves spontaneously in 1-2 weeks. Sometimes the lesions appear only one time, but usually, especially with HSV-2,

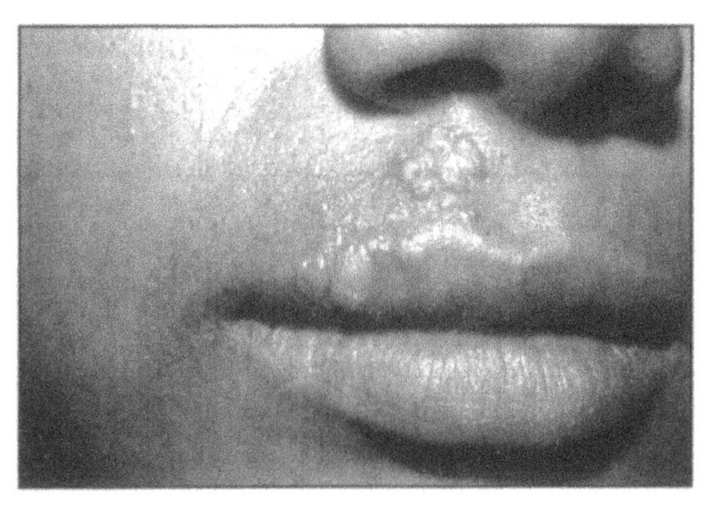

Fig. 53.1. Herpes simplex.

they occur 3-8 times the first year, 53% each month, 33% every 2-4 months and 14% more sporadically. The primary episode is the most serious, with erythema, edema and even necrosis. It lasts about 2-6 weeks, and it is accompanied by local and systemic symptoms like headache, malaise and fever. Labial herpes is generally localized at the margin between the skin and mucosa. It predominates on the lips or close to the mouth, but herpetic gingivostomatitis and geometric herpetic glossitis have been described. Genital herpes affects the glans or vulva (Fig. 53.2). Perianal and rectal herpes is seen in homosexuals. It is accompanied by tenesmus and anal discharge and is rarely complicated by urinary retention. Herpetic whitlow in children affects fingers by autoinoculation, almost always arising from an oral infection. In adults, it may be caused by HSV-2, and it follows digital-genital contact. There may be involvement of any part of the body such as cheeks, thighs, and buttocks. Neonatal herpes is uncommon. It is almost always due to HSV-2 (75%). It is disseminated and severe in 75% of cases. Sometimes it is confused with impetigo. There may be fever, generalized symptoms and systemic involvement (central nervous system, kidney, spleen or respiratory tract). It occurs more often in premature babies at the time of vaginal delivery in mothers with genital HSV infection by early membrane rupture or by contamination from nursing personnel. The risk of neonatal herpes decreases with caesarian section with intact membranes or within 4 h of membrane rupture. Herpetic eczema, or Kaposi's varicelliform-like eruption, is a severe, disseminated, vesicular and pustular form that may be lethal. Individuals with atopic dermatitis, congenital icthyosiform erythroderma, or Darier's disease seem to be the most frequently affected. In patients with HIV, skin ulcerations are extensive, especially in the anogenital area, and may extend to deep tissues. As a rule, in the presence of AIDS all chronic ulcers

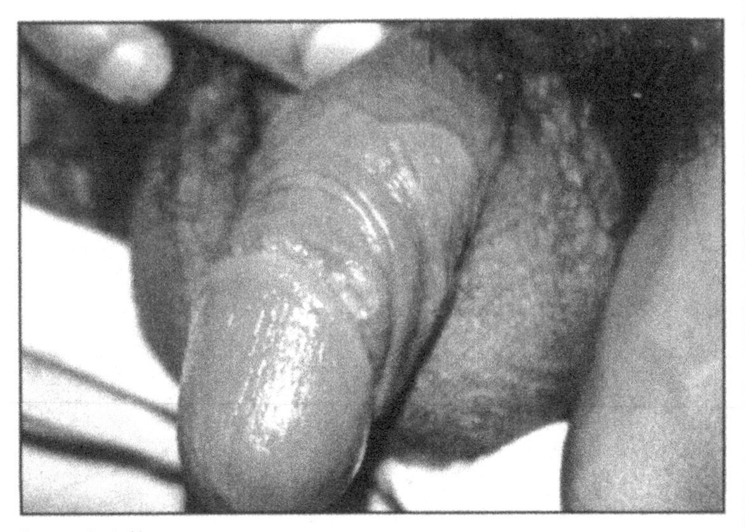

Fig. 53.2. Genital herpes.

are herpetic until proven otherwise. Viscera may be involved in the forms of esophagitis, pneumonia, hepatitis or encephalitis. Herpetic meningoencephalitis is more frequent than it was thought to occur, but it is rarely diagnosed.

LABORATORY DATA

A biopsy may be performed in immuncompromised individuals and in individuals with chronic ulcers. It is best to place the specimen in Bouin solution. An intraepidermal vesicle with ballooning and reticular degeneration is found as well as viral and nuclear changes. The Tzanck's cytologic smear shows giant, multinucleolated cells with nuclear inclusion bodies (Fig. 53.3). Serum antibody and antigen levels may be determined, as well as viral cultures, nucleic hybridization studies and the polymerase chain reaction (*J Am Acad Dermatol* 1995; 32(5 Pt 1):730-33). The direct immunofluorescence is a simple technique, sensible, specific and of low cost to identify the herpes virus. The studies must be done in samples of early lesions. Now under investigation is infrared thermography to count activity during the prodromal stage.

TREATMENT

No treatment completely eradicates latent infection. Treatment is symptomatic and prophylactic. In some cases the psychological factor is important. A placebo can decrease the frequency of recurrences 50%. Some medications decrease the

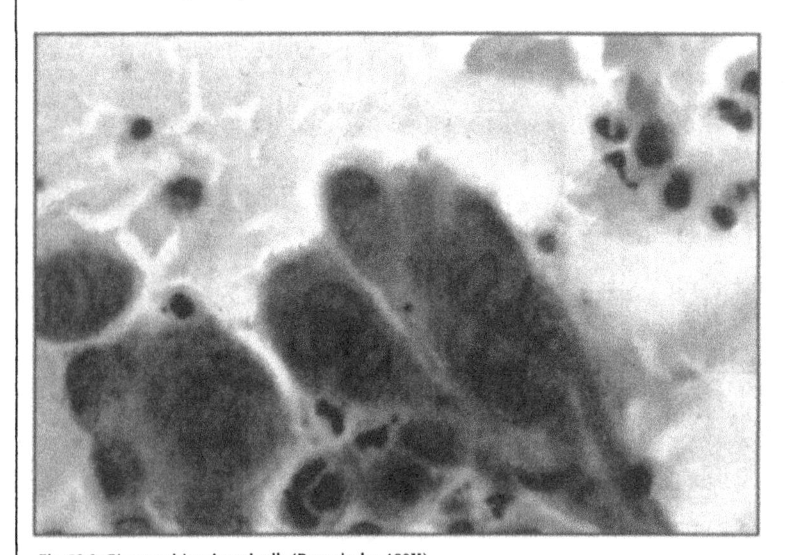

Fig. 53.3. Giant multinucleated cells (Papanicolau 100X).

course, but they do not reduce the frequency of recurrences except when administered continuously. On the skin, compresses of chamomile tea are suggested, and also lead or aluminum acetate, or other drying antiseptics. Drying powders like zinc oxide can be employed. Acetylsalicylic acid or indomethacin are indicated for pain. To shorten the course, vesicles may be open and ether or alcohol 60% may be applied. The following medications sometimes give acceptable results: iodoxiuridine (IDU), adenosine arabinoside, ribavirin, methisoprinol, vitamin C, 2-deoxi-D-glucose, zinc sulfate in 4% solution, topical or oral acyclovi, photodynamic therapy, human gammaglobulin, and levamizole (2-5 mg/kg/day in weekly dosage for six months), n-docosonal in cream 10%. Acyclovi is most effective when administered during the prodrome or the first 48-72 h. It decreases symptoms, contagiousness and healing time. It is recommended in the primary infection or if there are more than three episodes per year. The oral dosage is 200 mg 5 times a day for 10 days in primary infection and for 5 days in recurrences. For prophylaxis acyclovi 400 mg twice a day for a minimum of one year is recommended. In patients with HIV, 400 mg 5 times a day for 10 days and then 3-4 times a day for a prolonged time is recommended. Intravenous administration is reserved for severe infections in hospitalized patients. Also recommended is valacyclovir 1000 mg bid for primary infection or 500 mg bid for recurrences. Both regimens for 5 days and once a day for prophylaxis are recommended. In individuals resistant to acyclovi and immune compromise, foscarnet may be administered. Glucocorticoids are contraindicated, especially in herpetic ophthalmitis which can be treated with IDU in 0.1% ointment or vidarabine. For active genital herpes, abstinence or protected sex is recommended. Spermicide foam is viricidal in vitro. Individuals with genital herpes suffer from fear of rejection, low self-steem and guilt. Psychosexual treatment is important. In Germany, vaccines for HSV-1 and HSV-2 have been synthezised and heat-inactived (Lupidon-H and G).

Aphthous Ulcers

Recurrent aphthous stomatitis are oral ulcers, single or multiple, painful and recurrent, 2-3 mm to more than 1 cm in diameter. They are classified in larger, smaller, and herpetiform. Prevalence in the general population is 5-66% (average, 20%). They are frequent in North America. The larger and smaller aphthous ulcers occur equally in both sexes; they begin at 10-19 years of age. Herpetiform ulcers begin at 20-29 years of age, and they predominate in women. The etiology is unknown. Hereditary factors has been proposed (there may be family antecedents, but the relation with HLA is not clear), dietary, nutritional (zinc, folic acid, and vitamin B12 deficiency), endocrine, infectious, immune (atopy) and traumatic, but most frequently they are idiopathic. Aphthous ulcers may be a manifestation of an inflammatory, autoimmune disease (Sweet syndrome, cyclic neutropenia, periodic febril syndrome and pharyngitis) and they might even be an incomplete form of Behçet's disease. Aphthous ulcers begin during childhood or adolescence. They present as one or several round or oval, painful ulcers. They may be covered with a pseudomembrane and surrounded by erythema. They recur with intervals from days to months. Small apthous ulcers are more frequent;

one to five lesions involve the lips and tongue, and they do not scar (Fig. 53.4). Large aphthous ulcers are rare, but serious. There may be one to ten up to 3 cm in diameter. They may affect the pharynx and palate and may leave a scar. They are chronic and may persist for more than 20 years. Herpetiform aphthous ulcers appear in groups of 10-1000, are 1-2 mm, and coalesce. Small aphthous ulcers last from 4-14 days and recur in 1-4 months; other aphthous ulcers last more than one month and they generally recur in months. Histopathology reveals an inflammatory mononuclear infiltrate ($CD4^+$) during the pre-eruptive phase; in the ulcerative phase $CD8^+$ T cells appear and are replaced by $CD4^+$ cells during the healing stage. There may be leukocytes and an elevation of IgG and complement. There is no adequate treatment; it is just symptomatic, but it decreases recurrences. The following treatment is used: local glucocorticoids, antimicrobials like tetracycline, oral rinse with chlorhexidine 0.2-1%, or benzidamine cream with 5-aminosalicylic acid 5% three times a day. In severe cases, systemic glucocorticoids, colchicine 1-2 mg/day and thalidomide 100-200 mg per day are administered. Recently pentoxifylline 400 mg tid for 3 months has been used (*J Am Acad Dermatol* 1995; 33:680-682). Levamizole, transfer chromatic factor, gammaglobulin, sodium chromate and diaminodiphenylsulfone have also been employed with variable results.

Selected Readings

1 Biagioni PA, Lamley PJ. Electronic infrared thermography as a method af assessing herpes labialis infection. Acta Derm Venereol 1995; 75:264-68.

2 Nahass GT, Mandel MJ, Cook S et al. Detection of herpes simplex and varicella-zoster infection from cutaneous lesions in different clinical syages with the polymerase chain reaction. J Am Acad Dermatol 1995; 32(5 Pt 1): 730-33.

3 Wahba-Yahav AV. Pentoxifylline in intractable recurrent aphthous stomatitis: An open trial. J Am Acad Dermatol 1995; 33:680-82.

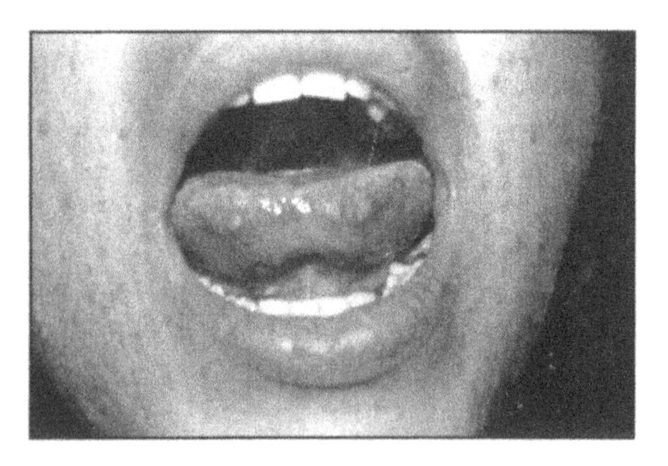

Fig. 53.4. Aphthous ulcers.

Varicella/Zoster

Roberto Arenas

Varicella is a primary infection caused by varicella-zoster virus (VZV). It is very contagious and self-limited. It is characterized by a centripetal eruption of vesicles with an erythematous base. The lesions evolve into umbilicated pustules and crusts that can leave small pits. Herpes zoster (shingles) is caused by reactivation of the varicella-zoster virus. It is manifested by hyperesthesia and pain with groups of vesicles over an erythematous base. It is self-limited and in adults and elderly can cause postherpetic neuralgia.

GEOGRAPHIC DISTRIBUTION

Varicella is endemic, but it generally occurs in an epidemic form, mainly at the end of winter or at the beginning of spring. In the United States 3 million cases a year are reported affecting both sexes equally. Ninety percent of cases occur in children 2-10 years of age. Varicella in adolescents and adults has a higher morbidity. When varicella is complicated by pneumonia, mortality is 10-30% and even higher in immune-compromised individuals. The risk in pregnant women is low because 90-95% of the population has antibodies to VZV. Yet congenital varicella occurs in 1.6 per 100,000 births. Zoster infection may occur sporadically all year long. The annual frequency is 0.74-3.4 cases per 1,000 habitants. In the United States, 3,000-4,000 patients a year are hospitalized, and there are 100 deaths. It affects all races, and there is a slightly higher incidence in men. It is rare in children less than two years of age, and the rate of maternal-fetal contagion is 2.5%. It occurs in 2-25% of immune-compromised individuals.

ETIOLOGY

VZV is acquired by inhalation or direct contact, and the initial areas of infection are the conjunctivae or respiratory mucosae. Cellular immunity controls VZV and leads to latency, but a decrease of T lymphocytes (neoplasm, transplantation, AIDS, aging) is related to reactivation. There are two cycles of replication, first in ganglia and then in the liver, spleen and other organs. It is highly contagious in the prodromal and vesicular stages, especially five days before or after the beginning of the exanthem. In herpes zoster (reactivation) there is a stage of replication that causes adenitis and transitory viremia with an acute inflammatory response and neuronal necrosis that causes neuralgia. The virus is then released causing similar cellular changes as seen in varicella. Fetuses of pregnant women with varicella in the first trimester may have congenital varicella syndrome. Neonatal varicella occurs when maternal varicella onset is 5 days before to 2 days after delivery. Neonatal mortality is 30%. If the varicella appears in the second trimester there is no embryopathy, the virus remains latent in the fetus, and it may give rise to herpes zoster before the child is 2 years old. On contact with individuals with herpes zoster or varicella, non-immune individuals may develop varicella but not zoster.

CLINICAL FEATURES

In varicella the incubation period is 10-20 days with an average of two weeks. It begins like a mild maculopapular exanthem, almost always followed by fever and malaise that lasts 24-48 h. In 4-5 days a centripetal, pruritic eruption appears on the trunk, face, hairy skin and extremities (Fig. 54.1). It is characterized by isolated vesicles over an erythematous base that give the impression of "a drop of dew over a rose petal". The vesicles soon become pustular and in 2-3 days form crusts, and when they fall off, may leave a small depression. Varicella may affect conjunctivae and oral and vulvar mucosae. In adults varicella is more serious than it is in children. In undernourished and immune-compromised individuals , hemorrhagic blisters and high fever may occur. Complications and reactivation in patients with HIV depend on the nature of the severity of the immune defect. Complications are rare in healthy children, but there may be associated cutaneous streptoccocal or staphyloccocal infections, thrombocytopenic purpura, pleuritis, pneumonia (1 per 400), encephalitis (1 per 1000), septicemia and Reye's syndrome related to acetylsalicylic acid.

Herpes zoster is localized to a single dermatome, rarely to two. It frequently involves intercostal (53%) and superior lumbar (T3-L2) nerves segments. It generally begins with hyperesthesia, pain or a burning sensation along the distribution of the nerve involved; 2-4 days later lesions appear that do not affect the midline (Fig. 54.2). In 12-24 h there are few or abundant vesicles 2-3 mm in diameter over an erythematous base. Most coalesce, dry-up, become pustular and leave golden yellow crusting (7-10 days). Blisters, purpuric lesions, necrosis and gangrenous scars may occur, which will heal slowly and leave hypo- or hyperpigmented scarring (Fig. 54.3). Sometimes there are no cutaneous lesions, "zoster without herpes". Usually there is regional adenopathy and generalized symptoms such as asthenia, headache and fever, especially in children. It affects the trigeminal nerve in 10-17% of cases, particularly the ophthalmic branch. When the nasociliary branch is involved, vesicles appear on the tip of the nose (Hutchinson sign). Ocular complications occur in 75% of cases, e.g., keratitis (50%), anterior uveitis, iridocyclitis, and panophthalmitis. Involvement of the facial and vestibulocochlear nerves may cause Ramsay-Hunt's syndrome (Bell's palsy, temporomandibular arthralgia and vertigo). In children mild pain presents early on but disappears rapidly. However, in individuals older than 40 there is an acute phase, "zoster's

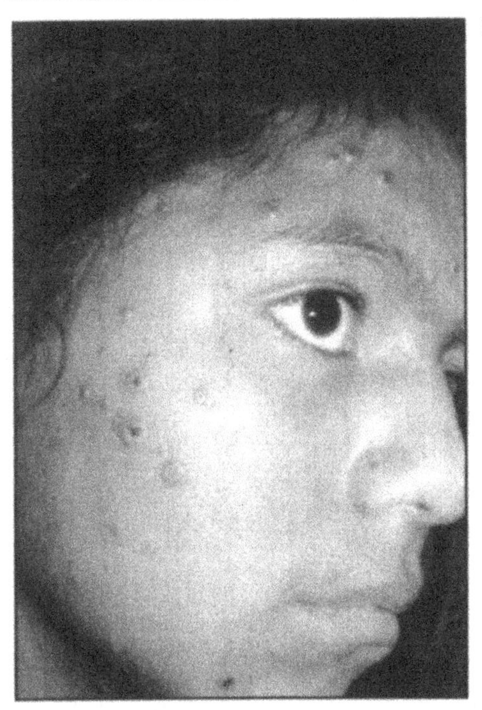

Fig. 54.1. Varicella.

algesia", marked by intense and persistent pain (50%) or postherpetic neuritis (10-15%). The pre-eruptive pain may simulate myocardial infarction, pleuritis, cholecystitis, appendicitis or other painful illnesses. The evolution of the dermatosis is acute or subacute and it lasts 2-3 weeks. It may recur (1%), especially in immune compromised individuals. Under these circumstances the generalized form may occur with atypical dissemination of zoster or the zoster varicelliform (40%) with ocular or neurological complications (Fig. 54.3). The congenital varicella syndrome may cause hypoplasia of the extremities, scars along the course of involved dermatomes, ocular and central nervous system damage (encephalitis and mental retardation) and death.

LABORATORY DATA

Laboratory examination is not required. On histological examination there are intraepidermal vesicles formed by ballooning degeneration, with giant multinucleated cells and epidermal cells with intranuclear eosinophilic inclusions. Tzanck's cytologic smear shows large multinucleated cells with nuclear inclusion bodies. On electron microscopy viral particles are observed.

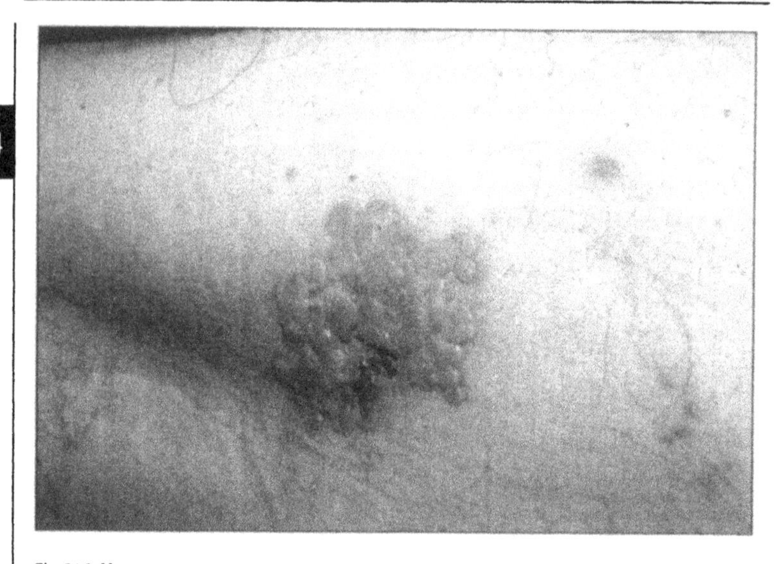

Fig. 54.2. Herpes zoster.

TREATMENT

Treatment of varicella is symptomatic. Standard measures include baby powder, zinc oxide, starch, oat's bath and baths without soap. Antihistamines or acetaminophen may be administered. In individuals in poor general health, antivirals or human gammaglobulin 16.5% may be administered, 0.12-0.22 ml/kg (1.5 ml/10 kg IV q 5d). In immune-compromised individuals or in neonatal varicella, acyclovir 5-10 mg/kg/d every 8 h for 5 days is recommended. In immune-compromised children 2-13 years old, 20 mg/kg of acyclovir may be given—not to exceed 800 mg—qid for 5 days, and in adults 800 mg 5 times a day for 7-10 days or valaciclovir 1000 mg bid for 5 days. A vaccine of attenuated live virus is available, but indications for use are not yet clear. Treatment is not necessary in children and youngsters when zoster is present. If there is additional infection, a mild antiseptic is recommended. The following is recommended to alleviate the symptoms: compresses with lead subacetate and equal parts of distilled water, Burow's solution (calcium acetate and aluminum sulfate) or chamomile tea, and unscented powder. Analgesics like acetylsalicylic acid 500 mg are suggested, or the combination with propoxyphene and caffeine bid. The use of glucocorticoids for treating neuritis is moot, e.g., during the first 5-10 days methylprednisolone 40-60 mg/day for 8 days may be administered and tapered progressively over three weeks. Also neuritis improves with the following: carbamazepine 200 mg/day in 2-3 divided doses, as well as amitryptyline, fluphenacine, or colloidal solution (protamide), 13.6 mg IM per day for 5 days. With severe pain, nerve blocks, alcohol instillation of the nerve may be performed. For

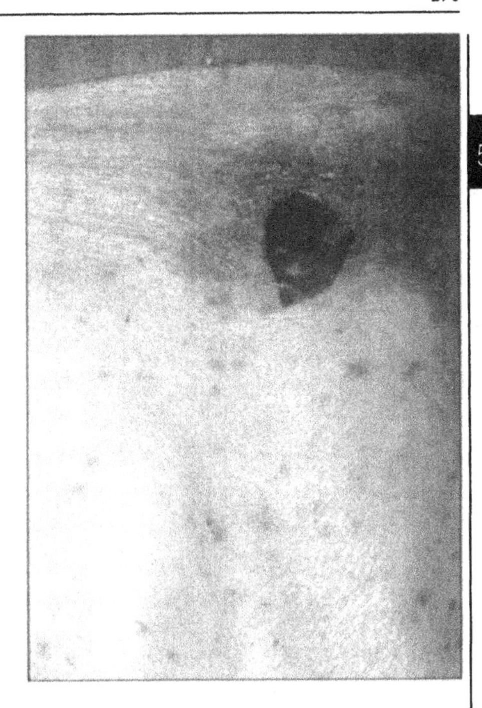

Fig. 54.3. Zoster and varicelliform eruption.

serious illness or in the presence of immunosuppression the following antivirals may be used: ribavirin, 400 mg po and cream 7.5% tid for six days, adenine arabinoside (vidarabine) 10 mg/kg/day intravenously (0.5 mg/ml IV over 12 h for 7 days requires hospitalization), acyclovir 800 mg po q4h for 7-10 days or in immunosuppressed individuals 500 mg/m^2 body surface IV q8h for 7-10 days, or 10 mg/kg IV over 1 h q 8h for 7 days. The use of 5% acyclovir cream as adjuvant is controversial. Sometimes acyclovir is combined with prednisone. The following have been recently used: valacyclovir 500-1000 mg PO tid for 7-10 days (it is becomes acyclovir after administration at therapeutic levels similar to those obtained with intravenous acyclovir, with few side effects) (*J Med Virol* 1993; (suppl 1): 150-53), and famcyclovir (pencyclovir), 250-500 mg tid for 7 days. Transforming growth factor has been used, but it is not yet available. The following are ineffective: vitamin B12, emetine, griseofulvin, imipramine, cimetidine and dihydroergotamine. If there are ocular lesions, they must be treated by an ophthalmologist. Mydriatics and acyclovir are recommended, but there is controversy regarding the use of glucocorticoids. Sorivudin 40 mg qd is under investigation in immunocompromised individuals (*J Infect Dis* 1996; 174(2): 249-55), and the use of brivudine PO 125 mg q6 h for 5 days (*J Med Virol* 1995; 46(3): 252-7) and a varicella-zoster live attenuated vaccine has been reported.

Selected Readings

1 Enders G, Miller E, Cradock-Watson J et al. Consequences of varicella and herpes zoster in pregnancy: Prospective study of 1739 cases. The Lancet 343:1547-50.
2 Memar OM, Tyring SK et al. Antiviral agents in dermatology: Current status and future prospects. Int J Dermatol 1995; 34(9):597-606.
3 Rockley PF, Tyring Sk. Pathophysiology and clinical manifestations of varicella zoster virus infections. Int J Dermatol 1994; 33(44):227-32.

Viral Warts/Focal Epithelial Hyperplasia

Roberto Arenas

Viral warts are benign, epidermal tumors. They have a low risk of transmission and are caused by human papillomavirus (HPV). They are characterized by raised, warty or vegetative lesions and are classified as flat (verruca plana), common (verruca vulgaris), plantar or acuminata. They are self-limited and remit spontaneously without leaving a scar.

GEOGRAPHIC DISTRIBUTION

These warts affect all races, both sexes and all ages. They are among the most common dermatoses, and have been found in 7.2% of students in Germany. Common warts, 70%, and flat warts, 3.5%, predominate in children and adolescents. They decrease with age. Warts can also be an occupational disease, .e.g., in butchers (HPV-2 and -7). Plantar warts occur in 6.5-24% of cases. Condyloma acuminata occur in adolescents and adults. In the United States they are considered the most common sexually transmitted disease. They affect 15-20% of people 15-49 years of age. In AIDS they are present in 5-27% of cases.

ETIOLOGY

Human papillomavirus is a double-stranded DNA virus with an icosahedral capsid of 72 capsomeres and of 50-55 nm that belongs to the Papovaviridae family, Papova group and to the papilloma subgroup. More than 55 types of HPV have been recognized (Table 55.1). The virus only replicates in well-differentiated keratinocytes. It causes epithelial proliferation. After infection it can become latent and then reactivate. Incubation varies from weeks to a year. The warts are manifested by pleomorphic tumors of varied form in skin and mucosa depending upon the location (Fig. 55.1). The lesions are benign and auto-inoculable. At the site of trauma they can present in a linear array in accordance with the Koebner phenomenon. They are transmitted directly from one individual to another or indirectly. Condyloma acuminata is transmitted by sexual contact, but the transmission of anogenital condyloma in children is difficult to determine. The virus

Table 55.1. Correlation between clinical lesions and human papillomavirus (HPV) types (modification of Lowy DR, Androphy E. Warts. In: Fitzpatrick TB, Eisen AL, Wolff K et al. Dermatology in General Medicine. 4th ed. 1993:2611-2621).

Clinical lesions	HPV type
Common warts	2, 1, 3, 4, 7, 10, 16, 29, 41, 60, 65
Flat warts	3, 10 (1 an 4?)
Plantar warts	1, 2, 3, 4, 10
Acuminated condylomata	6, 11
Laryngeal papilloma	6, 11
Oral leukoplakia	16, 80
Atypical lesions in the immunocompromised	20, 27, 49

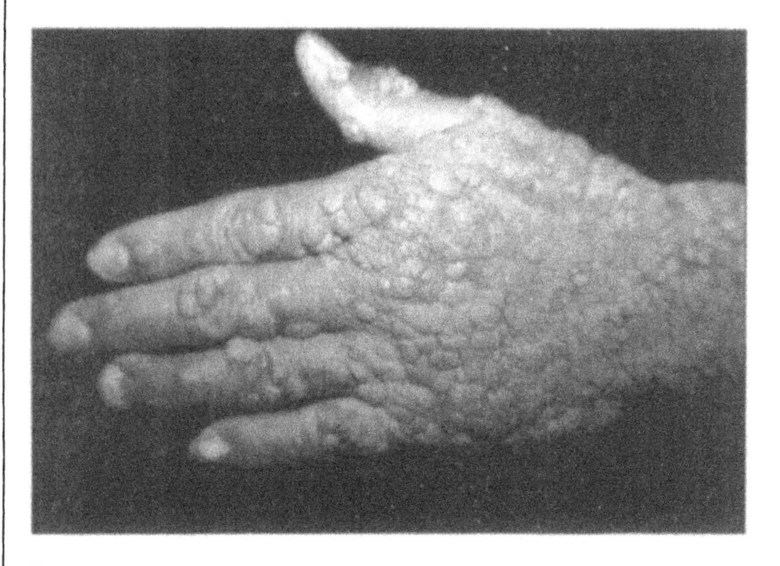

Fig. 55.1. Verruca vulgaris.

enters through superficial abrasions in the skin, and heat and humidity favor its growth. Only in a very small percentage of cases do lesions become dysplastic or neoplastic depending upon the type of HPV and genetic and environmental factors. The host defense against HPV is not well understood. Perhaps it depends on cellular immunity, because in the presence of decreased cellular immunity, the incidence of warts is greater, involvement is more extensive and there is a higher risk of malignant transformation.

CLINICAL PICTURE

The course of all the clinical varieties is chronic and unpredictable. Lesions can last months or years. In the absence of trauma, they usually resolve spontaneously without a scar. In the presence of lymphoma, in transplant recipients (15-87%) (HPV-2 and -4 or -3 and -5) and in persons with AIDS, warts are more numerous, generalized, exuberant and resistant to treatment. They can simulate verruciform epidermodysplasia and become malignant.

COMMON WARTS (VERUCCA VULGARIS)

They can occur on any part of the skin but predominate in exposed areas, mainly the face, forearm and back of the hand. They are single or multiple, isolated or confluent, hemispheric, well-limited and 3-5 mm up to 1 cm, rough, dry, skin-colored or gray and asymptomatic. They involute within an average of two years (Fig. 55.1). Generally they are sessile, but on the eyelid they are filiform and can be associated with conjuntivitis and keratitis. When they occur on the edge of a fingernail, they cause ungual dystrophy and are painful. If they affect genitals, they are more keratosic and less vegetative than condylomas. When they are localized on the inner surface of the lips or the oral mucosae, they are called papillomas (HPV-6 and -11), are vegetative and single or multiple. They can grow rapidly and constitute a picture known as florid oral papillomatosis.

FLAT OR JUVENILE WARTS (VERRUCA PLANA)

They are small, round or polygonal, 1-4 mm lesions confused with papules. They are slightly elevated, skin-colored, gray or mildly erythematous or pigmented, and are asymptomatic. They predominate on the face and back of the hands.

They can be few or abundant, and sometimes, because of the Koebner phenomenon, they have a linear configuration (Fig. 55.2). When involuting, they are more prominent due to inflammation, i.e., erythema and pruritus, and sometimes have halos of hypopigmention. They can be related to common warts and laryngeal papilloma.

PLANTAR WARTS

They are localized on the soles or toes. They are single or multiple, with up to 40-50 grouped lesions (HPV-2) or in a mosaic array; individual lesions are 0.5-1 cm in diameter. They are white-yellow with some dark or hemorrhagic areas and tender because the wart acts like a foreign body (Fig. 55.3). In 30-60% of cases, lesions involute within one year, faster in children.

CONDYLOMA ACUMINATA

They predominate on the genitals, in men on the balanoprepucial area and in women on the labia majora; they also can appear on the anus, rectum and urinary meatus. They are gray, vegetative lesions with a moist surface. They can appear like a cauliflower, mainly in women (Fig. 55.4). They are chronic and are sexually transmitted in up to 75% of cases. They are best seen with the application of acetic acid or a combination of methylene blue and toluidine blue.

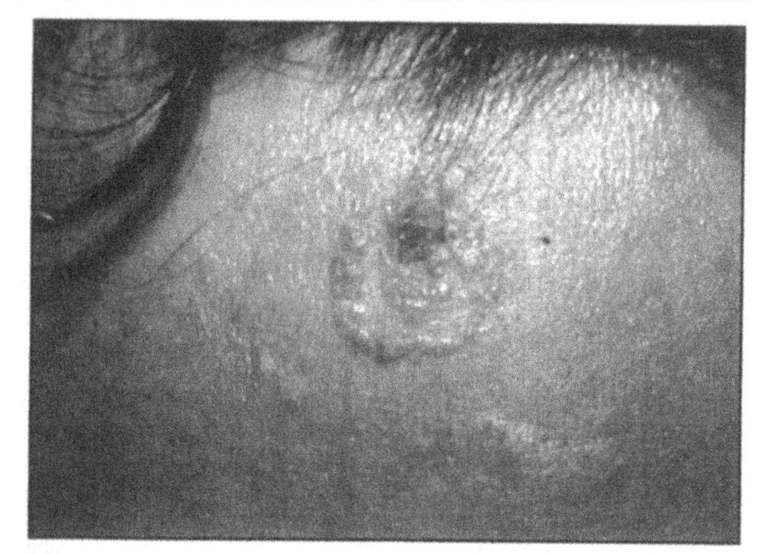

Fig. 55.2. Verruca plana.

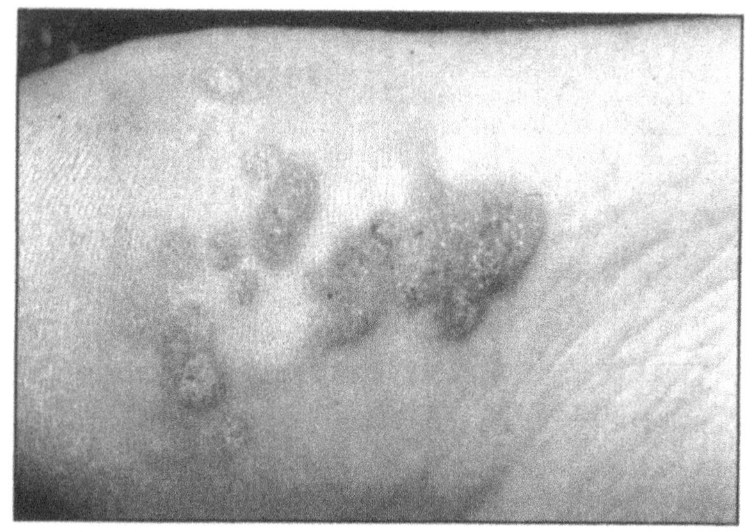

Fig. 55.3. Plantar warts.

LABORATORY DATA

Biopsy is not necessary. There are varying degrees of epidermal hyperplasia, e.g., hyperkeratosis with parakeratotic areas (especially in plantar forms) and acanthosis and granulomatous reaction. Cytopathic changes manifest as ballooning

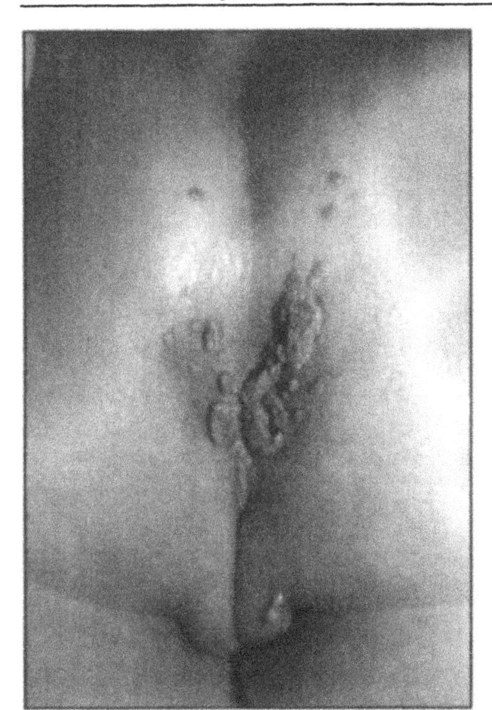

Fig. 55.4. Condyloma acuminata.

or reticular degeneration of epidermal cells with intracytoplasmic inclusions. There can be papillomatous (except in plantar warts). In condylomas there can be pseudoepitheliomatotous hyperplasia. Clinical and histophatologic examinations are adequate for diagnosis, but in experimental studies or for medical-legal reasons, electronic microscopy, ELISA, DNA hybridization and polymerase chain reaction (PCR) can be useful.

TREATMENT

There is no single, definitive treatment. Even placebos yield good results, and hypnosis has been used successfully. Surgery or electrocautery is reserved for single lesions and/or for areas not easily visible because these treatments leave scars. A keratolytic such as salicylic acid ointment 1-4% can be applied simply or can be applied combined with equal parts of lactic acid in four parts of colodion. In select cases salicylic acid 10-40% or topical retinoic acid can be used. With caution, cryotherapy, with liquid nitrogen for 10-15 sec, can be employed with special applicators or simply with a swab. In general, three or four sessions are necessary. It is very painful. Some people advocate injecting intralesional tetracycline with xylocaine. Podophillin

20-50% and podophilotoxin 0.5% are useful for anogenital warts, as are monoacetic and trichloroacetic acids and silver nitrate. Bleomycin, 1 mg/1 ml intralesionally in one or two applications of 0.2-1 ml, have been used. The successful use of dinitrochlorobenzene (DNCB) has been reported as the cantharidin under an occlusive dressing for 24-48 h or applied in acetone and colodion for a few hours. Treatment with 5% cream of 5-fluorouracil applied in an occlusive dressing, CO_2 laser, alpha interferon or oral retinoid has been reported. Antivirals or oral cimetidine and griseofulvin have been employed.

Focal Epithelial Hyperplasia

Heck's syndrome is a multifocal epithelial hyperplasia caused by HPV (HPV-13 or 32) characterized by multiple, circumscribed, papillomatous lesions of the mucosae that occur most often in malnourished individuals or in persons with a genetic predisposition. The syndrome has been described mainly in Eskimos and American Indians as well as in chimpanzees and rabbits. It is more frequent in children and adolescents, 97%, especially around the age of 11. It predominates in females with a rate of 2.2:1. Familiar cases account for 28%.

CLINICAL PICTURE

Oral lesions are abundant and are localized mainly in the movil mucosa, and non-keratinizing tissue such as lips and buccal mucosae. On the tongue, lesions appear in the lateral edges. It does not affect the soft palate or pharyngeal mucosa. The lesions are circumscribed, vegetative and asymptomatic (Fig. 55.5). In 50% they interfere with eating, and patients bite themselves frequently. Papulonodular and papillomatoid types have been described.

LABORATORY DATA

Biopsy reveals acanthosis, hyperkeratosis and cytopathic changes characteristic of cells with balonoide degeneration. There are numerous mitotic figures in the nuclei, and connective tissue is abundant.

TREATMENT

There is no specific treatment. Cryosurgery and vitamin E are recommended. Recently, vaccination with BCG has been used with success.

Selected Readings

1 Carlos R, Sedano HO et al. Multifocal papilloma virus epithelial hyperplasia. Oral Surg Oral Pathol 1994; 77:631-5.

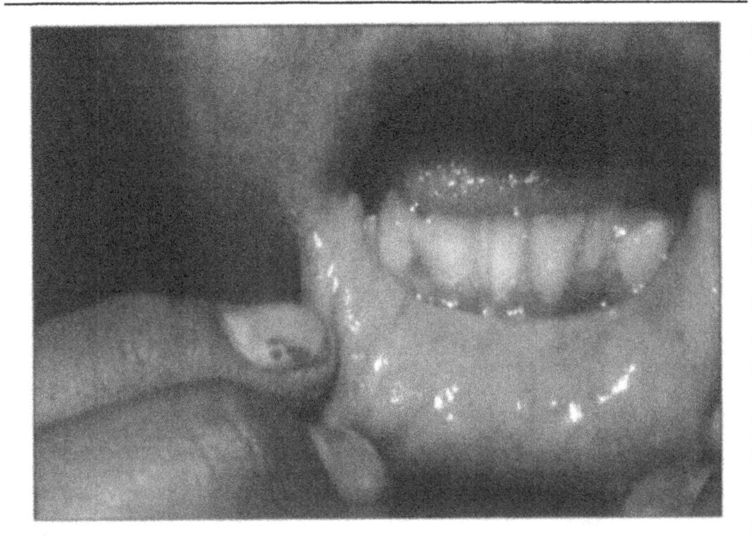

Fig. 55.5. Focal epithelial hyperplasia.

2 Fang BS, Guedes AC, Munoz LC, Villa LL et al. Human papillomavirus type 16 variants isolated from vulvar Bowenoid papulosis. J Med Virol 1993; 41(1):49-54.

3 Phelps WC, Alexander KA et al. Antiviral therapy for human papillomaviruses: Rationale and prospects. Ann Intern Med 1995; 123:368-82.

4 Sichiazza L, Occella C, Bleidi DE, Rampini E et al. Sul trattamento delle verruche volgari. Giorn It Derm Vener 1985; 121(3):43-45.

Molluscum Contagiosum

Roberto Arenas

This is a benign dermatosis with a viral origin. It is spread by autoinoculation, by direct contact or fomites. It is characterized by 2-3 mm, umbilicated, isolated or abundant lesions.

GEOGRAPHIC DISTRIBUTION

This disease occurs worldwide. It affects any race, age and sex. It predominates in children 10-12 years of age, sexually active adults and immune-compromised individuals. It is observed in 5-18% of individuals with AIDS. It is a sexually transmitted disease and is more common in humid and hot climates and where hygiene is poor. In some places it appears in 35% of families.

ETIOLOGY

It is produced by one of the biggest pox viruses (150-300 mm). The pox virus contains double-stranded DNA . The virus has two subtypes, MCV-1 and MCV-2. The lesions produced by both are indistinguishable. MCV-1 accounts for 76-96% of cases. MCV-2 is sexually transmitted, predominates in adult men and HIV positive individuals. The MCV-1:MCV-2 ratio in children is of 43:1, in women 11:1 and in men 1.3:1. In many patients a T lymphocyte deficiency or other immune abnormality is present. The role of Langerhans cells in infection is unknown. Viral transformation occurs on the deep portion of the Malpighi layer. In the granular layer there is a change of color, from eosinophilic to basophilic. Electron microscopy and immunohistochemistry reveal electrodense regions in the granular cells that contain tricohialin and electrolucid regions that contain filagrin (*Int J Dermatol* 1996:35(2):106-108). The first perhaps demonstrates the keratins related to hyperproliferacion (K6/K16) and the second keratins of differenciated epidermis (K1/K10).

CLINICAL FEATURES

The incubation period varies from several weeks to 50 days. In children lesions can occur on any part of the skin, usually on the face, trunk, and limbs; in adults they occur on the lower abdomen, thighs, pubis, penis, and perianal region.

Tropical Dermatology, edited by Roberto Arenas and Roberto Estrada. ©2001 Landes Bioscience.

Lesions rarely appear on palms of the hands, soles, mouth and around the eyes. Generally lesions are 1-3 mm and up to 1 cm, but are rarely larger. They are semi-spherical, hard, skin-colored or white-yellow, translucid and umbilicated. When squeezed they discharge a grumous material (Fig. 56.1). Usually, there are less than 30, but hundreds occur and can even form plaques. Sometimes they are inflammatory, and in 10% a perilesional eczematoid reaction has been observed. The course is chronic and asymptomatic and can persist for years. There can be mild pruritus. The scratching causes autoinoculation. Rarely lesions resolve spontaneously but recur. In individuals with AIDS, the number and size of lesions increase rapidly and they are confused with basal cell carcinoma, atypical mycobacteriosis, cryptococcis and cutaneous histoplasmosis.

LABORATORY DATA

Biopsy shows epidermed lobules with a stretch pore. There is a keratinous crater, with basophilic or amphophilic molluscum bodies (of more than 35 mm diameter). Also there are intracytoplasmic inclusions within keratinocytes. In 17% of cases there is an inflammatory reaction.

TREATMENT

When the lesions are limited, surgical excision with a needle or cautery and curettage can be employed with application of EMLA cream (a eutectic mixture of lidocaine and prilocaine), 1-2 h before the procedure. Similarly, cryosurgery

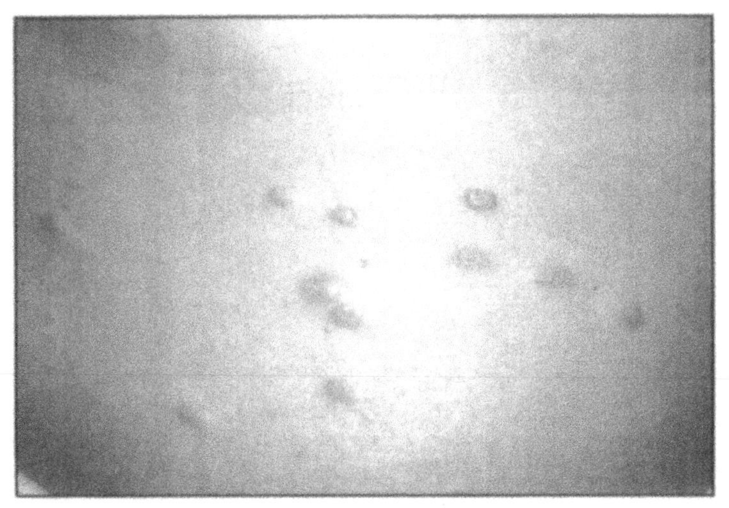

Fig. 56.1. Molluscum contagiosum.

can be performed following application of trichloractic acid. The effectiveness of orally administered griseofulvin or topical 5-fluorouracil is not very well demonstrated. The following is also recommended: cantharidin 0.9% in a solution of acetone and colodion applied under an occlusive dressing for 6-10 h, or tretinoin 0.05-0.1% topically. Podophillin, silver nitrate, iodine tincture, phenol and methiosazone are less effective.

Selected Readings

1 Gottlieb SL, Myskowski PL et al. Molluscum contagiosum. Int J Dermatol 1994; 33(7):453-461.
2 Koopman RJJ, Van Merrienboer FCJ, Vreden SG, Dolmans WMV et al. Molluscum contagiosum: A marker for advanced HIV infection. Br J Dermatol 1992; 126:528-29.
3 Uemure T, Kawashima M et al. Molecular epidemiologic analysis of Japanese patients with molluscum contagiosum. Int J Dermatol 1996; 35(2):99-105.

Dengue

Roberto Arenas

Epidemic disease caused by a arbovirus of the Flaviviridae family; it is transmitted by mosquitos such as *Aedes aegypti* and *A. albopictus*. The clinical manifestations vary according to the characteristics of the host. The virus has four serotypes. The infection can be asymptomatic or can manifest with fever, (dengue fever), malaise, arthralgia and a cutaneous eruption; sometimes a severe form is present (febrile hemorrhagic dengue).

GEOGRAPHIC DISTRIBUTION

Dengue occurs worldwide but is most common in the tropics and subtropics due to the favorable conditions for the vector. In the last 15 years the incidence of dengue has increased worldwide due to uncontrolled urbanization, population movement and poor control of the vectors. The disease is endemic in the East and West Africa, India, temperate zones of North America, Mexico (Chiapas, Yucatan), and also in others Latin American countries (El Salvador, Caribbean Islands, Brazil, Argentina, Paraguay and Bolivia), the Mediterranean, North Australia, Southeast Asia, China and the Pacific Islands. Hemorrhagic fever has been reported in 12 Asian countries, Pacific Islands, Cuba, Venezuela and Brazil. There is a possibility of an epidemic in the Americas. The virus can be carried by people visiting endemic areas. There are cyclic patterns of transmission, but in endemic areas dengue generally occurs year round. Mortality, especially in children, is 5%.

ETIOLOGY

The virus is spherical with single-stranded RNA. It has four serotypes (DEN1- DEN4) that manifest in the same clinical picture without cross immunity. Generally, the viruses in a specific geographic area share a topotype, although more than one topotype is sometimes found. The vectors that transmit the virus from person-to-person are mosquitoes of the gender *Aedes*, especially *A.* (*Stegomya*) *aegypti*, that have domestic habits and are found from sea level to 1,700 m and reproduce in standing water in urban and suburban areas. *A. polynesiensis* and *A. scutellaris* can act as vectors as well. In the United States, northern Mexico, Santo Domingo, Brazil and Guatemala, *A. albopictus* recently has been reported to be a vector. Transmission is via the female which bites during the day and the virus multiplies in salivaray glands. There can be transovaric and vertical transmission. Man is generally the most important reservoir although in some places the monkey is a resrevoir. After the bite, the virus multiplies in mononuclear cells of lymphatics and disseminates through the lymphatics and blood throughout the reticuloendothelial system and mononuclear cells of Peyer's plaques, timo and skin. After infection, there is lifelong homotypic immunity and heteropic immunity that lasts several months. The classic fever occurs in non-immune populations. Hemorrhagic dengue generally occurs under 2,000 ft in hot and rainy seasons. It occurs at 3-5 year intervals in populations where two or more serotypes are endemic/epidemic. It appears during the second infection with a heterotypic virus as a result of variations in virulence, interaction of the virus with environmental or host factors or by a combination of various risk factors. Hemostatic abnormalities are characterized by increased capillary permeability, thrombocytopenia and coagulapathy. Experimental studies demonstrate that once the virus reaches the lymphatics it disseminates throughout the reticuloendotelial system and then enters the bloodstream. There is endothelial damage with infiltrates and perivascular edema. In serious cases there is extravasation of blood. The factors that favor the presence of the mosquito are accumulation, lack of piped water, inadequate waste disposal and poor home design.

CLINICAL PICTURE

Dengue manifests as classic dengue or hemorrhagic fever. Classic dengue appears in two forms: infantile and adult. The infantile dengue is a febrile illness more or less indistinguishable of other viral diseases. It is characterized by rhinopharyngitis and on some occasions by gastrointestinal symptoms. Adult dengue requires incubation of 2-7 days. It presents with high fever (39-40°C) headache, myalgia, arthralgia and back pain (break-bone fever). In the following 2-6 days, anorexia, nausea, vomiting, colicky abdominal pain, constipation, taste disturbs, cutaneous hyperesthesia and palmar edema develop. This is accompanied by the dengue facies characterized by flushing, palpebral edema and conjunctival injection, retroocular pain, and photophobia. As the symptoms abate a morbiliform or scarlatiniform rash appears. This generalized pruritic rash begins on the face, neck and trunk; it does not involve the hands or soles. Sometimes there is a positive tourniquet test, hemorrhagic complications and petechiae on the back of hands, feet, arms and legs. It can be accompanied by a biphasic temperature pattern, lymphadenopathy, hair loss, bradycardia, weakness and prostration. Rarely there is encephalitis and Reye's syndrome. Hemorrhagic dengue or dengue shock syndrome occurs in people with previous infection or in children with transplacentally transmitted antibodies. It predominates in children 2-15 years old. The disease lasts 7-10 days. Initially there is abrupt fever, throat pain, malaise, headache, nausea, vomiting and, 2-5 days later, diaphoresis, tachypnea, tachycardia and hypotension. Temperature is 40-41°C, and convulsions can occur. It is accompanied by petechiae on the face, axillae, trunk and extremities, ecchymoses and sometimes gastrointestinal hemorrhage, lymphadenopathy, hepatomegaly, splenomegaly and renal failure (*Salud Publica Mex* 1995; 37:S29-44). In severe cases, patients deteriorate, suffer stroke, metabolic acidosis, severe bleeding and death.

LABORATORY DATA

With serologic studies an increase in the antibodies by methods of hemaglutination, complement fixation, neutralization, antibodies anti-dengue IgM and IgG, ELISA, and capture of IgM antibodies (MAC-ELISA) gives sensibility of 78-98%. Serotypes are identified by immunofluorescence with monoclonal antibodies. Early on the virus can be isolated by culture in mosquito cells. In hemorrhagic dengue there is leukopenia with lymphocytosis, hemoceoncentration and thrombocytopenia. There can be hypoproteinemia, hypoalbuminemia, hyponatremia and moderate elevation of alanine aminotranferase/aspartate.

TREATMENT

There is no specific treatment. Fluid and electrolyte losses must be replaced. In hemorrhagic dengue plasma or blood transfusion may be necessary. Acetominophen can be administrated and coagulation defects must be corrected. Aspirin and salicylates must not be used. The value of the corticosteroids or heparin has not been proven. A tetravalent vaccine of attenuated live virus is being developed. Prevention is oriented to the control of the vector and epidemiological vigilance. Given the domestic habits of the mosquito, elimination of stagnant waters, insecticides and larvicides are recommended. In endemic areas, vertical transmission and the risk of hemorrhagic infection in newborns should be considered.

Selected Readings

1 Kusner DJ et al. Dengue. In: Mahmoud AAF, ed. Tropical and geographical medicine. Companion Handbook. New York: McGraw-Hill. 2nd ed. 1993:204-7.
2 Martinez-Torrez E. Dengue and hemorrhagic dengue: The clinical aspects. Salud Publica Mex 1995; 37:S29-44.
3 Narro-Robles J, Gomez-Dantes H et al. El dengue en Mexico: Un problema prioritario de salud publica (The Dengue in Mexico: A prioritized problem of public health). Salud Publica Mex 1995; 37(Supl):12-20.
4 Nimmannitya S. Dengue and Dengue Haemorrhagic fever. In: Cook GC, ed. Manson's tropical diseases. Philadelphia: Saunder. 2nd ed. 1996: 721-29.
5 Ramirez-Rondo CH, Garcia CD. Dengue in Western hemisphere. Infect Dis Clin North Am 1994; 8(1): 107-28.

Acquired Immune Deficiency Syndrome (AIDS)

Roberto Arenas
Rodolfo Vick

This is an acquired immune deficiency of $CD4^+$ T cells. It is caused by the human immunodeficiency virus (HIV). It affects people of any age or sex, and it is manifested by a wasting syndrome, opportunistic infections and neoplasias.

GEOGRAPHIC DISTRIBUTION

AIDS is a serious public health problem worldwide. It has been reported in more than 136 countries. In 1983 the Centers for Disease Control in the U.S. had been informed about 3,000 cases. By the end of 1989, 200,000 cases had been reported to the WHO, but they estimated that the pandemic was from 5-10 million. In 1993, 75% (371,086) of the cases that were reported worldwide were from the Americas, and of these 289,320 were in the United States and 36,481 were in Brazil. In Mexico in 1996 there were 26,651 cases (57% dead); 38,083 cases have been reported overall, and this disease occupies the 19th place in causes of mortality. Conservatively, taking into account that the epidemic has lessened, the number of people infected by HIV for each AIDS case is estimated to be 4:1. It predominates in men. In America and Europe this disease is seen in homosexuals who are 20-35 years old and in bisexuals. In the United States 10% of the adult men population is homosexual. In Africa the homosexual:heterosexual ratio is 1:1, in this Continent and in Asia heterosexual transmission predominates. It affects all social classes.

There are several epidemiologic patterns of transmission. Pattern I includes homosexual and bisexual men, intravenous drug users, hemophiliacs, transfusion recipients and heterosexual couples and children of people in these groups. Pattern II comprises the active young heterosexual population, transfusion recipients and children of affected women. Pattern I is observed in the United States, Western Europe, Asia and Australia. Pattern II is seen in the non-industrialized countries of Africa and the Caribbean where the distribution in men and women is almost the same. In Latin America the pattern is changing from I to II, and now it is considered as I/II. Pattern III

Tropical Dermatology, edited by Roberto Arenas and Roberto Estrada. ©2001 Landes Bioscience.

appears in places where AIDS is only recently introduced such as Eastern Europe, North Africa, the Middle East, and in general, Asia and the Pacific. In Mexico AIDS is transmitted by sexual contact in 70-88% (35% homosexuals, 27.5% bisexual, 21.1% heterosexual), by blood and blood derivatives in 10.6-14.9% (transfusion 5.9%, renumerated donors 2.5%, hemophiliacs 1.3%), intravenous drug users in 0.4-0.9%, perinatal transmission in 1.1%, and occupational in less than 1%. In women transmission occurs via blood products in 53.6% and by heterosexual transmission in 43%. In children the transmission is perinatal in 54.9%, by transfusion in 25.5% (17.5% in hemophiliacs) and by sexual abuse in 2.1%.

ETIOLOGY

The human immunodeficiency virus (HIV-1), as with HTLV III or LAV, is a retrovirus. Its capsule contains several glycoproteins, among them gp120 (recognized by $CD4^+$ T cells) and gp41. The nucleocapside contains four proteins: p24, p17, p9 and p7. Within three months infected individuals produce antibodies against gp120, gp41 and p24. The virus loses its capsule when it penetrates the cell, and through a reverse transcriptase a copy of the DNA is made from the RNA viral genome. Shortly after the primary infection, 1 of 100 T cells contain virus. Immune defense diminishes when the number of $CD4^+$ lymphocytes falls below 600/ml. Transmission is influenced by the size and antigenicity of the viral source, the condition of the host and the genetic susceptibility. The virus is cytotoxic. It infects $CD4^+$ lymphocytes, monocytes/macrophages and Langerhans cells. Cell-mediated immunity is impaired; accordingly susceptibility to infection and neoplasia is enhanced. In the central nervous system, microglia are affected. Seroconversion occurs 4-6 months after infection. Early on, there is polyclonal expansion of B cells and high titers of IgG, IgM and IgA in the blood. Late in the disease, there is an increase is serum immunoglobulins. After the infection, there are transient clinical signs in 10-15% of cases, but patients either remain asymptomatic for prolonged periods or suffer generalized, persistent lymphadenopathy. Patients can become symptomatic the first 3 years of disease in 10-15% of cases; in 2-20% in six years and in 50% in ten years. The natural course can be modified by the following factors: sexually transmitted diseases (STD), intravenous drug use, pregnancy, childhood, cellular activation by another virus or antigen, and various factors such as malnutrition, immunosuppression and genetic susceptibility. Some histocompatibility antigens are involved in disease progression. For example, Kaposi's sarcoma is related to HLA-DR5 and lymphadenopathy to HLA-B35. In West Africa a related virus has been found, HIV-2. This has an epidemiological pattern similar to HIV-1. The infection is transmitted by sexual contact, blood and transplacentary transfusion. There is no proof of transmission by casual contact.

CLASSIFICATION

Originally HIV infection was classified in the following way:
a) asymptomatic;
b) syndrome of chronic lymphadenopathy (pre-AIDS, gay syndrome);
c) AIDS-related complex;
d) complete AIDS.

In 1993 a revised classification was proposed for adolescents and adults according to the number of $CD4^+$ lymphocytes (Table 58.1).

CLINICAL PICTURE

Latency varies from several months to more than 10 years. The initial phase is characterized by fever, diarrhea, weight loss (wasting syndrome in 40-68% of cases), adenopathy, pruritus and other manifestations depending upon the level of cell-mediated immunosuppression. Many sexually transmitted diseases accompany AIDS: gonorrhea, syphilis, hepatitis B, A, and non-A, non-B, pharyngitis, and chlamydial proctitis. In AIDS-related complex (ARC), neurologic or systemic illnesses appear, opportunist infections like oral candidiasis, multidermatomal herpes zoster, hairy cell leukoplasia, seborrheic dermatitis and retinal spots in cotton branches. Pre-existing dermatoses, like atopic dermatitis, psoriasis and Reiter's syndrome, deteriorate as the acquired immunodeficiency syndrome becomes increasingly severe. When AIDS is established, the following can appear: Kaposi's sarcoma (7-50%) (Fig.58.1), Hodgkin's disease and non-Hodgkin-lymphoma, *Pneumocystis carinii* pneumonia (74%) that presents on chest x-ray as bilateral infiltrates, atypical mycobacterioses and tuberculosis (17.1%), meningeal cryptococosis, cerebral toxoplasmosis, intestinal infections caused by *Cryptosporidium* that are manifested by diarrhea, sometimes with hematochezia, epigastric pain, nausea, vomiting and anorexia with loss of fluids of up to 3 L per day and a tendency to the chronicity, serious mucosal candidiasis, systemic infections caused by cytomegalovirus (CMV) and Epstein-Barr virus, ulcerated herpes simplex,

Table 58.1. Classification by clinical data and lymphocyte T cell count (modification of CDC. Update on acquired immune deficiency virus syndrome (AIDS) MMWR. 1992; 41:421).

Lymphocytes T CD4+*	Asymptomatic, acute (primary) or generalized and persistent lymphadenopaty	Symptomatic	AIDS
>500/ml	A1	B1	C1
200 a 499/ml	A2	B2	C2
<200/ml	A3	B3	C3

*AIDS indicator T cell count

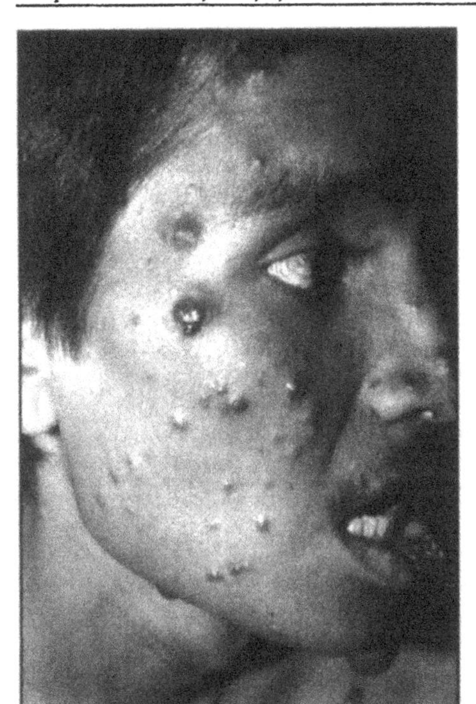

Fig. 58.1. Kaposi's sarcoma in AIDS.

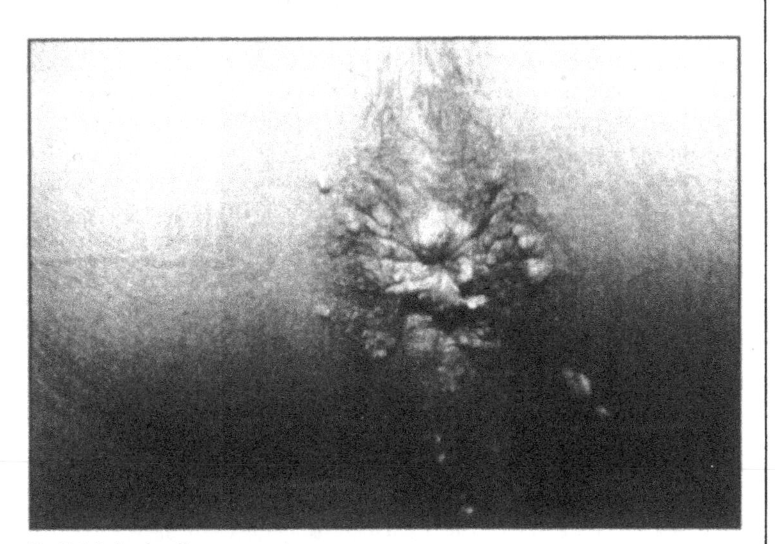

Fig. 58.2. Perianal molluscum contagiosum.

aspergillosis and multifocal progressive leukoencephalopathy caused by papovavirus.

In more than 90% of patients, cutaneous manifestations have been reported (Figs 58.2, 58.3). Kaposi's sarcoma appears suddenly. Onset is rapid and asymptomatic. It is manifested by lesions of 1 mm to several centimeters in diameter that follow the dermatomes (Fig. 58.1) and predominate on the trunk and head. There are macular and papular or tumoral lesions with an angiomatous appearance. A maculopapular exanthem has been described which lasts 1-2 weeks. It is related to CMV and Epstein-Barr virus. Hairy cell leukoplasia predominates on the lateral margins of the tongue. Pruriginous papular eruptions, 2-5 mm, affect head, neck, trunk and extremities. Bacillary epithelioid angiomatosis originates from a bacillus similar to the one that causes cat scratch fever. The lesions have a papulonodular aspect, are red-violet, scant or abundant, and can precede a systemic infection. They resolve with antibiotics. Sometimes cloacogenic carcinoma develops. Genital and perianal herpes is characterized by prolonged change and big ulcerations (37.9%). Varicella can be accompanied by pneumonia, encephalitis and visceral lesions. It is severe and disseminated, with necrosis and neuritis (8%). Molluscum contagiosum predominates on the face and genitals. Lesions are large and usually numerous (Fig. 58.2). In 10-25% of CMV there is a cutaneous lesion, i.e., maculopapular eruptions, purpurae, ulcers or vesicles. Candidiasis is seen in 50% and affects the buccopharyngeal mucosa and esophagus. Dermatophytosis is manifested by palmo-plantar keratoderma and atypical tinea corporis or cruris usually caused by *T. rubrum* and superficial white onychomycosis or proximal subungual white onychomycosis. Cryptococcosis is manifested by encephalitis and meningitis. On the skin there can be cellulitis, ulcers, vasculitis, abscesses and papulopustular lesions. Histoplasmosis occurs in 11% of cases. There can be papulopustular eruptions, cellulitis or erythema nodosum. In syphilis the period of latency is decreased. Neurosyphilis can occur within two years, and the secondary manifestations are exuberant with large papules, often hyperkeratotic (Fig. 58.3). Mycobacterioses have nonspecific cutaneous manifestations, e.g., papules, pustules, ecthyma and cellulitis. *M. avium-intracellulare, M. marinum* and *M. fortuitum* are frequently isolated. Scabies with atypical or crusted lesions are present. There can be infections by Demodex and Acanthamoeba. Malnutrition is a consequence of the chronic illness, associated infections and diarrhea. It is manifested by xerosis, pellagra-like changes, manifestations of zinc deficiency, finger nail fragility, hair loss and color change of the hair. Skin manifestations in drug eruptions are common; most frequently to trimethoprim-sulfamethoxazole. There is maculopapular eruption with peripheral cytopenia, erythema multiforme, Stevens-Johnson syndrome and toxic epidermal necrolysis, as well as pigmentation. In addition, lesions similar to seborrheic dermatitis, sometimes severe and extensive, have been described. Pruritus, yellow discoloration of the fingernails, eosinophilic folliculitis and trichomegalia of the eyelashes have been reported (Tables 58.2, 58.3).

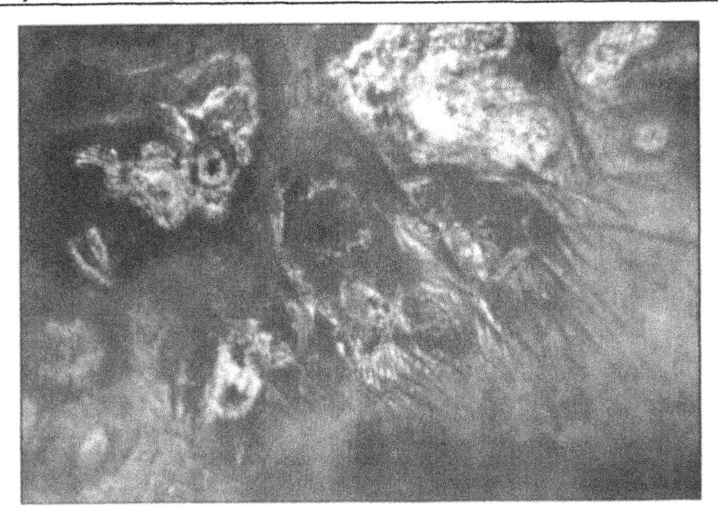

Fig. 58.3. Secondary, hyperkeratotic papules.

LABORATORY DATA

The immunodeficiency is manifested by lymphopenia, cutaneous anergy, and inversion of the T cell helper/suppressor ratio. Humoral immunity is normal or there is increased IgA. Antibodies are detected by ELISA, Western blot and RIPA (radio-immune-precipitation assay). In children less than 18 months old, a positive test is not definitive due to the fact that the antibodies can be of maternal origin. Definitive diagnosis requires Western blot, viral culture, determination of the viral p24 and PCR (polymerase chain reaction). Indicators of disease progression are $CD4^+$ T cell count, tests of cutaneous anergy or an altered mitogen/antigen response. VDRL and FTA can be negative in patients with syphilis or, conversely, titers can be high. When fungal infection is suspected, mycological studies must be done. The cryptosporidiosis is diagnosed when identifying the oocystes in feces by direct exam, with Giemsa or with modified Kinyoun. Biopsy in Kaposi's sarcoma shows vascular neoplasia with capillary neoformation and proliferation of perivascular connective tissue. Pearl's stain demonstrates hemosiderin. With bacillary angiomatosis the picture can be similar to Kaposi's sarcoma, but Warthin-Starry reveals the causative agent. In hairy cell leukoplasia there is hyperkeratosis and parakeratosis with acanthosis and ballooning cells; the virus and Candida have been isolated in culture. In papular pruritic eruptions the biopsy shows inflammatory perivascular infiltrate with eosinophils. There can be an dermo-epidermal dermatitis that simulates drug reactions. There is necrotic vacuolization of the basal cells and keratinocytes without eosinophils or polymorphonuclears.

Table 58.2. Clinical manifestations of AIDS

Neoplasia	**Infections (see Table 58.3)**	**Primary dermatosis**	**Nutritional**	**Others**
Kaposi's Sarcoma	Viral	Eosinophilic folliculitis	Acrodermatitis	Seborrheic dermatitis
Lymphoma	Protozoa	Exanthem	Enterophatica	Drug reactions
Bacillar angiomatosis	Fungi	Pruriginous papular		Vasculitis
Basal cell carcinoma	Mycobacteria	Eruptions		Thrombocytopenic purpura
Squamous cell carcinoma		Hairy leukoplakia		Icthyosiform skin or eczema Telangiectasia
				Thrombophlebitis syndome
				Aphthous ulcer
				Onychodystrophia
				Granuloma annulare

Table 58.3. Infectious diseases and etiological agents in AIDS (modification of Penny NS. Skin manifestations of AIDS. Philadelphia: Lippincott. 1990).

Bacteria:
S. aureus, Haemophilus, Branhamella, *P. aeruginosa*, Salmonella, syphilis, chanchroid, lymphogranuloma, bacillar angiomatosis, botryomycosis, tuberculosis and other mycobacteria
Virus:
Herpes simplex and zoster, varicella, CMV, molluscum contagiosum, common warts, acuminated condylomata, hairy leukoplakia
Fungi:
dermatophytosis, candidiasis, pityriasis versicolor, *cryptocococcis*, histoplasmosis, coccidiodomycosis, sporotrichosis, penicilliosis, *P. carnii*
Protozoa and others:
amibiasis, scabies, leishmaniasis

Prevention

Prevention is a priority, as is the development of curative treatment since vaccines are not yet available. Education is important. Monogamy should be emphasized, and in the case of new partners, safe sex with the use of condoms is essential. The notification and evaluation of sexual contacts is indispensable to interrupt the chain of transmission. Also, intravenous drug use, the selection of blood donors, the administration of blood products, and the proper use of dialysis equipment and other blood product delivery devices must be carefully monitored to minimize exposure to HIV.

TREATMENT

General measures are essential: good nutrition, treatment of opportunistic infections, chemotherapy of neoplasms, change of sexual habits, enrollment of the patient in STD and AIDS programs, a program of physical activity and psychological and family support. Treatment with nucleoside antiretroviral analogs such as zidovudine (azidotimidine, AZT), 500 mg/day has been effective. If started early it delays progression of the disease; the p24 antigen and the incidence of associated HIV dementia decrease. AZT can cause bone marrow suppression, nausea, headaches, insomnia, fatigue and myopathy. In patients intolerant of AZT and in whom there is clinical and immunologic deterioration, didanosine (DDI) 200 mg/12 h can be administered. It can cause pancreatitis, peripheral neuropathy and diarrhea. In advanced infection, deoxicitidine (zalcitabine, DDC) 0.75 mg/8 h is recommended; in 30% of patients it causes peripheral neuropathy. Also, d4T 20-40 mg/12 h, 3TC 150 mg/12 h, saquinavir 600 mg/8 h, ritonavir 600 mg/12 h and indinavir 800 mg/8 h can be used.

Selected Readings

1 Abraham A et al. AIDS in Tropics. In:Parish LCH, Millikan LE, eds. Dermatologic Clinics. Philadelphia:Saunders, 1994; 12(4):747-752.
2 Brostoff J, Scadding GK, Male D, Roitt IM et al. Clinical Immunology. London: Gower Medical,1991; 24.1-41.14.
3 Roitt I Brostoff J, Male D et al. Immunology. 3rd ed. London:Mosby, 1993.
4 Sande PA, Volberding PA et al. Manejo Medico del SIDA (The medical management of AIDS). Mexico:Interamericana/McGraw-Hill, 1992.
5 Soberon-Acevedo G. El SIDA in Mexico y el mundo: una vision integral(AIDS in Mexico and the world: An integral vision). Gac Med Mex 1996; 132 (Supl 1): 1-138.
6 Vick R, Vargas AS, Castro G, Cordova H, Fabian MG, Arenas R. Manifestatciones dermatologicas en el sindrome de inmunodeficiencia adquirida. Estudio de 46 pacientes(Dermatological manifestations in the acquired immunodeficiency syndrome). Dermatologia Rev Mex 1995; 39(4):197-201.

K. Malnutrition

Pellagra/Kwashiorkor

Pellagra/Kwashiorkor

Roberto Arenas and Guadalupe Chavez

Pellagra, also named rose's disease or Gaspar Casal's disease, is a systemic disease caused by niacin deficiency, with a predominant involvement of the skin, mucosae and central nervous system. The dermatosis is bilateral and symmetric, with a preference for sunlight exposed areas. It is characterized by erythema, scaling, blisters and pigmentation. It is known as the disease of four "Ds": dermatitis, diarrhea, dementia and death.

GEOGRAPHIC DISTRIBUTION

Pellagra is endemic in nonindustrialized countries with a high rate of malnutrition where corn constitutes the dietary basis. It predominates in farmers. Epidemics have been described in refugees of Mozambique and Cuba. In 1990 an epidemic in Malawi, affecting approximately 900,000 refugees, was the most extensive since World War II (*Int J Epidemiol* 1993; 22(3):504-11). The low frequency of pellagra in Mesoamerica may be due, to corn processing with calcium hydroxide before it is consumed in the form of "tortillas." This process promotes the liberation of niacin from certain conjugated forms. In the United States pellagra has practically disappeared although it can be observed in chronic alcoholics, the elderly, retarded persons, diabetics, in intestinal malabsorption or people with carcinoid tumors.

ETIOLOGY

Pellagra is due to deficiency of niacin or its precursor, tryptophan. Yet the clinical manifestations are usually due to deficiency of several vitamins. It is also caused by the presence of niacin antagonists such as isoniazid, 6-mercaptopurine, 5-fluorouracil (5-FU), and chloramphenicol. The first three inhibit the synthesis of coenzymes; isoniazid also acts by competing with nicotinamide, and 5-FU inhibits the conversion of tryptophan to niacin. It has also been related to anticonvulsants such as phenytoin, etosuximide and valproic acid. Niacin deficiency can be due to a combination of several causes, e.g., inadequate diet or poor intestinal absorption, increased metabolic requirements (pregnancy and lactation), interference with transport and utilization (chronic alcoholism) or the administration of the aforementioned antagonists. The existence of an anti-niacin in corn has been proposed, but it has not been proven effective. Corn has an incomplete protein with little tryptophan, the amino acid required for niacin production. It is also believed that the high amounts of leucine in sorghum and other cereals can cause pellagra by interfering with tryptophan and niacin metabolism. Manifestations of pellagra are observed in carcinoid syndrome wherein there is abnormal conversion of tryptophan to serotonin. Niacin is a component of two active co-enzymes: nicotinamide adenine dinucleotide phosphate (NADP) and nicotine adenine dinucleotide (NAD), also known as coenzymes I and II, respectively, which are essential for energy transport and cellular metabolism. In niacin deficiency cell repair is impaired in high turnover tissues such as skin and the digestive tract or in organs with large energy requirements like the brain. Photosensitivity (phototoxicity) can be related to a deficiency of coenzymes I and II. Both are essential in the repair of epidermis damaged by ultraviolet radiation.

CLINICAL PICTURE

The dermatitis of pellagra is bilateral and symmetric. It involves sunlight exposed skin, such as the face, neck, dorsum of the hands and feet. It is characterized by intense erythema, edema, burning, blisters and subsequently scaling, skin thickening and dark brown pigmentation. Lesions about the lower neck form a collarette of dermatitis, and are called "Casal's necklace" (Fig. 59.1). The dermatitis can be generalized. There can be maceration in folds and ulcerative areas. Follicular hyperkeratosis is observed in seborrheic areas such as sides of the nose and cheeks. The mouth may be sore with angular cheilitis and glossitis. This last originates as an inflammatory process with atrophy of the filliform papillae. There is pharynxitis and painful esophagitis with odynophagia as well as vomiting and diarrhea. Anemia, amenorrhea, miscarriages and weight loss may occur. Central nervous system symptoms include irritability, asthenia, anorexia, headache, insomnia, anterograde and retrograde amnesia as well as behavior disturbances that can progress to psychosis and dementia. Also, there are peripheral nerve lesions with motor and sensory disturbances. In children there is a suppression of growth, edema, diffuse alopecia with loss of brightness of the hair and bands of discoloration. The differential diagnosis includes Hartnup's disease which is characterized by symptoms similar to those of pellagra, cerebellar ataxia and sometimes mental retardation. In adolescents the mental changes are severe. In old people the lesions leave a certain degree of atrophy, and there usually are ungual changes like fragility, striae and leukonychia. There are incomplete forms in individuals with chronic malnutrition manifested by dry and scaly skin. When there are no cutaneous lesions it is called pellagra without pellagra. Individuals with AIDS present with a picture similar to classic pellagra, i.e., cachexia, dementia, diarrhea and dry skin. In patients with advanced cancer, cachexia and hypoalbuminemia are common, but chemotherapy can also cause vitamin deficiencies and malnutrition by causing anorexia, stomatitis and alimentary disorders.

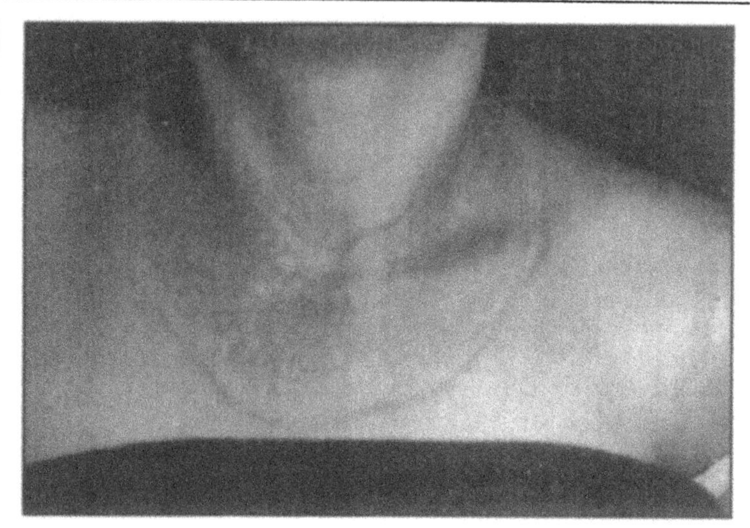

Fig. 59.1. Pellagra, Casal's necklace.

LABORATORY DATA

Biopsy is not relevant. Histologic changes are nonspecific. It may show intraepidermal or subepidermal vesicles or bullae and chronic inflammatory infiltrate. In old lesions there is hyperkeratosis with areas of parakeratosis, moderate acanthosis and hyalinization of collagen.

Kwashiorkor is an illness of protein-energy malnutrition (PEM), characterized by a diet poor in protein and rich in caloric energy in the form of starch and sugar. In the weaning period, nutrients are needed that satisfy the high demands required for normal growth. PEM is either primary, due to inadequate protein intake, or secondary, due to inefficient nutrient utilization or poor intestinal absorption. In underdeveloped countries there is a scarcity of protein. It is substituted by carbohydrates creating the so-called "sugar babies" because mothers feed their babies with carbonated drinks and bread of processed grains that are low in protein. Kwashiorkor is characterized by weight loss, neurologic findings, peripheral edema, diarrhea, hypoalbuminemia, mucosal and adnexal changes (Fig. 59.2). Skin manifestations first appear on areas exposed to rubbing or to pressure like the groin, knees, gluteus and elbows. In severe cases any part of the body can be affected. In children growth is delayed and there is a characteristic pot-belly. Skin color varies from an opaque to dark gray or takes on a yellowish hue so that hair appears blond, thin and fine. Fingernail thickness decreases, and dry skin alternates with areas covered with shiny scales called "flaky paint" (enamel paint areas). These lesions in undernourished children with edema are pathognomonic for kwashiorkor. They are not seen in marasmus. Also there is loss of fat, muscle atrophy and diarrhea. Signs of malnutrition manifest first on the skin and may vary, most

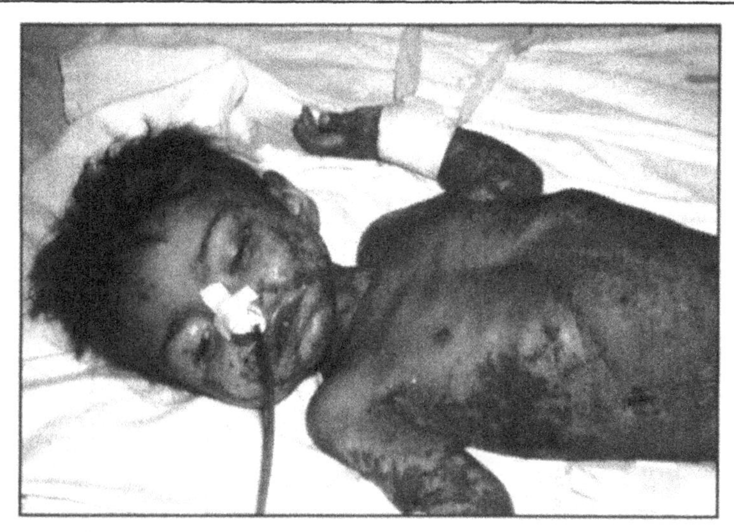

Fig. 59.2. Malnutrition, kwashiorkor.

commonly pallor, alopecia or hypertrichosis and cutaneous hyperpigmentation. Secondary skin infections such as pyoderma and candidiasis are common. And the cardiovascular, neuromuscular and gastrointestinal alterations are frequently fatal.

TREATMENT

Balanced and adequate nutrition is essential. A soft, high protein diet including meats, fish, milk, eggs, seeds and green vegetables, all high in tryptophan and niacin, are recommended. Supplemental nicotinamide, 50-100 mg bid or tid, is indicated. It is administered along with B complex.

Selected Readings

1 Elmore JG, Feinstein A et al. Joseph Goldberger: An unsung hero of American clinical epidermiology. Ann Intern Med 1994; 121:372-75.
2 Gould JW, Mercurio MG, Elmets CA et al. Cutaneous photosensitivity diseases induced by exogenous agents. J Am Adad Dermatol 1995; 33:551-73.
3 Malfait P, Moren A, Dillon JC et al. An outbreak of pellagra related to changes in dietary niacin among mozambican refugees in Malawi. Int J Epidemiol 1993; 22(3):504-11.

L. Sweaty Syndromes

Miliaria/Dyshidrosis

Caloric Intertrigo

Miliaria/Dyshidrosis

Roberto Arenas
Roberto Estrada

Miliaria is a dermatosis of sweat retention (sudamina) characterized by numerous pruritic pearled or erythematous papules and vesicles that predominate on the trunk and extremities.

GEOGRAPHIC DATA

It is observed at all ages and in both sexes equally. It predominates in children and is most frequent in the tropics.

ETIOLOGY

Overhydration of the stratum corneum in a humid environment causes terminal obstruction of the eccrine sweat and secondary anhidrosis. This occlusion can occur when the ambient temperature is high, and occlusive clothing is used, such as disposable diapers. Perspiration acts as an irritant because of its high sodium chloride concentration. This may cause pruritus. A quantitative relationship between bacterial flora and the suppression of perspiration has been demonstrated experimentally, and on histology a material that stains with PAS and is resistant to astringents has been observed. This material is produced by *S. epidermidis*; it has been identified as an extracellular polysaccharide (*J Am Acad Dermatol* 1995; 33:729-33. *J Am Acad Dermatol* 1995; 33:729-33)

CLINICAL PICTURE

Miliaria is a disseminated dermatosis involving the neck, thorax, folds, extremities and face. It is composed of papules or pearled/erythematous papulovesicles of 1-3 mm that appear as outbreaks. They produce fine scaling and resolve completely. It is called crystalline or sudamina miliaria when the vesicles are clear or have a pearly appearance. They are punctate, superficial and fragile. They break with only slight friction, as when bathing, and leave a fine scaling surface. This form is more common in newborns (Fig. 60.1) The bullous form, or miliaria rubra, is very pruritic. The vesicles or papulovesicles are deep and

Tropical Dermatology, edited by Roberto Arenas and Roberto Estrada. ©2001 Landes Bioscience.

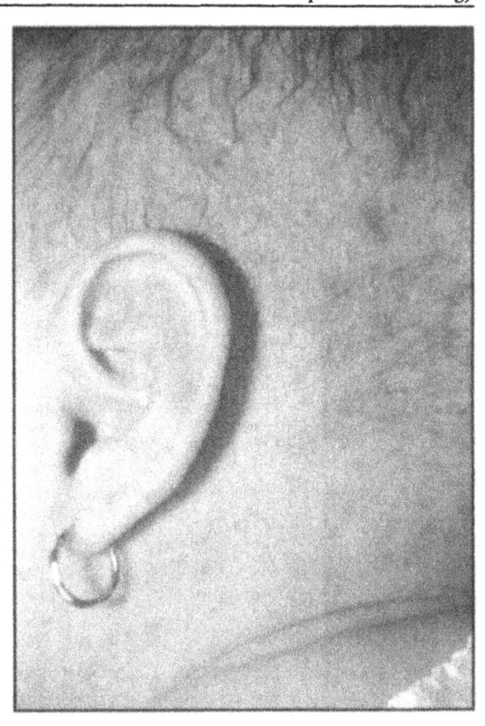

Fig. 60.1. Miliaria.

inflamed. They worsen with perspiration. Lesions are localized on the trunk and extremities and are associated with anhidrosis or local hypohidrosis. In severe cases, anhidrotic tropical asthenia accompanied by fatigue, nausea, drowsiness, palpitations, tachycardia and malaise can develop. Periporitis is the secondary infection of vesicles. It is characterized by micropustules, presents especially in nursing babies and predominate on forehead, bridge of the nose and hairy skin. Pustular miliaria appears after remission of some other dermatosis such as eczema, and it predominates in folds and on flexure surfaces. Apocrine miliaria or Fox-Fordyce disease occurs almost exclusively in women. It begins as transparent, pruritic vesicles in the axilla and perineum. It may appear by psychoaffective or genital stimulus and is non-thermodynamic.

LABORATORY DATA

Biopsy is not indicated. An intraepidermal subcorneal vesicle is seen. In rubra, or deep malaria, the stratum corneum is parakeratotic and hyperkeratotic and contains a homogeneous eosinophilic PAS-positive material, and a leukocytic or lymphocytic, periductal infiltrate. There can be spongiosis and exocytosis.

TREATMENT

The treatment is immersion baths and vigorous, frequent washing and clean, nonocclusive clothing. Strenuous physical activity and tight clothing of synthetic material should be avoided. Drying powders such as talcum, starch or zinc oxide, or simple lotions, can be applied. If there are signs of secondary infection, topical or systemic antibiotics, or antiseptics such as copper sulfate 1/1000 or clioquinol decrease bacterial populations. Topical corticosteroids have a limited benefit. In the deep form the anhidrous lanolin is helpful. In severe cases the use of aromatic retinoids such as isotretinoin, 0.5 mg/kg with the limitations already known for acne.

DYSHIDROSIS

This is also called pompholyx and dishydrotic eczema and cheiropompholyx. Dyshidrosis is a dermatosis of unknown etiology with disturbances of perspiration and atopy. It is characterized by a recurrent vesicular eruption of palms and soles.

GEOGRAPHIC DISTRIBUTION

It is more frequent in young people and in adults 30-40 years old. In the United States this dermatosis is observed in 0.2-0.9%, and up to 2.8%, of people with skin lesions. It predominates in men 3:1. It is observed especially in rainy or hot seasons.

ETIOLOGY

It is related to disturbances of perspiration (hyperhidrosis) and seasonal changes, tinea pedis, contact dermatitis (30%) or nickel ingestion. Sensitivity to nickel sulfate has been found (28%); a relationship to chrome (20%) and cobalt (16%) has also been suggested. IgE is elevated in 37% of patients. Emotional stress has been suggested as an etiologic factor.

CLINICAL PICTURE

The dermatosis occurs only on the palms, soles and interdigitally. It is characterized by nonerythematous vesicles of sudden appearance, encased in the epidermis that gives the impression of "sago grains." If they coalesce, they form blisters (Fig. 60.2). In general, they do not break; they occur as outbreaks and relapses. They leave a collarete of scales of 1-3 mm that give rise to erythemato-squamous lesions with lichenification and sometimes pruritus. The appearance of pustules indicates secondary infection.

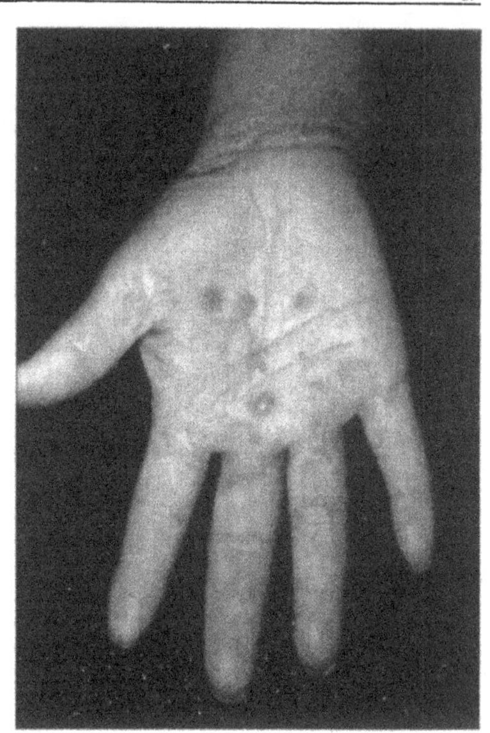

Fig. 60.2. Dyshidrosis.

LABORATORY DATA

Biopsy does not show sweat ducts involvement. Intraepidermal vesicles with thickening of the stratum corneum, usually without inflammatory changes are seen.

TREATMENT

The cause should be eliminated when it is identified. Control of stress favors improvement. Compresses of lead subacetate with equal parts of distilled or purified water, Burow's solution, oatmeal water or water with tobacco, drying powders like talcum and zinc oxide or a cream of neutral glycerite in starch are indicated. Erythromycin has been used as well as short courses of oral or topical glucocorticoids. PUVA is recommended in resistant cases, and pentoxyphillin, disodic chromoglycate, as well as the use of iontophoresis have been suggested. A diet without nickel is recommended so food should not be cooked in pans containing it. The diet should omit canned foods, asparagus, beans, mushrooms, onions, corn, spinach, tomatoes, peas, pears, wheat bread, tea, cocoa, chocolate and baking powder. Although it generates many side effects in patients with confirmed sensitivity to nickel, tetraethyltiuran (Antabuse) can be administered orally 50-200 mg/day, but alcohol must be avoided.

Selected Readings

1 Bose S, Ortonne JP et al. Diseases affected by heat. In: Rarish LCH, Millikan LE et al, eds. Global Dermatology. Diagnosis and Management According to Geography, Climate and Culture. New York: Springer-Verlag, 1994; 83-92.

2 Christen MM, McGinley KJ, Foglia A et al. The role of extracellular polysaccharide substance produced by Staphylococcus epidermidis in miliaria. J Am Acad Dermatol 1995; 33:729-33.

3 Kirk JF, Wilson BB, Chun W, Cooper PH. Miliaria profunda. J Am Acad Dermatol 1996; 35:854-6.

4 Moward CM, McGinley KJ, Foglia A, Leyden JL et al. The role of extracellular polysaccharide substance produced by Staphylococcus epidermidis in Miliaria. J Am Acad Dermatol 1995; 33:729-33.

5 Odia S, Vocks E, Rakoski J, Ring J et al. Successful treatment of dyshidrotic hand eczema using tap water Iontophoresis with pulsed direct current. Acta Derm Venereol 1996; 76:472-74.

6 Yokoseki H, Katayama I, Nishioda K et al. The role of metal alergy and local hyperhidrosis in the pathogenesis of pompholyx. J Dermatol 1992; 19(12):964-967.

Caloric Intertrigo

Roberto Estrada

This is a disease of cutaneous folds caused by heat, friction and moisture combining to irritate and macerate the skin.

GEOGRAPHIC DISTRIBUTION

This dermatosis occurs worldwide, but it predominates in tropical and subtropical areas.

ETIOLOGY

Obese individuals with deep skin folds are most commonly affected. Babies with short necks and naturally accentuated skin folds where sweat is retained, people with accelerated metabolism who copiously perspire in the heat, workers exposed to ovens, hot water boilers, foundries, weldings, sports people with prolonged exposure to the sun and individuals with poor hygiene are frequently affected. Other risk factors include: poorly ventilated houses, shelter constructed of tin, asbestos or tar with cardboard roofing, and those exposed to direct sun most of the day. Similarly, automobiles parked in the sun without air conditioning, and in teething babies, the accumulation of powder residue, saliva and food irritate the skin of the neck, especially if it is short and the chin constantly rests on the chest. The use of synthetic (acrylic) or wool underclothing that does not absorb moisture, which mothers traditionally use to cover newborns, even in hot climates, also are risk factors. Disposable diapers, which are waterproof do not allow any ventilation of the skin, predispose to intertrigo.

CLINICAL PICTURE

Involvement of large skin folds frequently precedes caloric intertrigo. Initially there is marked erythema which is followed, within hours, by the appearance of vesicles, eczema and maceration depending on the intensity. During the ensuing days satellite vesicles or dryness and scaling with hypochromic spots appear around the plaques. The most commonly affected areas are the neck, axillae, groin, intergluteal fold, submammary folds, popliteal fossae and

Tropical Dermatology, edited by Roberto Arenas and Roberto Estrada. ©2001 Landes Bioscience.

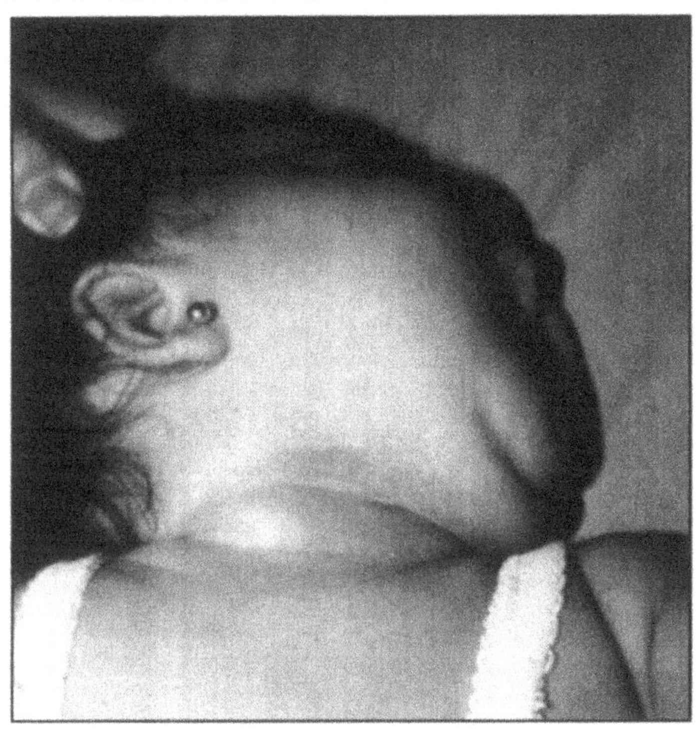

Fig. 61.1. Caloric intertrigo.

antecubital fossae. In babies and obese individuals, secondary folds exist on the extremities and abdomen and even in the retroauricular area. There is itching which patients describe as a sensation of "pins-and needles" which increases with elevated temperature. This is accompanied by irritability and anxiety. Heat and humidity so influence the course of the disease that relocation to a temperate, dry climate causes the clinical picture to improve within hours. Even after lesions resolve there is residual leukoderma. Symptoms sometimes become so intense that patients or their parents apply irritant products such as alcohol, flour or popular medications that end up worsening the picture. The intense scratching and the lack of hygiene may lead to impetigo, folliculitis or furunculosis. The differential diagnosis includes atopic dermatitis, candida intertrigo, contact dermatitis caused by deodorants and inverted psoriasis.

TREATMENT

The control of predisposing conditions is more important than the use of medications. Palliative measures include the use of cotton clothing which should be washed with soft soaps, improvement in housing ventilation, plants that provide shade and freshness, fans, and in the worst cases, air conditioning. Also it is important to avoid, at least temporarily, activities that stimulate perspiration, e.g., sources of intense heat and exposure to the sun. And, if all of the above is inadequate, the patient must be transported to a temperate climate which cures most cases. Medications used to improve the eczema and maceration are substitution of bath soaps, rinsing with astringent solutions such as Burow's solution 2-3 times a day and the use of a topical corticosteroid. If the eczematoid changes are not severe or the skin is not irritated after the rinsing, steroid can be applied alone. However, the patients should be informed that the steroid is not to be used longer than a week which is the time that required to clear most cases because there is the risk of this medication causing atrophic striae if it is used on skin folds for long periods of time. One must also treat associated mycotic (candidiasis) and bacterial infections.

Selected Readings

1 Arenas R. Dermatologia. Atlas, diagnostico and treatment (dermatology, atlas, diagnostic, and treatment). Mexico:Interamericana/McGraw-Hill. 1996:139, 315-17.

M. Tropical Ulcer

Tropical Ulcer

Tropical Ulcer

R. J. Hay

Tropical ulcer or tropical phagedenic ulcer is a synergistic bacterial infection caused by a combination of organisms, one of which is a Fusobacterium species, known as *F. ulcerans*; other bacteria which have been implicated include spirochetes and other aerobic and anaerobic bacteria (*Br J Dermatol* 1987; 1616:31-37; *Int Rev Trop Med* 1963; 2: 267-291). The disease shares striking similarities with cancrum oris (noma) or the veterinary disease, foot rot of sheep. Tropical ulcer is mainly a disease of older children and young adults. It often occurs in clusters of cases.

GEOGRAPHICAL DISTRIBUTION

Tropical ulcer is common throughout the hot and humid tropical regions of the world, in some of which, e.g. northern Papua New Guinea (*J Hyg Trop Med* 1989; 92:215-220), it is the commonest skin disease. Notwithstanding its common distribution, it may be patchy leading to a pattern of disease in which endemic areas may occur close to other areas where the infection is seldom if ever seen. Endemic foci occur in both rural and urban populations and generally the climate associated with the disease is tropical and night temperatures remain high. Typically tropical ulcer is an infection of children and young adults. Males are generally more commonly infected than females. There are some striking differences in distribution. For instance, in one community in Papua New Guinea young adult women were mainly affected, a feature which correlated with their main activity—fishing by wading in a coastal lagoon.

ETIOLOGY

There is no clear evidence that host-predisposing factors play a role in the pathogenesis of this condition. Previously malnutrition was thought to be critical to the development of tropical ulcers, but it is likely that social factors such as overcrowding may be equally important. Recent studies have not shown a correlation between nutritional indices and the development of tropical ulcer (*Trans Roy Soc Med Hyg* 1985; 80:132137), although the possible role of deficiencies in micro-nutrients such as zinc is unproven (*Am J Clin Nutr* 1985; 41:43-51). The body of data would suggest therefore that this is usually a disease of otherwise healthy individuals. There is now considerable evidence to suggest that this disease in an infection (*Am J Trop Med Hyg* 1980; 29:291-297). The condition has been shown to be transmissible by inoculation of material from affected patients (*J Roy Coll Surg Edin* 1966; 11: 196-199). Recently, careful sampling of early ulcers and preulcerative papules has resulted in the isolation of *F. ulcerans*, a fusobacterial species unique to tropical ulcers. This is less often found in late stage ulcers. Other anaerobes are also sometimes isolated and a consistent histological finding is the presence of spiral bacteria, spirochetes, which have not been isolated in culture but which have the ultrastructural features of *Treponema* spp. In late-stage ulcers other bacteria including many aerobes such as *S. aureus* and *Pseudomonas* spp can be found in tropical ulcers. *F. ulcerans* causes destruction of tissue culture cells and can synergize with other bacteria to cause destructive skin lesions in experimentally infected animals. It is likely therefore that the infection is a synergistic anaerobic infection confined to regions of the humid tropics. While the reservoir of *F. ulcerans* is not known, it has been isolated from mud and stagnant water in endemic areas (*Int J Dermatol* 1988; 27: 49-53). The proposed mode of infection is through exposure to contaminated mud or stagnant water in pools or puddles and entry of organisms through a scratch. There are, however, a number of unresolved questions. For instance it is not known why the disease is uncommon in the New World. In addition the spirochetes associated with the disease have not been isolated in culture and their precise role and source is unknown.

CLINICAL FEATURES

Most tropical ulcers develop at exposed sites on the limbs, possibly areas subject to trauma such as a scratch or insect bite, and are therefore commonest on the lower legs and on the unshod foot. They may also occur on the arms. The ulcer goes through an evolutionary phase starting with a preulcerative papule on apparently healthy skin. This is an area of hyperpigmentation and swelling which may rarely blister (Fig. 62.1). A characteristic is its rapid breakdown to form an ulcer with an indurated edge. The sudden appearance of the ulcer from an area of discomfort on an exposed site together with subsequent pain and the regular punched out appearance are very typical (Fig. 62.2). The floor of the ulcer is covered by a foul-smelling, grayish, purulent slough. Pain is common and there may be fever and constitutional symptoms. There is usually no regional lymph node swelling. If the lesion is treated promptly with careful dressing, the spread is limited to a lesion, 2-4 cm in diameter, which heals slowly. In about 4-15% of cases, depending to some extent on the availability of primary health care, chronic ulcers may form and may involve deep structures such as tendons and periosteum (*Int J Dermatol* 1988; 27:49-53). Chronic tropical ulcers are more irregular than the acute forerunner and may extend over a wider area with an irregular margin.

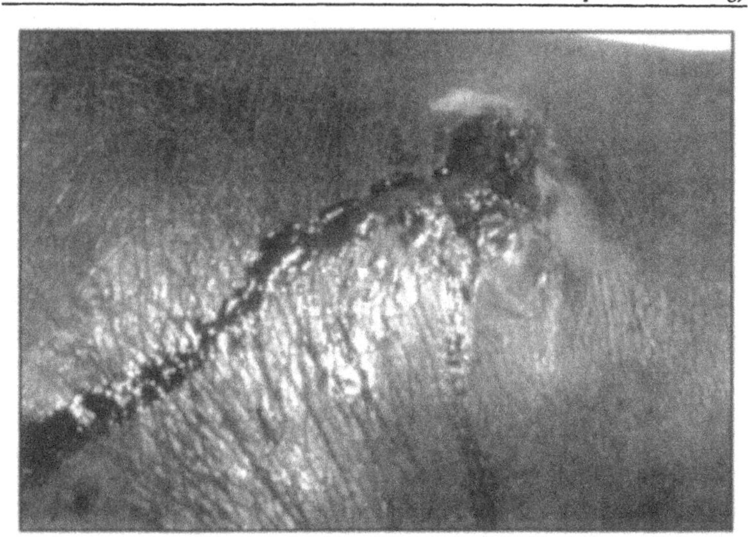

Fig. 62.1. Preulcerative papule, tropical ulcer.

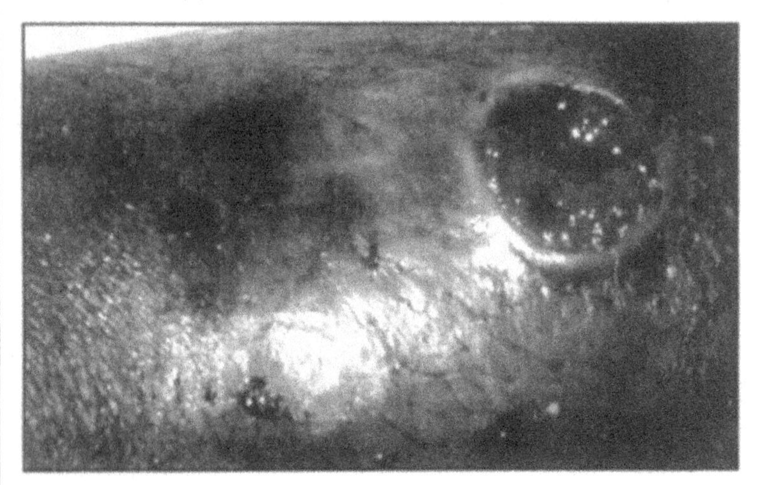

Fig. 62.2. Acute tropical ulcer.

These are difficult to distinguish from ulcers of other etiologies such as stasis ulcers. Squamous epithelioma develops after 10 years or more in some such cases. The main features are the young age group of those affected, the clusters of cases in a community, the rapid onset, pain and regular appearance of lesions. Sickle cell ulcers may also affect a young age group although they are seldom as regular as tropical ulcers. Other lesions that ulcerate include infected sores, cutaneous diphtheria and leishmaniasis; all should be distinguished. In some endemic areas in Africa and the West Pacific, yaws co-exists, although often the lesions of primary yaws are exophytic.

LABORATORY DATA

There are no specific laboratory tests for tropical ulcers.

TREATMENT

Rest, elevation of the limb and adequate diet are important. Any underlying chronic disease should be treated. During the early stages of the disease, penicillin or metronidazole are recommended, combined with topical antiseptics such as potassium permanganate solution. In resistant cases grafting may be done and is important in ensuring that the lesion does not become extensive and chronic (*J Hyg Trop Med* 1989; 92: 215-220).

Selected Readings

1. Adriaans B, Hay RJ, Drasar B et al. The infectious aetiology of tropical ulcer—a study of the role of anaerobic bacteria. Br J Dermatol 1987; 1616: 31-37.
2. Kuberski T, Koteka G. An epidemic of tropical ulcer in the Cook Islands. Am J Trop Med Hyg 1980; 29:291-297.
3. Morris GE, Hay RJ, Srinavasa A et al. The diagnosis and management of tropical ulcer in East Sepik Province of Papua New Guinea. J Hyg Trop Med 1989; 92:215-220.
4. Robinson DC, Hay RJ. Tropical ulcer in Zambia. Trans Roy Soc Med Hyg 1985; 80:132-137.
5. Robinson DC, Adriaans B, Hay RJ et al. The epidemiology and clinical features of tropical ulcer. Int J Dermatol 1988; 27:49-53.
6. Wilkinson M, Agett P, Cole TJ. Zinc and acute tropical ulcers in Gambian children and adolescents. Am J Clin Nutr 1985; 41:43-51.

N. Contact Reactions

Reactions to Contact with Marine Animals

Reactions to Contact with Marine Animals

Giancarlo Albanese

In order to limit such a wide-ranging subject, I shall limit my considerations here to the Coelenterates, a group including a large number of marine organisms responsible for a wide and recurring variety of reactions. Not to be neglected, however, and hence at least worth mentioning, are dermatitis caused by contact with Echinoderms and with several other animals (mollusks, arthropods, sponges, water worms, fish). Cutaneous pathology caused by Coelenterates is yet another example of how changing habits in the population can lead to subsequent changes in some reactions of series of dermatologic cases.

GEOGRAPHICAL DISTRIBUTION

In the Mediterranean area, cases of dermatitis caused by contact with Coelenterates used to be attributed to jellyfish and the like. Now, however, due to the increasing frequency of sea and diving holidays, we are beginning to see the consequences of contact with forms of life that are hard to find in the Mediterranean, such as corals, for example. The distribution of these species thus varies greatly according to the different areas considered, proving to be more important, for example, in coral reef areas.

ETIOLOGY

The phylum Coelenterata (Cnidaria) is a large group of invertebrates (9,000 components, 100 of which are harmful to man) which includes not only corals but also hydras, jelly fish and sea anemones. There are various species of Coelenterates (subdivided into three classes: Scyphozoa, Anthozoa and Hydrozoa), but they all use the same mechanism (nematocysts) to capture their prey and this is what affects humans who come into contact with them. Nematocysts are capsules containing hollow coiled threads which are violently ejected in reaction to mechanical or chemical stimuli sensed by an external receptor (cnidocil); these threads inject toxins (proteinic tetramethylammonium hydrate, catecholamine, adenine, 5-hydroxytryptamine, histamine, etc.). As far as these substances are concerned, it should be underlined that in some areas, such as the Mediterranean, such toxins cause quite moderate local reactions, while in other geographical areas, such as Australia and the Pacific islands, the toxic effect may be lethal. Schematically speaking, the most important clinical pictures, in terms of frequency or severity, are the result of contacts with jellyfish, Physalia and corals. A separate clinical picture, linked to the same mechanisms but with different clinical features, is seabather's eruption.

JELLYFISH

Jellyfish usually cause local reactions on the contact area (Fig. 63.1); these are quite evident but not serious (linear erythematous-edematous lesions or even urticarial blisters). The most frequent situations are then related to these local reactions (toxic reactions, angioedema, persistent and relapsing delayed allergic reactions, allergic contact dermatitis, papular urticaria) and to their sequelae (keloids, pigmentation, scars). Lethal toxic systemic reactions and anaphylaxis are rare. However, particularly serious cases linked to species present on the Atlantic and Pacific coasts have been reported. For example, the two species present on the costs of Australia: *Carukia barnesi* and *Chironex fleckeri* (*J Dermatol* 1996; 37 (Suppl 1):S23-226) are characterized by three toxic components: dermatonecrotic, hemolytic and myocardial. They thus cause papular erythematous lesions to the skin that ulcerate, become necrotic and are slow to heal; neuromuscular paralysis and cardiovascular collapse may be fatal within minutes of poisoning.

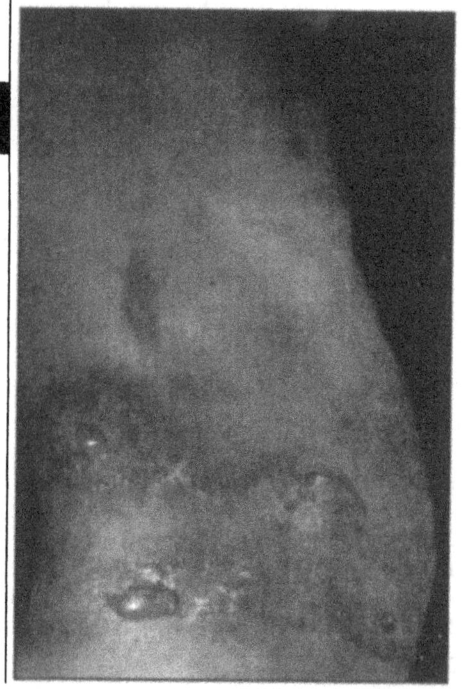

Fig. 63.1. Jellyfish contact dermatitis.

PHYSALIA

These coelenterates of the Hydrozoa class normally live in the tropical areas of the Pacific, Atlantic and Indian oceans. Floating in the current, with round bodies and tentacles several meters long, they have been known to reach the Atlantic coasts of Europe and the Mediterranean. The most widespread species is *Physalia physalis*, the so-called "Portuguese man of war." The venom released as a result of contact with the tentacles is used to capture the small fishes that they feed on (*Ann Emerg Med* 1989; 18:312-315). This substance has a paralyzing effect on fish. In man it causes intense pain, followed by the appearance of erythematous lesions, blisters and ulcers. After a few minutes, respiratory problems may arise, and may lead to respiratory arrest and cardiovascular collapse.

CORALS

A few species of coral are isolated, but most of them are organized in structures known as coral reefs, a frequent attraction for divers and swimmers. Corals consist of a skeletal framework of calcium carbonate crystals, secreted by the vital part of the coral polyp. The latter has structural features typical of the Coelenterates, including the offensive part designed to capture their prey, i.e. cnidoblasts. The dermatological consequences of contact with corals include continuous lesions caused by the calcareous skeleton; foreign body reactions and

Fig. 63.2. Red Sea coral contact dermatitis.

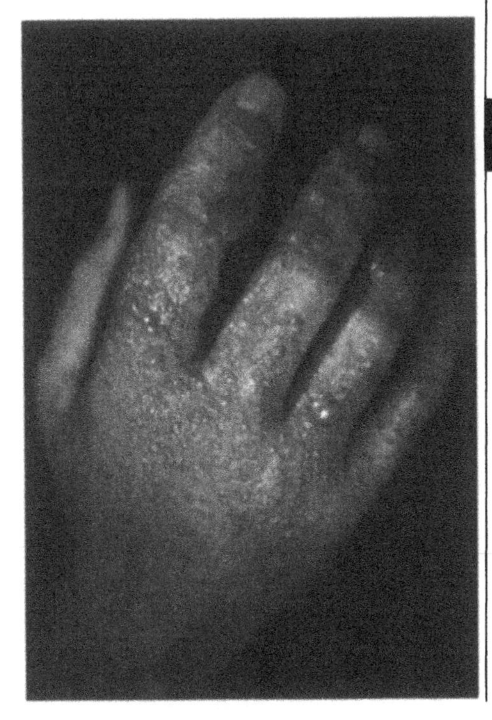

secondary infections; and dermatitis from contact with nematocysts. The most frequent situations involve the nematocysts (Fig. 63.2), exactly as described above, and thus linked to local reactions (toxic, angioedema, recurring, persistent, delayed allergic reactions, allergic contact dermatitis, papular urticaria) and their sequelae (pigmentation, scars, keloids). Fatal systemic toxic reactions and anaphylaxis are rare (*Contact Dermatitis* 1993; 29:285-286).

Seabather's Eruption

This is a very pruritic, papular eruption also known as marine dermatitis, which affects the areas of the body covered by the bathing suit; it has been reported especially on the southeast coat of Florida and sometimes in Cuba, in Mexico and in the Caribbean. It therefore occurs in geographical areas in which the range of agents thought to cause this clinical picture is widest (marine plants, hydroids, corals). Recently, however, cases have also been reported on Long Island, 1000 miles further north than the areas in which cases were previously reported (*N Engl J Med* 1993; 329:542-544). Recent studies have identified the urticant nematocysts of a cnidarian larvae (*Limuche unguiculata*) deriving from some jellyfish, as being responsible for the cases found in Florida (*J Am Acad Dermatol* 1994; 30:399-400). On Long Island, however, the cnidarian larva of *Edwardsiella lineata* has been identified as the culprit. In seabather's eruption, papulo-vesicular,

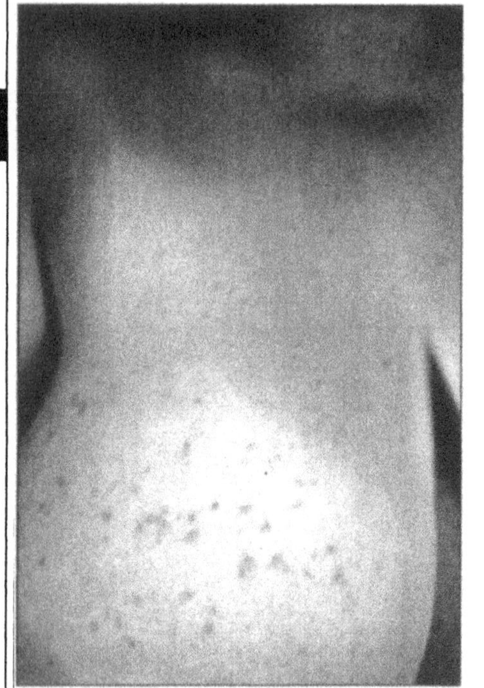

Fig. 63.3. Seabather's eruption.

erythematous and very itchy lesions (Fig. 63.3) usually appear a few hours after exposure and then last up to 2 weeks with spontaneous recovery. As already mentioned, these lesions only involve the areas covered by a bathing suit. This is because the larvae are trapped inside the costume, and the pressure stimulates the release of the urticant nematocysts and their venom. Seabather's eruption should be distinguished, in particular, from swimmer's itch, a geographically ubiquitous clinical picture which is located especially in exposed areas of the body and is caused by the cercariae of schistosomes penetrating the skin. This dermatosis affects people swimming in both fresh and sea water, in a cycle in which water rodents and birds are normally the primary hosts of the schistosome, and some types of mollusks act as intermediaries.

TREATMENT

With regard to jellyfish, treatment of the less serious forms involves inactivating the stinging threads with vinegar and treating the symptoms with topical steroids. The more serious forms may require more complex therapy (analgesia, cardiovascular and pulmonary resuscitation). In benign cases of *Physalia physalis*, slow-healing cutaneous lesions may be accompanied by persistent urticarial eruptions. Local treatment is exactly as described for contact with jellyfish; the patient often requires cardiopulmonary resuscitation. The clinical picture of seabather's eruption is self-limited, thus requiring only symptomatic treatment with topical steroids and oral antihistamines. It therefore differs from swimmer's itch from the therapeutic point of view too, in this case requiring anti-schistosome treatment (albendazole 400 mg per day po for 3 days).

SELECTED READINGS

1. Camarasa JG,Nogues Antich E, Serra-Baldrich E. Red Sea Coral contact dermatitis. Contact Dermatitis 1993; 29:285-286.
2. Freudenthal AR, Joseph PR. Seabather's eruption. N Engl J Med 1993; 329:542-544.
3. Holmes JL. Marine stingers in far north Queensland. Australas J Dermatol 1996; 37 (Suppl 1):S23-226.
4. Stein MR, Marraccini JV, Rothschild NE, Burnett JW. Fatal Portuguese man-o'-war (Physalia physalis) envenomation. Ann Emerg Med 1989; 18:312-315.
5. Wong DE, Meinking TL, Rosen LB, Taplin D, Hogan DJ, Burnett JW: Seabather's eruption. J Am Acad Dermatol 1994; 30:399-400.

O. Reaction to Arthropods

Reactions to Arthropods (Spiders)

Edmundo Velazquez-Gonzalez

More than 30,000 species of spiders are known, but less than 60 have medical implications (Wilson CD, King Le Jr. Spiders and Spider Bites. Dermatologic Clinics 1990; 8:277). Two of these species are reviewed here because of their importance: *Latrodectus mactans* (black widow) and *Loxoceles* (brown recluse spiders).

GEOGRAPHIC DISTRIBUTION

Latrodectus mactans and *Loxoceles spp.* are distributed worldwide, *L. reclusa* occurs in the Southeast United States and Mexico. *L. laeta* occurs in South America (Chile), Central America and the Southeast United States.

ETIOLOGY

In North America *Latrodectus mactans* is known as "black widow," in Argentina as "arana de lino," in Bolivia as "micomico," in Chile as "arana brava" and in Peru as "lucacaha." The poison is a neurotoxin (alpha-lathrotoxin) that depletes acetylcholine at motor nerve endings. Brown spiders are from the *Loxoceles* spp, brown recluse (*Loxoceles reclusa*) or South American brown reclusa (*Loxoceles laeta*). Their bite causes a distinctive cutaneous necrosis called necrotic arachnoidism (Fig. 64.1.1). Sphingomyelinase D has been identified as the toxic factor responsible of the cutaneous necrosis, hemolysis and systemic involvement.

CLINICAL PICTURE

The local reaction of *Latrodectus mactans* bite is often unremarkable or only mildly erythematous. But within 30 min extracutaneous manifestations ensue: lymphadenopathy, myalgias, abdominal wall muscle cramps that may simulate an acute abdomen, nausea, fever, tachypnea and hyperreflexia. Death occurs in 1% of cases (*Ann Emerg Med* 1987; 16:18). The dermatologic manifestations of brown recluse spider bites vary from mild irritation with two points of inoculation at the bite to full-thickness necrosis of the skin. There is pain, tumefaction, induration, edema, lymphangitis and erythema with vesicles and pustules (Fig.

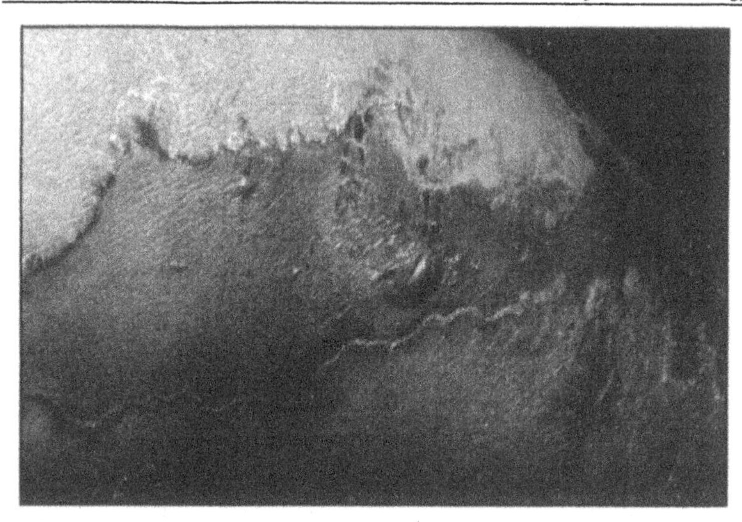

Fig. 64.1.1. Cutaneous loxocelism (Courtesy of Patricia Chang).

64.1.2). Systemic involvement occurs in 10% of cases associated with necrotic lesions: fever, chills, tachycardia, vomiting, diarrhea, icterus, arthralgia, hematuria, hemoglobinuria, general weakness, dry skin and convulsions, especially seen in children.

LABORATORY DATA

Laboratory tests are nonspecific. Leukocytosis, albuminuria and hemoglobinuria occur. In brown recluse bites there can be massive hemolysis causing the erythrocyte count in the peripheral blood to fall below 1 million per mm^3 with a corresponding reduction of hemoglobin.

TREATMENT

Treatment is aimed at decreasing pain and controlling muscle spasms. The principal agents used are narcotics, muscle relaxants and—the most effective—10% calcium gluconate 1-2 ml/kg without exceeding 10 ml, iv slowly. The dose may be repeated according to the clinical response. The mechanism of action is unknown. Small children and elderly cardiac patients might require hospitalization. There is an antidote isolated from horse serum, but there is a risk of anaphylaxis. This antidote is reserved only for high risk cases. In brown recluse bites the treatment includes rest, elevation of the extremity and application of ice compresses. The use of corticosteroids is controversial. They are usually reserved for

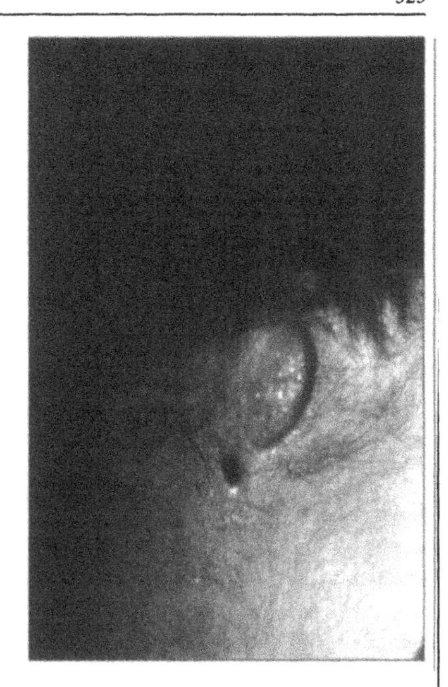

Fig. 64.1.2. Loxocelism with secondary ulceration.

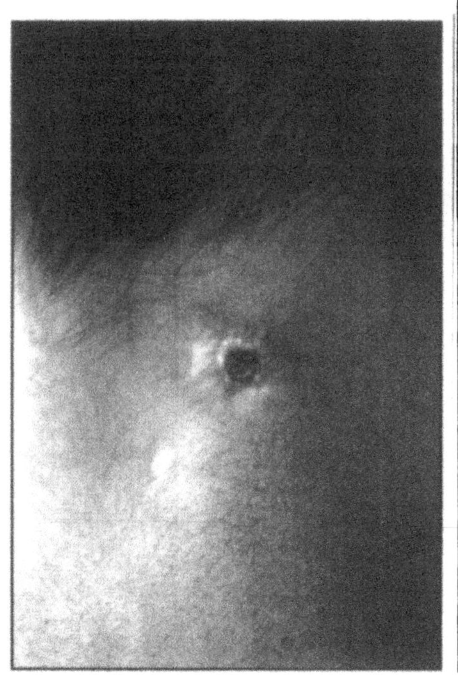

Fig. 64.1.3. Loxocelism after dapsone 1 mg/kg/day.

systemic involvement and for children. The dosage is 1 mg/kg/day. Early surgical excision is reserved for wounds that progress to full-thickness necrosis of the skin. Antibiotics are indicated if there is a secondary bacterial infection. Antitetanus prophylaxis is recommended. In serious cases of systemic illness, blood transfusion is indicated-especially in children less than 8 years of age. A specific antidote has been developed at Vanderbilt University in Tennessee (615-322-BITE). If the antidote is not available, dapsone 100 mg/day in adults and 1 mg/kg/day in children (*JAMA* 1983; 250:646) is indicated. The presence of glucose phosphate dehydrogenase, indicates iatrogenic hemolysis, methemoglobinemia and leukopenia due to dapsone (Fig. 64.1.3).

Selected Readings

1. Kenneth SW, Adel AFM. Tropical and Geographical Medicine, 2nd ed. New York: McGraw-Hill. 1990:562.
2. King LE Jr, Rees RS. Dapsone treatment of a brown recluse bite. JAMA 1983; 250:646.
3. Moss HS, Bindeh LS. A retrospective review of black widow spider envenomation. Ann Emerg Med 1987; 16:188.
4. Newcomer VD, Young EM Jr. Heridas especiales y urgencias. Clin Dermatol 1993; 11:747.
5. Wilson CD, King LE Jr. Spiders and spider bites. Dermatologic Clinics 1990; 8:277.

Reactions to Arthropods (Scorpions)

Edmundo Velazquez-Gonzalez

Scorpions are eight-legged arthropods that live mainly in warm climates. There are 650 species worldwide. The poisonous species are: in North Africa, *Androctonus australis*; in Turkey and North Africa, *A. crassicauda*; in the Mediterranean, *Buthus occitanus*; in Trinidad and Venezuela, *Tityus trinitatis*; in Brazil and Argentina, *T. serrulatus*; in California, New Mexico, Arizona, Baja California, *Centruroides sculpturatus*; and in Mexico, *C. suffusus* and *C. noxius*.

GEOGRAPHIC DISTRIBUTION

Worldwide.

ETIOLOGY

Bites of these species can be lethal: *A. australis*, *C. infamatus*, *C. noxius*, *T serrulatus*, *Leivrus quinquestriatus* (Egypt and Israel).

CLINICAL PICTURE

Scorpion bites produce a painful burning sensation or paresthesias localized to the wound, generally within the first hour. Systemic manifestations include tachycardia, hypertension, respiratory insufficiency, cardiovascular collapse, diaphoresis, pallor, sialorrhea, confusion, anxiety, nausea, abdominal wall muscle cramps and allergic reactions such as bronchial constriction.

LABORATORY DATA

There are no specific laboratory tests.

TREATMENT

Immersion of the affected extremity in cold water and application of a proximal tourniquet—both to reduce the displacement of the toxin. IV fluid replacement is crucial. In cases of vascular collapse, vasopressors and intravenous corticosteroids are indicated. Calcium gluconate may also be administered. Scorpion antidote may be obtained in the Arizona Poison Control Center (602-626-6016).

Selected Readings

1. Kenneth SW, Adel AFM. Tropical and Geographical Medicine, 2nd ed. New York: McGraw-Hill. 1990:562.
2. King LE Jr, Rees RS. Dapsone Treatment of a brown recluse bite. JAMA 1983; 250:646.
3. Moss HS, Bindeh LS, A retrospective review of black widow spider envenomation. Ann Emerg Med 1987; 16:188.
4. Newcomer VD, Young EM Jr. Heridas especiales y urgencias. Clin Dermatol 1993; 11:747.
5. Wilson CD, King LE Jr. Spiders and Spider Bites. Dermatologic Clinics 1990; 8:277.

P. Dyschromias

Ashy Dermatosis

Ashy Dermatosis

Roberto Arenas

Erythema dyschromicum perstans, also known as Oswaldo Ramirez's disease or figurative chronic erythema with melanodermia, is a pigmented dermatosis of unknown etiology. It is characterized by asymptomatic, blue-gray spots with marginal erythema. It is chronic dermatosis for which there is no specific treatment. It is perhaps influenced by inflammatory, hormonal or environmental factors.

GEOGRAPHIC DISTRIBUTION

This dermatitis occurs in America, Europe, and Asia following a limited fringe northward to Finland and southward to Ecuador. It predominates in Latin America. It is observed in people of brown skin (type IV), especially those from 20-40 years of age. It predominates in women and practically never affects Caucasians.

ETIOLOGY

The cause is unknown; perhaps there is a genetic basis. Accumulations of cytotoxic T lymphocytes interact with melanocytes to cause vacuolar degeneration of the dermoepidermal layer and incontinence of pigment. This disease is often confused with pigmented lichen planus or it is classified as a variety of the lichen planus. The influence of factors such as sunlight, climate, environmental contaminants, nutrition, occupation, hormones, inflammation, parasitosis and infections is unknown.

CLINICAL PICTURE

It presents as a disseminated dermatitis of the face, neck, trunk and proximal part of the extremities. It spares hairy skin, palms and soles, and it is rare on mucosae. Blue-gray or ashy-colored macules of 0.5 to several centimeters in diameter, with an elliptic form, appear in bands or irregularly. At the outset they are surrounded by an erythematous, circinated edge that usually disappears with time and leaves a hypochromic halo (Fig. 65.1). There are generally no symptoms except for mild pruritus early on. Many times there is dermographia. The disease

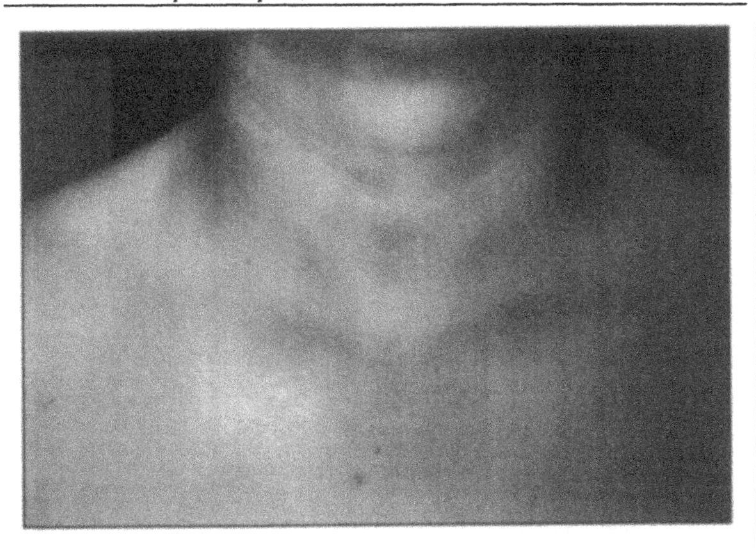

Fig. 65.1. Ashy dermatosis.

must be distinguished from pigmented lichen planus, a variety of lichen planus characterized by brown-gray or plain gray spots that generally affects the face, neck and trunk (*Actas Dermosifilogr* 1996; 87:38-40). This clinical form is difficult to distinguish from actinic lichen planus or lichenoid melanodermatitis that manifests as annular plaques of violaceous pigment on exposed skin which predominates in tropical climates (Fig. 65.2). Melanosis caused by friction or macular amyloidosis secondary to the use of nylon brushes is a dark melanodermia of reticular aspect that appears in thin people with brown skin, especially in Latins or Asians. This disease is due to mechanical repeated irritation of bony protuberances such as clavicle, scapula and vertebrae (Fig. 65.3).

LABORATORY DATA

The lesions are not obvious with a Wood's lamp. On biopsy there is orthokeratotic hyperkeratosis sometimes more evident at the follicules, vacuolization of the basal cells and pigment incontinence, with the presence of colloid or Civatte bodies. In the superficial dermis there is a mild perivascular lymphocytic infiltrate and abundant melanophages (Fig. 65.4).

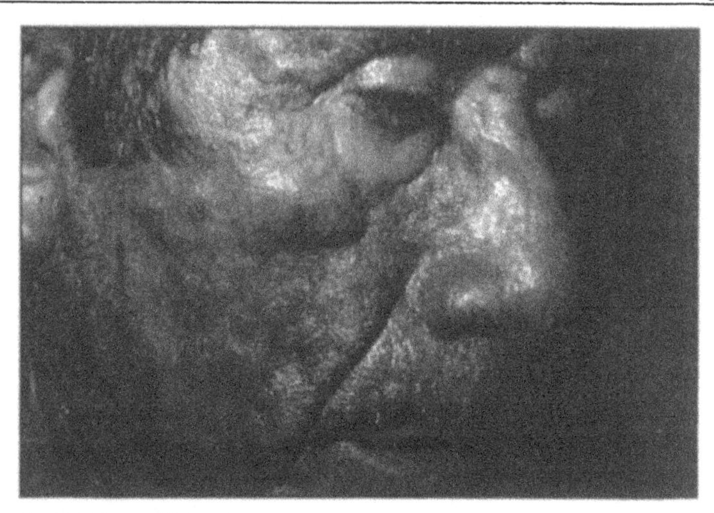

Fig. 65.2. Pigmented lichen planus.

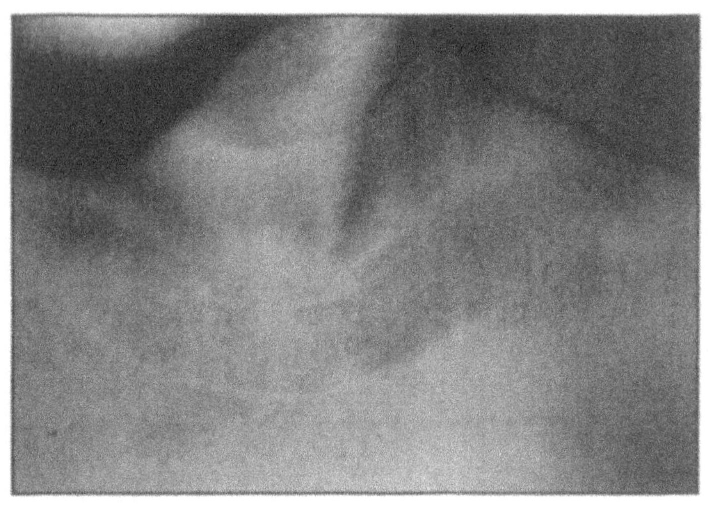

Fig. 65.3. Melanosis caused by friction.

TREATMENT

There is no effective treatment. The following have been used: Antihistamines, vitamins, glucocorticoids, antibiotics, diaminodiohenylsulfone, isoniazid, griseofulvin, chloroquine, estrogens and placebo. Clofazimine 50-100 mg 3 times a week for 3-5 months yields the best results (*Int J Dermatol* 1989; 28(3):198-200).

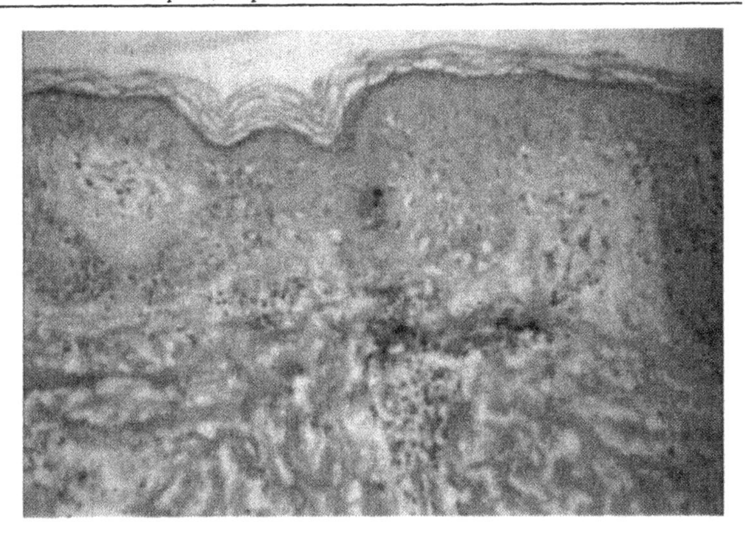

Fig. 65.4. Ashy dermatosis, histpathology (H.E. 20X).

Selected Readings

1 Arenas R, Bautista M et al. Dermatosis cenicienta estudio de 8 pacientes tratados con clofazimina. (Ashy Dermatosis: The study of 8 patients treated with clofazimine) Rev Soc Colombiana Dermatologia 1992; 1(3):103-105.

2 Baranda L, Torres-Alvarez B, Cortes-Franco R et al. Involvement of cell adhesion and activation molecules in the pathogenesis of erythema dyschromicum perstans (ashy dermatitis). Arch Dermatol 1997; 133:325-29.

3 Piquero-Martin J, Perez-Alfonzo R, Abrusci V et al. Clinical trial with clofazimine for treating erythema dyschromicum perstans. Int J Dermatol 1989; 28(3):198-200.

4 Sumitra S, Yesudian P. et al Friction amyloidosis: a variant or an etiologic factor in amyloidosis? Int J Dermatol 1993; 32:302-7.

5 Vega-Memije ME, Waxtein L, Arenas R, Hojyo MT, Dominguez-Soto L et al. Ashy dermatosis and lichen planus pigmentosus: A clinicophathologic study of 31 cases. Int J Dermatol 1992; 31(2):90-94.

Q. Others

Acne

Ainhum (Spontaneous Dactylosis)

Actinic Prurigo

Keloids

Verruga Peruana: An Infectious Endemic Angiomatosis

Communitary Dermatology (Epidermiology and Dermatology)

Acne

Roberto Cortes-Franco

Acne is perhaps the most common dermatological ailment in the world and the most frequent cause of dermatological consultations. It is an inflammatory disease of the pilosebaceous follicle that occurs in the adolescence although it persists with frequency to adult life. It commonly causes psychological problems at an age in which physical appearance is critical for the proper development of personality.

ETIOLOGY

The etiology is multifactorial. Testosterone and adrenal androgens, especially dehydroepiandrosterone, stimulate the development of the sebaceous glands and the production of sebum. As the production of sebum increases, levels of linoleic acid (an essential fatty acid that when deficient induces hyperkeratinization) decrease by dilution favoring an increase in the proliferation and cohesion of epithelial follicular cells. The result is obstruction of the follicular lumen by keratin, cellular debris and sebum to form micro-blackheads (closed-comedones). The blackheads are colonized by *Propionibacterium acnes*, a diphtheroid anaerobe gram-positive bacteria that secretes lipases that hydrolyze the triglycerides of the sebum to form free fatty acids that are pro-inflammatory and comodogenic. Moreover, *P. acnes* secretes chemotactic factors, proteolytic enzymes and hyaluronidase that contribute to the inflammatory process (development of papules and pustules). The wall of the blackheads eventially burst; its contents enter the dermis and enhance the inflammation.

CLINICAL PICTURE

Acne affects mainly the face, but it is also found on the neck, chest, back and shoulders. The lesions may or may not be inflamed. The non-inflammatory lesions, comedones, can be closed or open. The inflammatory lesions are the papules, pustules and abscesses. There are two rare and severe forms of acne:

1) conglobata in which inflammatory lesions such as abscesses, nodules, cysts and fistulae predominate and
2) fulminans characterized by inflammatory lesions, extensive necrosis of the affected skin and systemic complications such as fever, arthralgias and erythema nodosum (Fig. 66.1).

Any form of acne may leave erythematous or hyperpigmented spots that disappear with the time. Yet sometimes there is scarring, especially in severe forms that tend to be very difficult to treat. In some patients their acne worsens in warm and humid climates. This is due perhaps to overhydration of the corneocytes that exacerbates the plugging of the pilosebaceous ducts. Tropical or estival acne (Majorca's acne) has been described in people whose acne worsened after returning from a beach vacation probably because the application of tanning oils and sunscreens on the face may have a comodogenic effect. Similarly, patients who use retinoic acid or isotretinoin and expose themselves to the sun, develop erythema and severe skin irritation.

TREATMENT

The ideal treatment for acne is designed to control excessive production of sebum, to eliminate *P. acnes*, to normalize follicular keratinization and to decrease inflammation. Topical and systemic medications are available, and their use in combination for long periods is common.

Topical Agents

Retinoic acid (tretinoin) affects follicular keratinization, and because of this, it is an excellent comedolytic. In is available in cream, gel and solution in concentrations of 0.01% to 0.1%. It may be used as the sole treatment in cases of exclusively comedonic acne by applying it on the face twice a day. When comedones are inflamed, retinoic acid should be combined with topical antibiotics or benzoyl peroxide, and in severe cases, with systemic antibiotics. The most common side effects of retinoic acid are erythema, dryness, scaling and burning of the skin, affecting most of the individuals who use it. The side effects are more severe when the concentration is higher, but tend to be transient. They diminish or completely disappear within 3-4 months of continuous use. Although the systemic administration of retinoids is contradicted during pregnancy, the topic application of tretinoin seems to be safe. In some countries a topic formulation of isotretinoin is available in 0.05% cream. Its mechanism of action, dose, side effects, indications and efficacy are the same as for retinoic acid. Its safety during pregnancy is not yet demonstrated. Adapalene gel 0.1% is a new topical retinoid similar to others and with a potency equivalent to 0.025% retinoic acid. It is believed to cause less irritation than retinoic acid (*J Am Acad Dermatol* 1996; 34:482-3). Azelaic acid is a bicarboxylic aliphatic acid (nonanodioic acid); it is bactericidal, it decreases the proliferation and enhances the differentiation of the follicular keratinocytes, inhibits 5-(alpha)-reductase enzyme and decreases lipogenesis. Unfortunately, in vivo its efficacy is reduced, and in general, it is considered less effective than retinoic acid. It yields better results in light to moderate comedonic acne, and it is usually necessary to combine it with a topical or systemic antibiotic. It produces erythema and mild, transient scaling of the skin. Topical antibiotics are useful in the management of inflammatory acne, especially if combined with a topical retinoid. The most commonly used are clindamycin 1% in lotion, gel and solution and

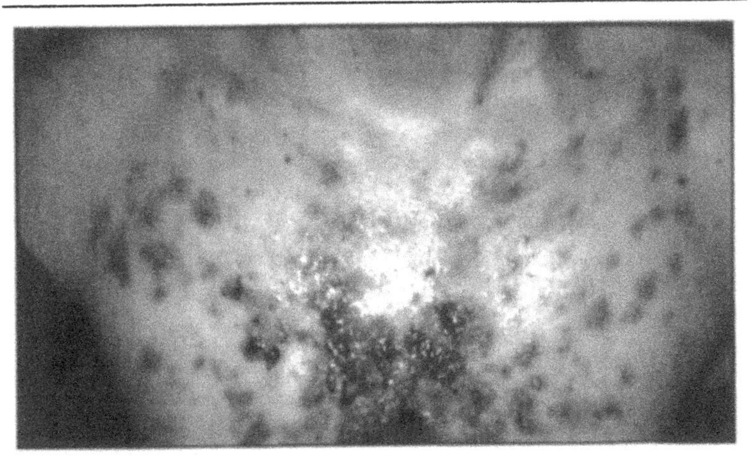

Fig. 66.1. Acne fulminans..

erythromycin 2-4% in gel or solution. The efficacy of both is similar. They reduce *P. acnes* and the follicular inflammatory phenomena caused by it. Generally they are combined with a topical retinoid in cases of mild acne with inflammatory lesions and with systemic antibiotics in moderate and severe cases. They are well tolerated and are practically free of side effects. A limitation of their continous use is the development of resistance. To avoid resistance, topical antibiotics are periodically omitted and substituted by benzoyl peroxide for a short time, e.g., after 3-4 months of treatment with topical erythromycin, it is stopped and benzoyl peroxide is substituted for one month. There is a preparation that combines erythromycin 2% with 5% benzoyl peroxide. Benzoyl peroxide, bacteriostatic and comedolytic, is available in gels, lotions and creams in 2.5-10% concentrations. It is usually applied once a day combined with a retinoid or a topical antibiotic. It is an irritant, especially in high concentrations, causing erythema, scaling, dryness and even contact dermatitis. Skin tolerance improves with a minimum of irritation with continuous use. It can bleach the color out of the clothing and is irreversible. It is recommended that benzoyl peroxide be started at low concentrations; the concentration can be increased, if necessary, as skin tolerance increases.

Systemic Antibiotics

As *P. acnes* populations are reduced, the inflammatory cascade that this bacteria provokes diminishes. Tetracyclines, which inhibit neutrophil chemotaxis, are indicated in moderate to severe cases with inflammatory lesions. Systemic antibiotics are generally used in combination with topical therapy. The most common are tetracycline, 250-2000 mg/day, minocyclin, 50-200 mg/day, and doxycycline 50-150 mg/day. Often they are administered for months or years without side effects. Every now and then there is gastric intolerance, drowsiness or vulvovaginal candidiasis. They should be taken on an empty stomach, especially tetracycline, because intestinal absorption decreases with foods that contain calcium or zinc. Also, exposure to the sun should be avoided because they are photosensitive medications, especially doxycycline. In children under 10 years of age and in pregnant women antibiotics are contraindicated because they cause retardation in bone growth and irreversible tooth pigmentation. In these situations, in allergic individuals or in those who cannot tolerate the tetracyclines, other antibiotics can be used such as erythromicin, 500-2000 mg/day, sulfamethoxazole/trimethoprim, 800/160 mg q.d. or b.i.d., or dapsone, 25-150 mg/day. Isotretinoin is sebostatic. It lessens the affects of follicular keratinization, inhibits the proliferation of *P. acnes* and is anti-inflammatory. It is useful for severe forms of acne (conglobata or fulminans) as well as for the moderate to severe forms with a significant inflammatory component. The dose is 0.5-1 mg/kg/day for 16-20 weeks. Generally, once treatment is suspended the acne remains in remission for long periods of time or indefinitely. When treatment begins, other topical or systemic anti-acne medication should be stopped. Isotretinoin causes dryness of the skin, lips and ocular and nasal mucosae that can be improved with moisturizing creams, lip protecters, artificial tears and petrolatum, respectively. It is highly teratogenic, so women in childbearing years who use it should use some kind of birth control one month before starting the treatment and up to one month after it is completed. Before beginning the medication hepatic function should be evaluated and cholesterol and serum triglycerides must be checked one month after the onset of treatment. If normal, it is not necessary to repeat them. But if abnormal they should be repeated monthly and for one month after treatment is complete. Some patients have significant elevations of lipids that justify the suspension of treatment. Side effects, high cost and its high teratogenicity limit their use in a great number of patients. Cyproterone acetate has antiandrogenic and sebostatic effects. It should only be used in women with severe acne who have not responded to conventional therapy. The best result is observed in those that have high serum levels of dehydroepiandrosterone sulfate (DHEA-S). If these patients also have high levels of free testosterone, they can benefit from low dose prednisone. Cyproterone acetate is available in coated pills that contain 2 mg of cyproterone acetate in combination with 0.035 mg of ethinylestradiol. It is administered beginning the first day of the menstrual cycle and for 21 consecutive days. Then treatment is omitted for 7 days, and a new cycle is resumed. It can be used for 6-24 months and its beneficial effect may be observed from the third cycle. It has a good contraceptive effect; accordingly it can be used in women who will also receive isotretinoin. The use of non-steroidal anti-inflammatory drugs (NSAIDs) such as ibuprofen 400-600 mg tid during the first months of usage accelerates the response of the systemic antibiotics in the inflammatory acne. NSAIDs can be discontinued after 2-3 months of continuous use.

Management of Sequelae

Scars and pigmented areas often remain after the inflammatory lesions disappear. The manipulation of lesions increases the risk of scarring. The compression and extrusion of comedones and intralesional infiltration (in cysts or abscesses) of steroids can be done occasionally. Yet the management of scars is difficult and often inadequate. Some deep puntiform scars or fibrous tracts can be treated surgically. Hypertrophic and keloidal scars will respond to the intralesional injection of triamcinolone. For moderately deep scars and for the lesions of atrophodermia, mechanical dermabrasion and "peelings" have been employed with variable success. The best therapeutic tool for the management of scars is perhaps the CO_2 laser.

Complementary Measures

Mild soap or soap substitutes during the first months of treatment are recommended. Most topical agents and systemic isotretinoin produce dryness of the skin that may be exacerbated by the detergent effect of the soaps. Similarly, exposure to sunlight increases irritation induced by the treatment. There are no special diets for the patients with acne. A relationship between diet and acne has never been demonstrated. But if patients insist that certain foods exacerbate their acne, those items should be avoided. Strict diet is difficult and has not been demonstrated to cause an improvement, but unfortunately no improvement has been reported.

Selected Readings

1 Berson DS, Shalita AR et al. The treatment of acne: The role of combination therapies. J Am Acad Dermatol 1995; 32:S31-417.
2 Shalita A, Weiss JS, Chalker DK et al. A comparisson of the efficacy and safety of adapalene gel 0.1% and tretinoin gel 0.024% in the treatment of acne vulgaris: A multicenter trial. J Am Acad Dermatol 1996; 34:482-7.
3 Sykes Jr NL, Webster GF. Acne. A review of optimum treatment. Drugs 1994; 48(1):59-70.

Ainhum (Spontaneous Dactylolysis)

Roberto Arenas

This disease is also known as serraro and quijilla in Brazil and in Africa as Banko-Kerende, Sukhapakla. It is the spontaneous amputation of a toe by a chronic, progressively constricting fibrous band.

GEOGRAPHIC DISTRIBUTION

This disease is almost exclusively limited to the in Black race. It predominates in Africa although it has been seen in Asia, America, especially in the Antilles, the Middle East and Oceania. In Nigeria the incidence is 1.7 per thousand. It is more common in men (2:1), is often familial, and it is more frequent in farmers 30-50 years old. It is occasionally seen in children.

ETIOLOGY

The etiology is unknown. It is due to a fibrotic reaction in genetically predisposed individuals. It is preceded by minimal trauma when walking barefoot. It has been related to inadequate posterior tibial artery circulation and to absence of the arterial plantar arch.

CLINICAL PICTURE

Constriction is caused by a fibrous band that affects the fifth toe bilaterally (75%), sometimes the fourth, and rarely the fingers. The band inserts at the digital-plantar junction. Progressive constriction leads to atrophy. The process is chronic and it is accompanied by little to moderate pain; sometimes there is secondary infection. The toe swells, becomes globous and cyanotic. Within approximately 5 years the toe spontaneously sloughs off. Pseudo-ainhum presents in much the same way but the cause is known, such as vascular, sensory, diabetes, scleroderma, syringomyelia, leprosy or a genodermatosis like diffuse plantar keratodermia, meleda disease or keratoderma hereditaria mutilans (Vohwinkel syndrome). It has even been seen in psoriasis or it is acquired by the wrapping with hairs or fibers (Fig. 67.1)

Tropical Dermatology, edited by Roberto Arenas and Roberto Estrada. ©2001 Landes Bioscience.

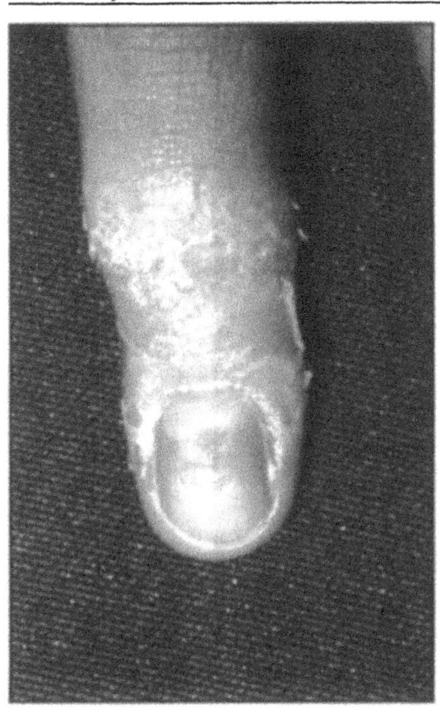

Fig. 67.1. Pseudo-ainhum.

LABORATORY DATA

X-ray reveals decreased osseous density with thinness of the cortex and decreased diameter of the phalanx. There may be absorption of the medial-distal part of the middle phalanx. Doppler demonstrates decreased blood flow in the posterior tibial artery. On biopsy there is hyperkeratosis and fibrosis of the connective tissue.

TREATMENT

There is no effective treatment. If there is infection, antibiotics are indicated. Sometimes intralesional corticosterods are useful. Surgery, with Z-plasty or amputation sometimes is necessary. In pseudo-ainhum oral retinoids, such as tretinoin 1 mg/kg/day (*Australas J Dermatol* 1992; 33(1):19-30), have been used and antifibrotic agents like tranilast have been tested (*J Hand Surg Br* 1995; 20(3): 338-41).

Selected Readings

1 Canizares O. Ainhum. In Canizares O, Harman R. Clinical tropical dermatology. Oxford:Blackwell, 2nd ed. 1975: 670-71.

2 Palungwachira P, Iwahara K, Ogawa H. Keratoderma hereditaria mutilans. Etretinate treatment and electron microspcope studies. Australas J Dermatol 1992; 33(1):19-30.

3 Pierard GE, Franchimont CF, Arrese-Estrada J. Hyperplasies conjoctives. In: Pierard GE, Caumes E, Franchimont C, Arrese-Estrada J et al, eds. Dermatologie tropicale. Bruxells:AUPELF, 1993:586-87.

4 Pisoht, Bhatia A, Oberlin C. Surgical correction of pseudo-ainhum in Vohwinkel syndrome. J Hand Surg Br 1995; 20(3):338-41.

Actinic Prurigo

Esther Guevara, Elisa Vega-Memije Ma Teresa Hojyo, Roberto Cortes-Franco, Luciano Dominguez-Soto

Actinic prurigo (AP) is a photodermatosis that has been called solar prurigo, polymorphous light eruption of prurigo type, and solar dermatitis.

GEOGRAPHIC DISTRIBUTION

Most patients live above 1000 meters altitude although there have been isolated cases in individuals living at the sea level. It affects mestizos of Latin America and American Indians whose skin phototypes are IV and V.

ETIOLOGY

The etiology is not known. Studies demonstrate:

a) the lesions can be caused by UVA light as well as UVB (*Dermatologia Rev Mex* 1993; 37:328);
b) increased T lymphocytes in pheripheral blood with a predominance of helper cells;
c) predominance of HLA-A-28 and HLA-DR4 subtype 0407; and
d) by immuno-histochemistry. B lymphocytes are found at the center of the lymphoid follicles and T lymphocytes in the periphery where no lymphoid follicles are formed both subtypes intermix as has been confirmed with studies of genetic rearrangement.

Elevated levels of tumor necrosis factor alpha and IgM have been found. These recent studies indicate the important role of cellular and humoral immunity in the physiopathology of the disease (Arrese-Estrada J, Guevara E, Alfaro G. First International Meeting SIDLA. Mexico).

CLINICAL PICTURE

Actinic prurigo can begin in infancy, but the highest incidence is 6-8 years of age, with a slightly more female cases. The dermatitis is generally disseminated, bilateral and symmetric. It affects sun exposed areas such as cheeks, dorsum of the

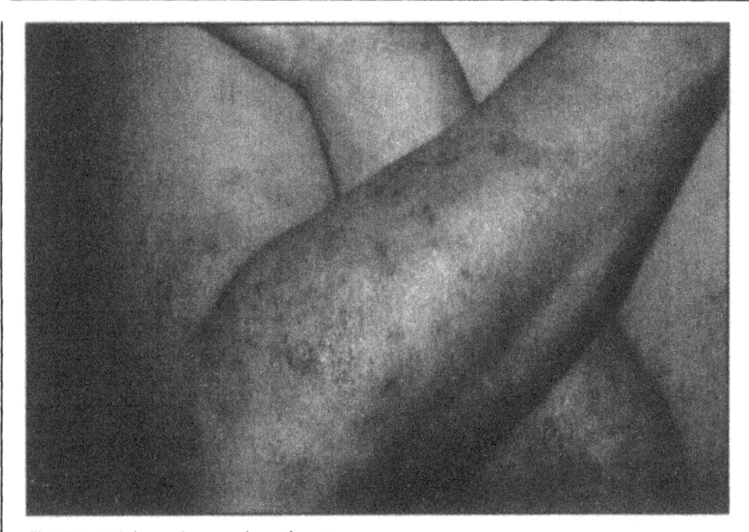

Fig. 68.1. Actinic prurigo, papules and crusts.

nose, forehead, chin, ear lobes, V of the neck and chest, the extensor surface of the arms and forearms, and the dorsum of the hands (Fig. 68.1). It can affect any area exposed to the sun. In 80% of cases the lips are involved and in 10% of cases the lips are the only site of involvement. In 45% of cases the conjunctivae are affected (Fig. 68.2). Lesions are erythematous papules, single or in very pruriginous groups and they can form large plaques. The lesions with hematic crusting, and since the ailment is chronic, lichenification is eventually seen. Chronic scratching of the face can produce pseudoalopecia of the eyebrows. The usual complications are contact dermatitis and impetigo. Lesions of the lip are manifested by cheilitis with edema, scales, fissures, ulceration, scabs and hyper-chromic spots. Conjunctival involvement is manifested by conjunctival hyperemia, epiphora, brown pigmentation, formation of pseudopterigium and photophobia. The differential diagnosis includes atopic photosensitivity dermatitis and chronic actinic dermatitis; dermatitis caused by chronic photocontact, and actinic reticuloid.

LABORATORY DATA

On biopsy there is hyperkeratosis, parakeratosis, regular acanthosis and thickening of the basal membrane. In the dermis a dense inflammatory infiltrate composed of lymphocytes can be seen. Biopsy of the lip reveals hyperkeratosis with parakeratosis, crusting, regular acanthosis, spongiosis, vacuolization of the basal layer and ulceration of the epithelial. In the chorion, there is edema of the stroma, vasodilatation and congestion of the blood vessels. But the most characteristic finding is a dense lymphoplasmocytic infiltrate that tends to form

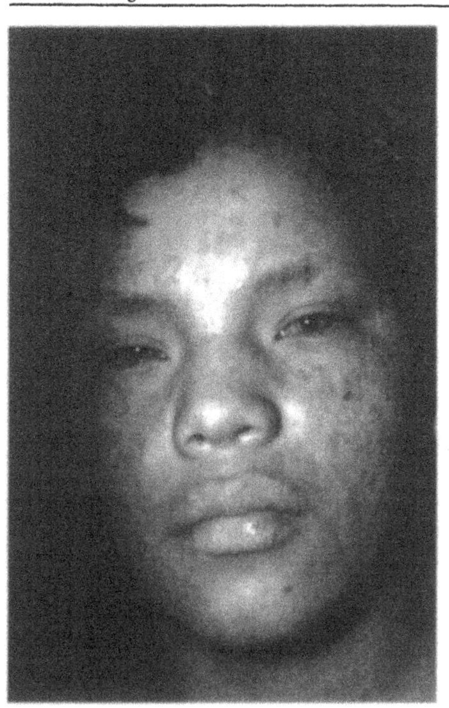

Fig. 68.2. Cutaneous lesions, conjunctival affection, and cheilitis.

lymphoid follicles. Many eosinophils and plasma cells can be observed. In conjunctivae the histology shows hyperplasia of the epithelium alternating with areas of atrophy and vacuolization of the basal layer. In the chorion, vasodilatation and an inflammatory infiltrate composed of lymphocytes are seen which also form lymphoid follicles. Also, there are eosinophils, and in most cases, incontinence of the pigment is found (*Dermatologia Rev Mex* 1993; 34:295-97).

TREATMENT

It is very important to explain the nature of the disease to patients, as they must accept the daily protection from the sun with hats and umbrellas. The cornerstone of pharmacologic treatment is thalidomide 100 mg/day with strict birth control of women in childbearing years in whom the medication is teratogenic. If there are complications, secondary infection or eczema, then compresses with Burow's solution and copper sulfate 1:1000 can be used. On some occasions steroids and topical antibiotics are indicated. Once the skin lesions remit, sunscreens should be used. Thalidomide can be gradually reduced; it can be reinstituted in case of relapse. Other medications have been used with less favorable results: antihistaminics, antimalarials, beta carotenes and PUVA.

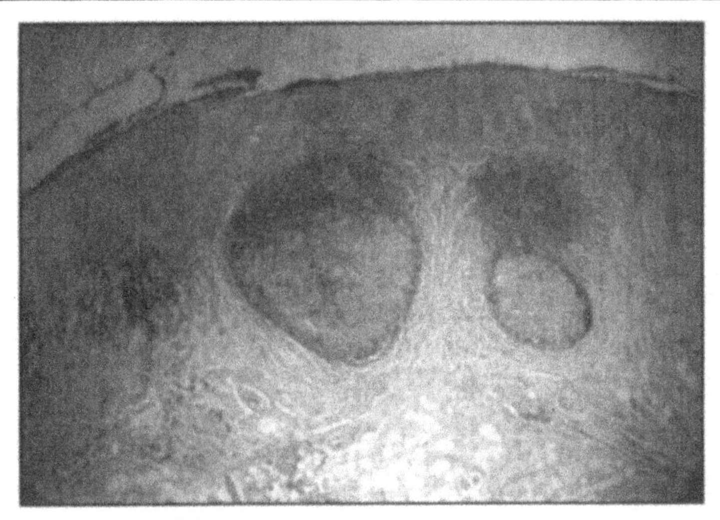

Fig. 68.3. Lymphoid follicles (H.E. 20X).

Selected Readings

1 Arenas R. Dermatologia. Atlas diagnostico y tratamiento. 2nd ed. (Atlas diagnostic and treatment) Mexico:Interamericana/McGraw-Hill. 1996: 72-73.

2 Hoyjo Tomoka M, Vega Memije E, Granados J et al. Actinic prurigo: An update. Int J Dermatol 1995; 34:380-84.

3 Hojyo MT, Vega ME, Cortes R et al. Prurigo actinico como modelo de fotodermatosis cronica en Latinoamerica. Med Cut ILA Lat Am 1996; XX: 265-77.

4 Novales J. Prurigo actinico. Caracteristicas clinicas. Dermatologia Rev Mex 1993; 37:293-94.

Keloids

Roberto Arenas and Josefina Carbajosa

Keloid is an exuberant and persistent scar that extends into nearby normal tissue and causes pruritus or burning.

GEOGRAPHIC DISTRIBUTION

The incidence is 4.5-16%. Keloids are seen in any race, age or sex. It predominates between 20-30 years of age and is more frequent in blacks (3.5-15:1) and in women. Its frequency increases during puberty and pregnancy.

ETIOLOGY

The etiology is unknown. There is a definitive genetic influence and its development is related to increased skin tension, growth factors and interleukins (*Int J Dermatol* 1994; 33(10) 681-91; *Int J Dermatol* 1994; 33(11):763-69). Lesions predominate in places with a high concentration of melanocytes. They are related to hypophyseal hormones, e.g., MSH, blood group A and cellular hypoxia. The degree of fibroplasia, cellular metabolism and the amount of intercellular substance vary. In fibroblastos there is an increase in the production of fibronectin,which continues for months or even years. The production of collagen and procollagen It is greater than in a neurotrophic scar, and collagen breakdown is diminished.

CLINICAL PICTURE

Keloids predominate in the deltoid region, sternal area, neck, back and legs (Fig. 69.1). They are related to wounds in areas with tension or folliculitis in the neck. Ear lobes are involved relatively frequently because of piercing required for earrings (Fig. 69.2). The lesions are exuberant, persistent, smooth, firm, hard, non-distensible, skin-colored, slightly pigmented or erythematous, and telangiectatic. They spread in a claw-like manner toward normal tissue, causing pruritus or burning. Rarely they are asymptomatic. The hypertrophic scar is confined to the site of the original lesion; it is less exophytic and symptomatic. It tends to disappear with time (about six months), or it flattens and softens. The clinical and microscopic distinction may be useful.

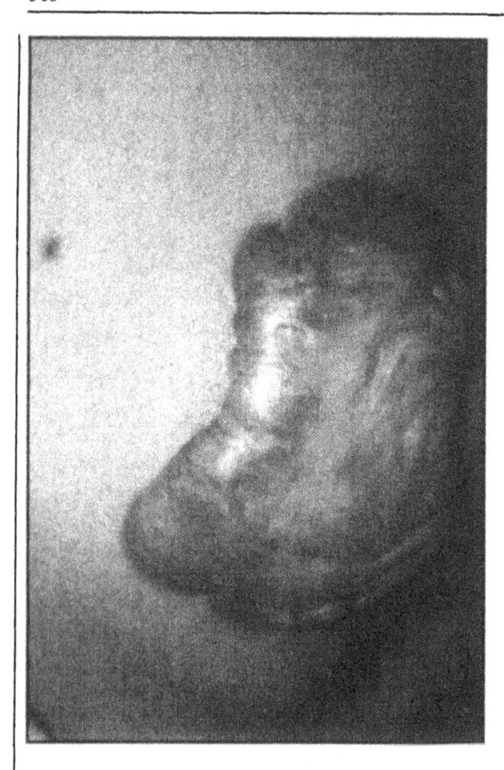

Fig. 69.1. Keloid after BCG-vaccination.

LABORATORY DATA

Biopsy is not recommended, to avoid the stimulation. In keloids the epidermis is flattened, the collagen fibers are thick and hyalinized. They are compact, wide, irregular and form nodules and whorls. There are fibroblasts, extracellular matrix and, in the superficial dermis, there are dilated vessels. In hypertrophic scars the fibrous tissue is immature and the fibroblasts show disposition at random. There is little or no extracellular matrix; the foreign body reaction is prominent and persistence of fibroblasts is frequent. The distinction is only possible by immunohistochemistry and electron microscopy which can demonstrate increased hydroxylase and proline collagenase.

TREATMENT

No treatment is completely effective. Recurrence following excision and cryosurgery is 100%. Excision and immediate application of topical or intralesional (triamcinolone) glucocorticoids or glucocorticoids combined with

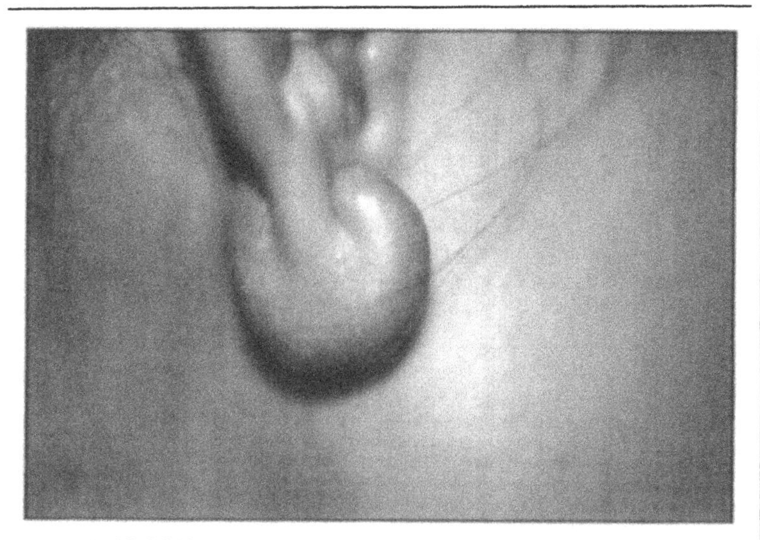

Fig. 69.2. Ear lobe keloid.

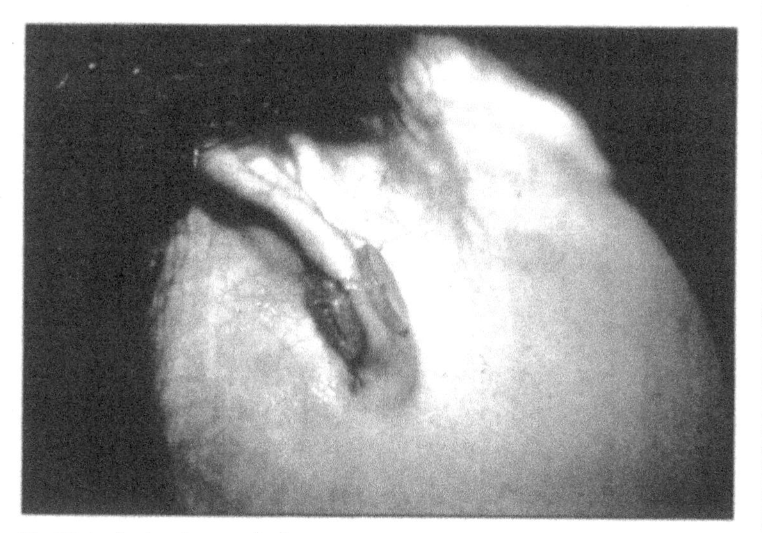

Fig. 69.3. Application of compression buttons.

dimethylsulfoxide (topical) has been employed. Colchicine 1-2 mg/day po and radiotherapy 1200 rad for 1-12 days (mean cumulative biologically effective dose is 28 Gy with efficacy evaluated at intervals of at least 2 years) have been described (*J Am Acad Dermatol* 1994: 31:225-31).

The Flashlamp-pumped pulse-dye laser, 585 nm, produces an improvement. Mechanical compression, involving bandages and specially designed elastic garments, has yielded acceptable results. On the ear lobe, excision followed by external compression with a mechanical prosthesis gives good results (*Ann Plast Surg* 1978; 1(6):579-81). Superior results in the ear lobe have been achieved by application of a plastic button secured with 2-0 Dermalon or thin metallic suture of non-rusting steel (*Br J Plast Surg* 1974; 24:186-87). The button is left in place for a minimum of two months. When removed, compressive ear clips are indicated (Fig. 69.3). Recently, in hypertrophic and keloid scars, occlusive silicon sheeting (polydimethylsiloxane) have been used with good results. Positive pressure is not required; they are placed for at least 12 h/day. Improvement begins after the first month. There are no significant side effects (*Dermatol Surg* 1995; 21:947-51; *Int J Dermatol* 1995; 34(7):506-509). The mechanism of action is not known, but it is probable that angiogenesis is inhibited, by a direct effect on fibroblasts and by hyperhydration of the cellular subcutaneous tissue.

Selected Readings

1. Berman B, Bieley HC. Adjunt therapies to surgical management of keloids. Dermatol Surg 1996; 22(2):126-30.
2. Carbajosa J. Queloides del lobulo de la oreja. Tratamiento con compresion externa. (Keloids of the ear lobe: Treatment with external compression) Dermatologia Rev Mex 1992; 36(6):366-68.
3. Datubo-Brown DD. Keloids: A review of the literature. Br J Plast Surg 1990; 43:70-7.
4. Rockwell WB, Cohen IK, Erlich HP. Keloids and hypertrophic scars. A Comprehensive review. Plast Reconstr Surg 1989; 84:827-37.

Verruga Peruana: An Infectious Endemic Angiomatosis

Hector Caceres-Rios

Verruga peruana (VP) is the eruptive phase of human bartonellosis (HB) or Carrion's disease, which follows an acute hematic phase known as Oroya Fever (OF). HB has been present in Peru since ancient times, as depicted in pre-Inca monoliths. In 1885 a medical student named Daniel A. Carrion decided to prove a common origin for OF and VP. He inoculated himself with the blood of verruga and twenty-one days later he developed OF and died. This experiment proved that OF and VP are two distinct clinical phases of the same illness.

GEOGRAPHIC DATA

Bartonellosis is an infectious endemic angiomatosis limited to the valley regions of the Andes Mountains in South America, including Colombia, Ecuador and more commonly, Peru. Several well-demarcated regions of the Andes situated from 500-3,000 meters above sea level, are the natural habitats of the hematophagous flies of the genus Lutzomyia, which transmit the disease.

ETIOLOGY

The etiological agent is *Bartonella bacilliformis* (Bb) a 3 μm highly polymorphic bacterium which can be found in bacillary or coccoid forms. Bb is closely related to *Rochalimaea quintana* and *R. hanselae*, the etiological agents of bacillary angiomatosis (BA). It is now clear that Rochalimaea and Bartonella belong to the same genus; therefore, the two genera have been merged under the Bartonella designation. *Bartonella henselae* causes cat scratch fever and many cases of BA, *Bartonella quintana* causes trench fever and some cases of BA; and Bb is responsible for HB. HB has four distinct and sequential clinical periods.

CLINICAL PICTURE

The incubation period lasts approximately 21 days. The second period known as OF varies from an oligosymptomatic to fatal course. Malaise, anorexia, arthralgia and high fever followed by rapidly progressive and severe anemia can occur

Tropical Dermatology, edited by Roberto Arenas and Roberto Estrada. ©2001 Landes Bioscience.

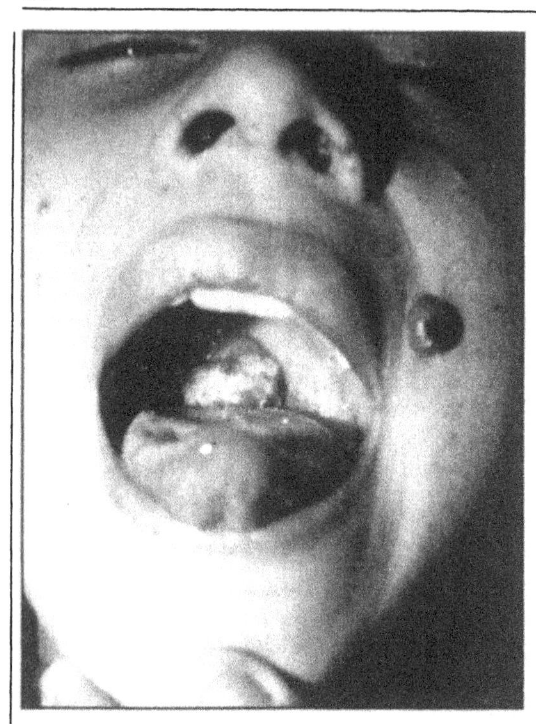

Fig. 70.1. Angiomatous tumors on the skin and mucous surfaces.

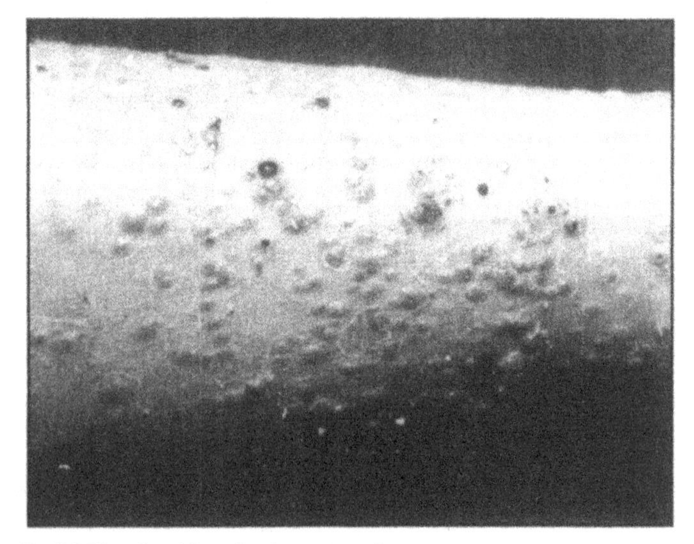

Fig. 70.2. Dissemianted form of angiomatous papules.

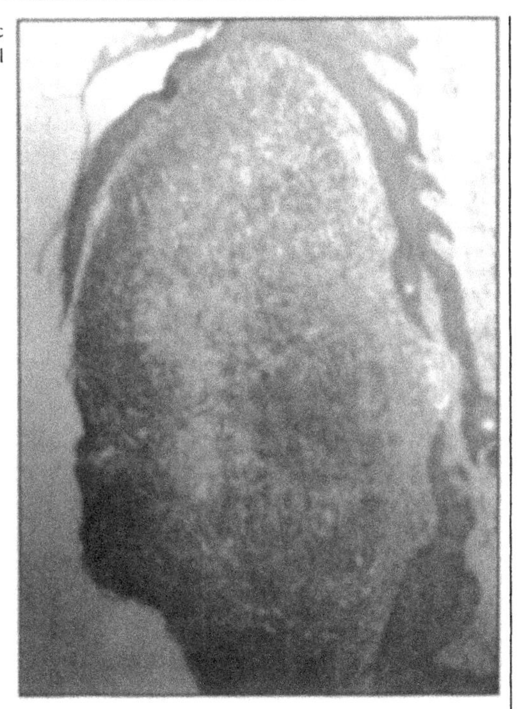

Fig. 70.3. Angioblastic and histiocytic proliferation limited by epidermal ridges (HE 20X).

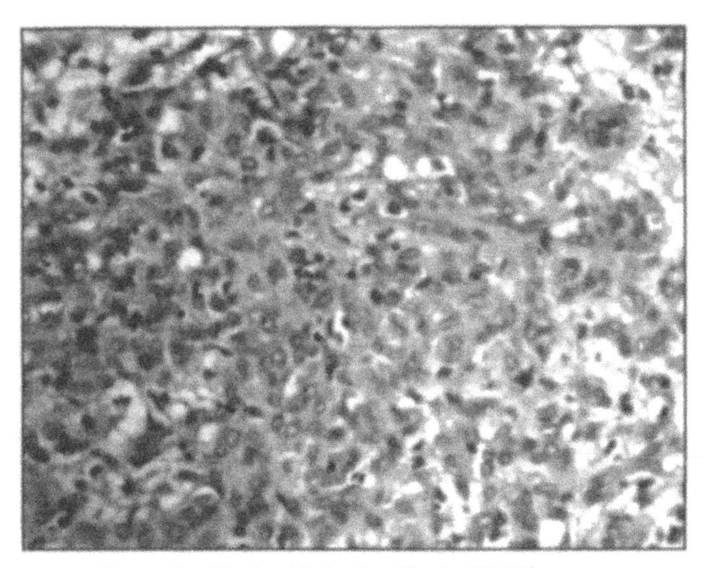

Fig. 70.4. Close-up of angioblastic and histiocytic proliferation (HE 40X).

during the next four weeks. A latency period follows and is characterized by immunobiological equilibrium between host and microbe. When the latency is broken, the eruptive or histioid phase known as VP results. Cutaneous lesions of VP are angiomatous, bleed easily and can be distinguished as papules, nodules or tumors of different sizes. Lesions can appear in the skin or mucous surfaces. Children are affected in about 50% of the cases, and lesions are mainly localized in the face and extremities, in isolated or disseminated forms. The varied morphology of the clinical lesions makes it easy to confuse them with pyogenic granulomas, histiocytomas, warts, ecthyma and folliculitis or neoplastic processes such as Kaposi's sarcoma. BA is another illness that resembles VP clinically, histopathologically and etiologically.

LABORATORY DATA

Histopathologic features of verruga peruana consist of angioblastic and histiocytic proliferation limited by epidermal ridges. Electron microscopy reveals coccoid forms in the interstitial spaces or intracellularly. Combined use of immunohistochemistry and electron microscopy shows that the verrucoma is primarily the product of the proliferation of two cell populations. The majority are positive for factor VIII, *Ulex europaeus* and Weibel-Palade bodies characteristic of endothelial cells, while the other stroma-like elements correspond to primitive reticular mesenchymal cells, histiocytes, or fibroblasts. It has been recognized that Dendrocytes and Langerhans cells have been observed among the histiocytes. Garcia has demonstrated that Bb produces an angiogenic factor that stimulates proliferation of endothelial cells through production of a tissue plasminogen activator (t-PA). Bb probably also induces the expression of soluble immune mediators such as TNF-α, BFGF, IL-8 or factor thirteen, all of which have angiogenic properties.

Selected Readings

1 Cockerell C, Tierno A, Friedman- Kien A et al. Clinical, histologic, microbiologic, and biochemical characterization of causative agent of bacillary (epithelioid) angiomatosis: A rickettsial illness with features of bartonellosis. J Invest Dermatol, 97:812-817 (1991).

2 Caceres-Rios H, Rodriguez-Tafur J, Bravo-Puccio F et al. Verruga peruana: An infectious endemic angiomatosis. Crit-Rev-Oncog 1995; 6(1): 47-56.

3 Garcia F, Wojta K, Broadley J et al. Bartonella bacilliformis stimulates endothelial cells in vitro and is angiogenic in vivo. Am J Pathol 136 (5):125-1135 (1990).

Symbols

A

B

C

N

O

P

U

V

W

Y